Fumes and Gases in the Welding Environment

A Research Report on Fumes and Gases Generated During Welding Operations

Research performed at Battelle-Columbus Laboratories under contract with the American Welding Society and supported by industry contributions

Under the direction of the AWS RESEARCH COMMITTEE ON SAFETY AND HEALTH

Edited by F. Y. Speight, Manager, Safety and Health, and
H. C. Campbell, Consultant

AMERICAN WELDING SOCIETY
550 N. LeJeune Road, Miami, Florida 33126

International Standard Book Number: 0-87171-174-5

American Welding Society, 550 N.W. LeJeune Road, P. O. Box 351040, Miami, Florida 33135

Printed in the United States of America
Reprinted January 1987

This report is published as a service and convenience to the welding industry and is the product of an independent contractor (Battelle-Columbus Laboratory) which is solely responsible for its contents. The materials in this report have not been independently reviewed or verified and are only offered as information. AWS assumes no responsibility for any claims that may arise from the use of this information. Users must make independent investigations to determine the applicability of this information for their purposes.

Contents

Personnel

The authors, all employed at Battelle-Columbus Laboratories:

R. M. Evans	R. G. Luce
L. J. Flanigan	D. C. Martin
D. G. Howden	H. E. Pattee
K. W. Lee	R. E. Robinson

Contributor to the experimental studies: W. H. Stefanov

Edited by: F. Y. Speight, American Welding Society
H. C. Campbell, Consultant

AWS Research Committee on Safety and Health

A. Lesnewich, Chairman	Airco Welding Products Division
F. Y. Speight, Secretary	American Welding Society
R. C. Becker	International Harvester Company
*K. L. Brown (Fumes and Gases)**	Lincoln Electric Company
R. M. Gage	Union Carbide, Linde Division
J. F. Hinrichs (Radiation)	A. O. Smith Corporation
*D. S. Janetka (Hazard Evaluation)**	International Harvester Company
*W. C. Janes (Hazard Evaluation)**	U.S. Steel Corporation
T. Kilmer (Analytical Methods)	Airco Welding Products
*J. O. Hayden (Thermal Spraying)**	Hayden Corporation
*W. B. Meyer (Thermal Spraying)**	St. Louis Metallizing Company
*R. W. Niles (Sampling Fumes and Gases)**	Eastman Kodak Company
*R. L. Peaslee (Brazing)**	Wall Colmonoy Corporation
A. W. Platt	Caterpillar Tractor Company
F. Carl Saacke (ex officio)	Airco, Incorporated
*N. B. Shankland (Noise)**	Sciaky Brothers, Incorporated
*P. Shaughnessy (Resistance Welding)**	Chrysler Corporation
*H. Trabbold (Ventilation)**	Aircair Company

*Consulting Member

Research Finance Committee

C. H. Renicke, Chairman 1975-77	Chemetron Corporation
E. B. Scripture, Chairman 1977-	Teledyne-McKay
R. E. Bruggeman	Caterpillar Tractor Company
A. W. Platt (resigned 1975)	Caterpillar Tractor Company
R. H. Schwegman	General Electric Company

Preface

The environment presented by welding, cutting, and similar operations is a concern to industry. Management seeks to provide a safe and healthy working environment for welders, and the Occupational Safety and Health Act of 1970 has added the force of law to this desire. Although fumes and gases associated with welding and cutting are the most obvious potential hazards, other factors—radiation, heat, and noise—must also be considered.

Prior research for industry and for the American Welding Society has scarcely scratched the surface of this problem. Some work conducted has raised more questions than it has answered. Typical of these questions are the following:

(1) Can fumes and gases be characterized on the basis of electrode classification?

(2) Can simplified procedures be developed to evaluate the hazards presented by airborne contaminants?

(3) What are typical noise levels associated with welding and cutting operations?

(4) How can ozone and nitrogen dioxides be detected accurately in the presence of dense fumes?

(5) Are fume concentrations affected in a predictable manner by the welding variables?

(6) Can the health and safety of workers be protected without affecting productivity?

Reliable data are clearly needed to demonstrate that welding operations are safe when carried out properly and do not constitute an unusual occupational hazard.

To help industry, AWS proposed a continuing investigation at Battelle Memorial Institute-Columbus Laboratories to study the various environments to which welders are exposed, to quantify them, to demonstrate means to cope with them in a safe manner, and to correlate laboratory and shop generated data.

The processes to be studied were arc welding, oxyfuel gas welding and cutting, brazing, and thermal spraying.

The investigation required funding at a rate of $150,000 per year for two years. Support was obtained from industry subscriptions, and progress reports were distributed to the sponsoring companies.

This report summarizes work completed to September 1978, except for studies of the noises and radiations generated in arc welding and cutting. A research report on the measurement of welding and cutting noise has been published separately as "Arc Welding and Cutting Noise" (Ref. 1)*. Two reports on radiation have been issued by the U.S. Army Environmental Hygiene Agency in cooperation with the American Welding Society and Battelle-Columbus Laboratories (Refs. 2 and 3) and a brief summary has been published in the "Welding Journal" (Ref. 4).

All measurements shown in the tables and text of this report are given in the system (U.S. customary or S.I.) in which they were made, followed by an approximate S.I. conversion in parentheses if the measurement was in U.S. units. No conversions are considered necessary for scientific measurements made directly in metric units.

*References are grouped together following the appendix.

Supporting Organizations

Airco Welding Products
A.O. Smith Corporation
Arcair Company
Arcos Corporation
Armco Steel Corporation
AWS Detroit Section
Babcock and Wilcox Company
Cabot Corporation (Stellite Division)
Caterpillar Tractor Company
Champion Commercial Industries, Incorporated
Chemetron Corporation
Combustion Engineering, Incorporated
Dover Corporation (Bernard Division)
Eastman Kodak Company
E.I. Dupont de Nemours
General Electric Company
The Heil Company
Hobart Brothers Company
Huntington Alloys, Incorporated
International Harvester Company
J.I. Case Company
Lincoln Electric Company
Miller Electric Company
Motor Vehicle Manufacturers Association
National Constructors Association
National Institute for Occupational Safety and Health (NIOSH)
Sandvik, Incorporated
Steel Plate Fabricators Association, Incorporated
Stoody Company
Sun Shipbuilding and Dry Dock Company
Teledync McKay
The Trane Company
Union Carbide Corporation, Linde Division
United States Steel
Victor Equipment Company
WABCO Construction and Mining Group (American Standard, Incorporated)
Westinghouse Electric Corporation

Contributing Organizations

Alabama Oxygen Company
AWS Central Texas Section
AWS Cleveland Section
AWS Houston Section
AWS Los Angeles Section
AWS New Jersey Section.
AWS Philadelphia Section
AWS Pittsburgh Section
Bel Welding Supply Company, Incorporated
Coastal Welding Supply
Industrial Welding Supplies, Incorporated (Alabama)
Iweco, Incorporated
Lake Welding Supply Company
Lex Company, Incorporated
McKenzie Repair, Incorporated
Mineweld Company of Indiana, Incorporated
Prest-O-Sales and Service, Incorporated
Rodgers Welding Supply Company
Suburban Welders Supply, Incorporated
United Association

In addition to the supporting and contributing organizations listed above, 119 other organizations contributed amounts up to $500 each to the program. All contributions are gratefully acknowledged.

Fumes and Gases in the Welding Environment

A Research Report on Fumes and Gases Generated During Welding Operations

Abstract

The environment surrounding many welding processes contains *fumes* (particulate matter) that may be harmful (toxic) or relatively harmless and *gases* that may have pulmonary or non-pulmonary effects. This report summarizes five experimental studies and several literature surveys (conducted by Battelle Memorial Institute-Columbus Laboratories for the American Welding Society) to evaluate the extent to which ventilation may control the exposure of the welder to these fumes and gases and to investigate the nature of the various fumes and gases generated in arc welding, in brazing with silver-based filler metals, in thermal spraying, and in oxyfuel gas cutting.

Comprehensive fume control requires exhaust flow rates adequate to reduce room contamination below critical levels, or cross-draft ventilation (or air-ventilated helmets) to remove fumes from the welder's breathing zone, or sometimes both.

Tables in this report show what fumes and gases are generated by fourteen types of covered electrodes for shielded metal arc welding, seven electrodes for flux cored arc welding, eleven gas metal arc solid electrodes, two BAg-class brazing filler metals, seven spraying and surfacing metals, and three thicknesses of carbon steel plate severed by oxyacetylene and oxymethane cutting under various operating conditions.

These data can be used in part to determine blower capacity and exhaust flow rates needed for ventilation.

Introduction

The American Welding Society has sponsored a two-year program, entitled "Improving the Welding Environment," conducted by Battelle's Columbus Laboratories. This book is a report of that research and consolidates the content of five topical reports on the program.

Overview of Fume Constituents

Fumes are particles formed by electrode (and base metal) constituents that are vaporized and subsequently condensed in the welding area. Because of their small size, fume particles may remain suspended in the aerosol form for long periods. Since the particles have mass and size and are affected by air movement, electrical fields, gravity, diffusional forces, and other external forces, they tend to agglomerate into clumps that gradually settle on the floor and other surfaces. While suspended, however, they are inhaled by all persons in the vicinity.

In addition to fume particles, there are also gases formed that have toxic effects. These include ozone, oxides of nitrogen, and carbon monoxide.

The presence of fumes and gases in the welding environment is a matter of concern to those responsible for the well-being of welding personnel. Some constitute a potential hazard to the health of the welder while others are merely a nuisance. Potential problems can be anticipated by estimating the concentrations of fume and gas constituents in the welding area and comparing these data to established exposure standards.

There are three basic ways to protect the welder from the fumes and gases generated in his or her vicinity:

(1) Isolate the welder from the fumes through the use of respirators and air supplied helmets.

(2) Remove the fumes from the welder's immediate environment by ventilation and exhaust ducts.

(3) Reduce the quantity of fumes at the source through changes in the welding process or the conditions of welding.

In the broad view, both the welder and those in the vicinity of the welding area must be protected from these fumes since no one should be exposed to accumulating concentrations of fumes and gases that are hazardous to safety and health. Accordingly, ventilation devices (method 2) are expected to become increasingly relied upon. Method 1 ignores the local environment, although it may be necessary to use it on particular tasks while keeping other personnel out of the contaminated area. Method 3 requires knowledge of the fume generating characteristics of all available processes and procedures and, in the long range, the development of fume-free processes.

The cost of operating fume control devices (method 2) will be an important factor as energy costs rise. Determining the most effective method of controlling fumes and gases generated during various welding processes will require iterative calculations of cost trade-offs that are a function of the concentration and characteristics. The level of control required will have a major effect on the cost, but in many cases more than one of the possible methods of control may need to be used in order to minimize costs.

Survey of Part I, Ventilation Studies

A major objective in studying the various welding processes was to generate data needed to design systems that will control the flow of fumes and gases. In order to meet this objective, it was necessary first to characterize the flow fields associated with fumes and gases generated by these welding processes. Then, theoretical and experimental studies were conducted to obtain the data used to define design curves for the performance of various local ventilation devices. Two series of experimental studies were conducted: local ventilation experiments were performed to define the effect that various ventilation conditions have on the welder's exposure to welding fumes, and local exhaust experiments were conducted to define the effect of various local exhaust devices on the welder's exposure to welding fumes and on the reduction in environmental contamination resulting from the collection of the fumes.

Certain welding processes were studied to provide basic information on the fume generation characteristics of commonly used arc welding electrodes as a function of welding process, electrode classification, and selected operating variables. This task emphasizes a comprehensive laboratory investigation of arc welding fumes to determine the rate at which they were produced with selected welding conditions in order to relate the fume generation rate to various electrode characteristics. More limited studies were also conducted to detect gases in the welding area, determine fume composition, and characterize fume particles in terms of size, crystalline structure, etc. Such data will be useful to industry in planning industrial hygiene programs and determining ventilation requirements.

The technical literature of the past two decades was reviewed for references to research conducted that sampled fumes and gases in the welding area, determined individual fume and gas constituents, and calculated concentrations of these constituents for comparison with recognized allowable limits. Many of these references were included in the AWS publication, *The Welding Environment* (Ref. 5). These investigations generally had limited objectives because they were undertaken to study the fumes associated with a particular welding application, to evaluate a specific electrode, to assess the performance of a ventilation system, or for some other definite purpose. The usefulness of some of this work is questionable today because supporting data on welding procedures, sampling methods, and analytical techniques are often incomplete or totally lacking. Nonetheless, these studies are important because the results form the basis for much of today's knowledge.

The design of fume control systems was not examined during this study. The American Welding Society has identified 84 welding processes by letter designations for diverse welding applications. It is not practical (from cost and time viewpoints) to study all possible combinations of factors and develop design data specifically for each combination. The approach taken in this study was limited to experimental and theoretical efforts to generate the information needed to design systems that will control the flow of fumes and gases generated during specific welding processes.

The experiments to define the effectiveness of the exhaust devices were conducted using a simulator gas instead of the arc welding fumes and gases. The simulator gas plumes duplicated the thermal and fluid mechanic characteristics of plumes from welding operations and enabled the experiments to be conducted continuously over a long period of time under reproducible conditions. Two subsets of experiments were conducted to ensure that the pertinent thermal and fluid mechanic characteristics of the simulator plume were identical to those of the weld plume. These consisted of tests to characterize the weld plumes for various welding processes and tests to establish the simulator operating conditions required to duplicate the actual welding plumes.

Theoretical analyses were conducted to define the overall room concentration level of welding fumes and gases in terms of the ventilation and weld fume generation rates.

Survey of Part II, Arc Welding Fumes and Gases

The key task in the AWS research program, "Fumes and Gases in the Welding Environment," was an investigation of the fumes and gases associated with conventional arc welding operations. This part of the program underscored the importance of these airborne contaminants and the need to develop a comprehensive understanding of factors governing their occurrence and behavior in and around welding areas. Fumes and gases are produced during all arc welding operations. Their potential effect on the well-being of welding personnel is highly dependent upon their amount and composition. More fumes, for example, are produced when carbon steels are welded with the shielded metal arc process than when austenitic stainless steels are welded in the same manner. On the basis of fume generation rate alone, the carbon steel welding operation appears to be potentially most harmful. Such may not be the case, however, because the fumes associated with the welding of 300-series stainless steels contain nickel and chromium compounds with threshold limit values (TLV's® [Ref. 6]) much lower than that of iron oxide, the principal constituent in the fumes produced during the welding of carbon steels.

In other words, the concentrations of the nickel and chromium compounds associated with the welding of stainless steel may exceed allowable limits even though the fume generation rate is low. In evaluating problems associated with the welding environment, the following must be taken into account:

(1) Welding process
(2) Welding consumables
(3) Welding conditions
(4) Base metal
(5) Characteristics of the welding area (size, height, air movement, etc.).

In addition, some knowledge concerning the relative hazards of specific fume and gas constituents is needed.

A limited number of welding processes was selected to represent the family of welding processes because of their common use and their wide range of fume generation rates. The selected processes included shielded metal arc welding (SMAW), flux cored arc welding (FCAW), gas metal arc welding (GMAW), and gas tungsten arc welding (GTAW) operations. Commercially-available electrodes were used in this study at manufacturer-specified arc voltage and current conditions.

A comprehensive matrix of tests was selected and followed to ensure versatility in the application of the results. Relationships were established between fume generation rate and various electrode characteristics (melting rate, metal deposition rate, etc.) to obtain data that could be used effectively by welding engineers, industrial hygienists, ventilation specialists, and others concerned with the health and safety aspects of arc welding. Research was also conducted to investigate the fume generation characteristics of selected electrodes as a function of various process variables such as current, voltage, and arc length.

During the course of this investigation, limited research on fume particle characterization, fume analysis, and gas detection during welding was included. No determination of the oxidation state of chromium in the fumes was attempted, however. Although fume characterization was an objective of the original program, budget limitations (especially in view of the expense of sophisticated analytical research) precluded a thorough analytical program, even though the need is recognized. A NIOSH criterion document issued in 1975 and the recent indictment of hexavalent chromium by OSHA identify certain compounds of chromium of value +6 as being carcinogenic (cancer-causing). It remains for some future program to determine the oxidation state of chromium in fumes from welding stainless steel. The concentrations of Cr given in this report must be identified as total Cr. Possibly two-thirds of the total Cr may be hexavalent, of which 95 percent may be water-soluble and, therefore, noncarcinogenic; the remaining 5 percent of that two-thirds would be presumed to be insoluble and carcinogenic.

Analytical research coupled with fume and gas studies could improve our ability to reduce the quantity of fumes and gases generated at the source by varying the process conditions. A limited amount of such research was included in this program. The results are sufficiently useful to indicate the value of further study.

Survey of Part III, Brazing Fumes and Gases

Another purpose of the American Welding Society research program was to examine the fumes generated during brazing. The broad objective of this task was to show experimentally the relationship of the fumes given off to the brazing filler metal used.

A literature search on brazing references was conducted to learn the extent and usefulness of the bibliographic data available. Study of this literature revealed the adequacies and deficiencies of the available information.

The experimental research program was developed without benefit or knowledge of any published prior experience by other investigators. Its magnitude and scope were restrained by the availability of funds assignable for brazing fume studies.

The investigation was limited to the BAg-class filler metals which are predominantly silver but contain copper, zinc, and cadmium. It was directed toward determining the fume generation characteristics of two filler metals having widely different cadmium contents: Ag-15Cu-16Zn-24Cd (BAg-1) and Ag-22Cu-20Zn-7Cd-1Sn. These filler metals have similar melting characteristics and are used extensively in industry for torch brazing.

The research plan was designed to minimize the variables caused by changes in flame shape, character, and position that occur during torch brazing. Therefore, procedures similar to those used during welding fume studies were suggested. Total fume samples collected in the AWS-Battelle welding fume collection chamber were used to find the influence of brazing variables on the fume generation rates and quantities. The fume samples were analyzed by atomic absorption methods to determine how brazing variables affect fume compositions. Unfortunately, it became increasingly evident, as the program progressed, that the procedures chosen did not provide the precision required for accurate measurement of brazing filler metal fumes.

In the brazing studies, the small quantities of fume produced and its apparent extremely small particle size presented problems in filter choice, chamber design, sample weight-analysis balance, and sampling environments. To minimize these problems, a fume collection chamber having the least volume possible, containing an atmosphere that is inert or readily controllable, and which can be made part of a closed system is advised. Filters or fume collection media must be capable of collecting very fine particles ($<1\mu m$) and should be small enough to avoid masking of actual fume weights by large blank filter weights. Continuous sampling through an atomic absorption type of analytical system appears worthy of consideration instead of filter sampling systems.

Attention to the details of actual filler metal melting and fluxing practices is also required. Melt sizes, quantities of flux used, exposed melt surface areas, and heating and cooling rates need attention. Melt size, particularly with respect to exposed areas, may alter fume generation rates because of flux cover variation. Crucibles should be a metal that is wet by the filler metal to avoid flux collection around the melt. This implies combined weight determination that includes the crucible. Weighing exercises should include checks of flux losses in addition to filler metal losses. Ideally, all tests should begin in a cool chamber and heating furnace to minimize differences in heating and cooling rates. The closed chamber already suggested should eliminate any influence of humidity in the test area.

Analytical procedures for fume analysis should be validated for the tests being conducted. An early phase of any analytical work should be directed toward determining the chemical character of the fume (metal, oxides, soluble, insoluble, etc.). Early analytical work should include complete analyses. Subsequent analyses requirements may be relaxed when the whole story is known.

Survey of Part IV, Thermal Spraying Fumes and Gases

As another part of the program, Battelle investigated the fumes and gases produced by thermal spraying operations and the manner in which the nature of the spraying process and the type of metal being sprayed affect their characteristics and the rate at which they are generated.

Thermal spraying processes are used by industry to repair or build up worn parts and to produce coatings that resist oxidation, corrosion, wear, erosion, abrasion, or impact. In practice, finely divided metallic or nonmetallic particles in the molten or semimolten state are sprayed on a suitable substrate to produce a dense, adherent deposit or coating. Thermal spraying encompasses three distinct processes that are differentiated mainly by the manner in which the particles are produced and heated: (a) gas combustion spraying, (b) electric arc spraying, and (c) plasma arc spraying. These processes are characterized by relatively low deposition efficiencies, moderate-to-high fume levels, and moderate-to-high noise levels.

In the area of thermal spraying, research efforts were directed toward a study of the fumes associated with the most commonly used processes: gas combustion spraying and electric arc spraying. In each instance, the metal being sprayed was in the form of wire. The metals included in the investigation were mutually agreed upon by the AWS Research Committee and Battelle.

Fume collection studies for thermal spraying were limited to those conducted at recommended spraying conditions. No work was done on detonation spraying, which is also included among the thermal spraying processes; it is a proprietary process and little is known about its characteristics.

Survey of Part V, Fumes and Gases from Oxygen Cutting

Finally, the fumes associated with oxygen cutting operations were investigated. These fumes occur as the result of the chemical reaction between iron and high purity oxygen and the process through which molten oxides and unoxidized metal are removed from the kerf by the cutting jet. Fume quantities are dependent upon the cutting process variables, base metal thickness, base metal composition, surface contaminants, and other factors. Fume compositions are governed by the composition of the metals being cut and any coating that may be on the surface of the metal.

To study the fumes associated with oxygen cutting operations, total fume samples were collected to accumulate data for use in determining fume generation rates and quantities relating fume generation rates to important process variables.

Research was directed toward a study of the fumes produced during oxyacetylene and oxynatural gas (methane) cutting of carbon steel plate in thicknesses of 1/2, 1, and 2 in. (12.7, 25.4, and 50.8 mm) at conditions recommended by the cutting torch manufacturer. Also included were studies of the effect of selected process variables on fume formation rates.

Although the potential problems presented by oxygen cutting fumes have long been noted, there are few references to this subject in the technical literature on safety and health. There is little information on the rate at which fumes are produced during oxygen cutting operations. The data obtained in the course of this investigation only partially fill this void.

Additional research would be needed to more fully characterize cutting fumes. Such research should include studies to determine the distribution of fume particles by size and examination of fume deposits with the scanning electron microscope.

In this study, concentrations of major elements could not be determined accurately because of difficulty in interpreting the spectrographic record and because of the lack of suitable standards. As a result, concentrations of such elements are expressed in a range of weight percentages instead of discrete weight percentages. If more accurate compositional data are needed, optical emission spectroscopy should be supplemented by other analytical techniques such as atomic absorption analyses or wet chemistry methods.

No study of arc cutting was made, although that was an original aim of the research program: Suitable equipment was not available at Battelle, and alternative locations at which tests could run could not be identified by the Committee.

Organization of This Report

Each of the five sections of the body of this report is organized in the following format:

Introduction
Scope
Limitations
Background (Part II only)
Program
Equipment, Materials, Procedure
Description of Work (Organized under Section A, Section B, etc.)
Discussion
Summary

Appendixes have been included at the end of the report which present the tabular data for Parts II, III, IV, and V in sequence.

The references cited in the Preface, Introduction, and Parts I, II, III, and V have been grouped together following the appendixes.

An index completes this compilation.

Part I. Ventilation Studies

Introduction

The research performed by Battelle-Columbus Laboratories for this portion of the American Welding Society program obtained ventilation design data for use in controlling welding plumes. The study was conducted over a 6-month period that began in May, 1977. The plume control study was monitored by a committee appointed by the AWS Research Committee.

About 250 tests were carried out in the course of this study. The results of the study show that various ventilation methods can be used effectively to reduce the welder's exposure to fumes and, conversely, that certain ventilation conditions can result in higher welder fume exposures than those experienced with no ventilation. Operating conditions for various local exhaust devices are specified in terms of a welder exposure factor and a room contamination factor which make them easy to use.

The report is divided into three major subsections paralleling the three main areas of investigation: "General Room Ventilation," "Local Ventilation," and "Local Exhaust Studies." Included in each subsection, where appropriate, is a description of the test apparatus and procedures and a discussion of test results. These subsections are followed by a summary. The references are found following the appendixes at the end of this report compilation.

An example problem is included in the report to show how the study results can be used. The example outlines one possible application of the results; it should be kept in mind that the results have a wide scope of application toward designing devices that can be used to control welding fumes and gases. Familiarity with the various experiments conducted during this study and the associated results will enable the reader to apply the results to a variety of welding problems in order to achieve cost-effective and environmentally acceptable control of welding plumes.

Scope

Theoretical and experimental studies were combined to generate data on the effect of ventilation air currents on fume concentrations in the welder's breathing zone and to define the effectiveness of exhaust devices in reducing environmental contamination.

One major series of experiments defined the fume concentration level in the breathing zone as a function of ventilation velocity and relative orientation of the ventilation source with respect to the welder using a large flow area source (4 x 4 ft. [1.2 x 1.2 m]).

A second major test series used a simulator gas instead of the arc welding fumes and gas. The simulator gas plumes were identical to the weld plumes in thermal and fluid mechanic characteristics and enabled the experiments to be conducted continuously over a long period of time under reproducible conditions. Two subsets of experiments were conducted to ensure that the pertinent thermal and fluid mechanic characteristics of the simulator plume were identical to those of the weld plume: tests to characterize the weld plumes for various welding processes, and tests to establish the simulator operating conditions required to duplicate the actual welding plumes.

Theoretical analyses were conducted to define the overall room concentration level of welding fumes and gases in terms of the ventilation and weld fume generation rates. The results of these analyses combined with the results of the experimental studies provide the information needed to design effective systems for control of welding fumes and gases.

Limitations

The design of fume control systems was not examined during this study.

Previous application of fume control methods to specific welding processes has been limited because sufficient data are not available. Most of the quantitative information in the literature is very general, having been obtained at a time when the need to conserve energy and to control the welder and plant environments was not so acute as it is today. However, in view of the large number of recognizable welding processes and variations of them, it was not practical for this study to consider more than a few representative combinations of factors which may influence the ultimate design of ventilation equipment. A limited number of welding processes was selected as being representative of the family of welding processes because of their common use and their wide range of fume generation rates. The selected processes included shielded metal arc welding (SMAW), flux cored arc welding (FCAW), and gas metal arc welding (GMAW) operations. Commercially-available electrodes were used in this study at manufacturer-specified arc voltage and current conditions.

Program

Ventilation provides a means for reducing to an acceptable level the hazardous and noxious fumes and gases to which a welder may be exposed. Many techniques are available for controlling welding fumes and gases. These range from systems that employ natural convection to localized devices such as air-ventilated helmets. Examples of methods for controlling the welder's exposure to fumes and gases include:

(1) General room ventilation
(2) Overhead exhaust hoods
(3) Portable local exhaust devices
(4) Down-draft tables
(5) Cross-draft tables
(6) Extractors built into the welding equipment
(7) Air-ventilated helmets

Fluid dynamics were applied to the theory of the welding plume. The governing differential equation for the contamination level in a closed room was integrated to obtain an expression for the room contamination level as a function of time. Steady-state fume concentration levels were calculated for rooms of all sizes. Transient concentration levels were calculated for times prior to achieving steady-state conditions.

Experiments were performed to determine the effect of air currents on these theoretical fume concentrations in the welder's breathing zone (local ventilation). Apparatus was developed to measure the reduction in operator exposure to contaminants, using two methods to measure the fume concentration level: weighing a filter and counting condensation nuclei. Sampling probes inside the welding helmet collected fumes at the level of the welder's mouth. Sampling efficiency of the probe was measured over a range of particle diameters.

An air current generator was devised. Data were then obtained with no ventilation and with variations in air speed and direction.

Local exhaust studies evaluated the effect of collector configuration, collector location, and exhaust flow rate. Three types of experiments were conducted:

(1) Welding fume characterizations
(2) Simulator characterizations
(3) Inlet configuration studies

Equipment, Materials, Procedure

Experimental apparatus

The experimental apparatus used to measure the reduction in operator exposure to contaminants resulting from the use of local ventilation included fume sampling equipment, an air current generator, and welding equipment consisting of a power supply, arc voltage and current instrumentation, and a welding table. The apparatus was set up in an area having a high ceiling and no natural drafts. Figure 1.1 is a schematic diagram of the experimental setup viewed from above the apparatus.

Fume Sampling Equipment. Two methods were used to measure the fume concentration level in the welder's breathing zone: the filter weighing method, and the condensation nuclei counter method.

The fume sampling system used in the filter weighing method of analysis is shown schematically in Fig. 1.2. The sampling technique was specified by the American Welding Society (AWS) as a standard procedure when using the filter weighing method (Ref. 1.1). The sampling system consisted of sampling probes, a filter and filter holder, a critical flow orifice, a pressure gage, and a vacuum pump. Essential features of the present sampling system are described below.

Two sampling probes were installed inside the welding helmet, positioned at the level of the welder's mouth 1-1/2 in. (38 mm) on either side of the vertical centerline of the helmet. Two probes were used to obtain a less biased sample from the welder's breathing zone by halving the sample flow rate, thus reducing interference effects caused by the welder's breathing. Figure 1.3 is a photograph of the helmet showing the sampling ports. These sampling probes were connected to the filter holder mounted on the outside of the helmet using a standard Y connector. In order to minimize any possible particulate losses through the sampling system, the inlet end of the probes was sharply machined, and relatively large tubing (1/4 in. [6.4 mm diameter]) was used. The distance between the inlet ends of the probes and the filter holder was about 4 in. (100 mm). Measurements of the sampling efficiency of this system showed it to be essentially 100 percent efficient when collecting particles below approximately 0.05 micron (μm) in diameter and 97.0 percent efficient with 2 μm diameter particles.

The filter holder used was a standard 25 mm in-line holder (Model 1109, Gelman Instrument). Millipore

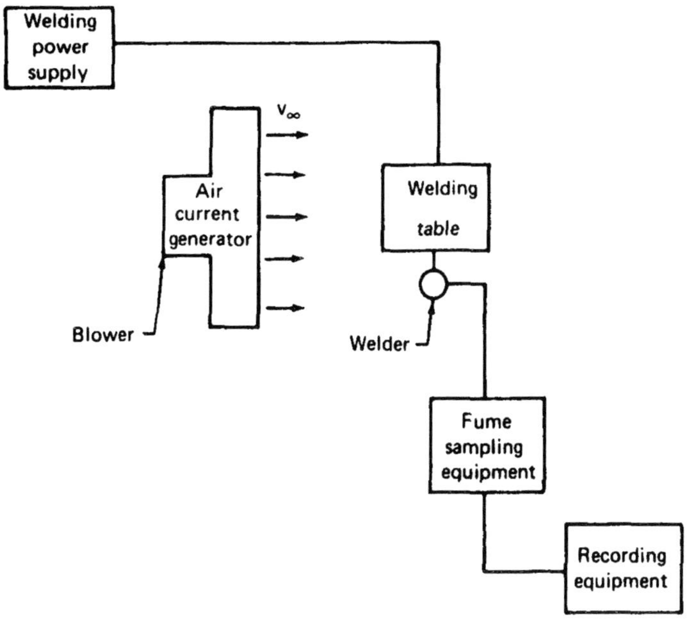

Fig. 1.1–Schematic of experimental setup

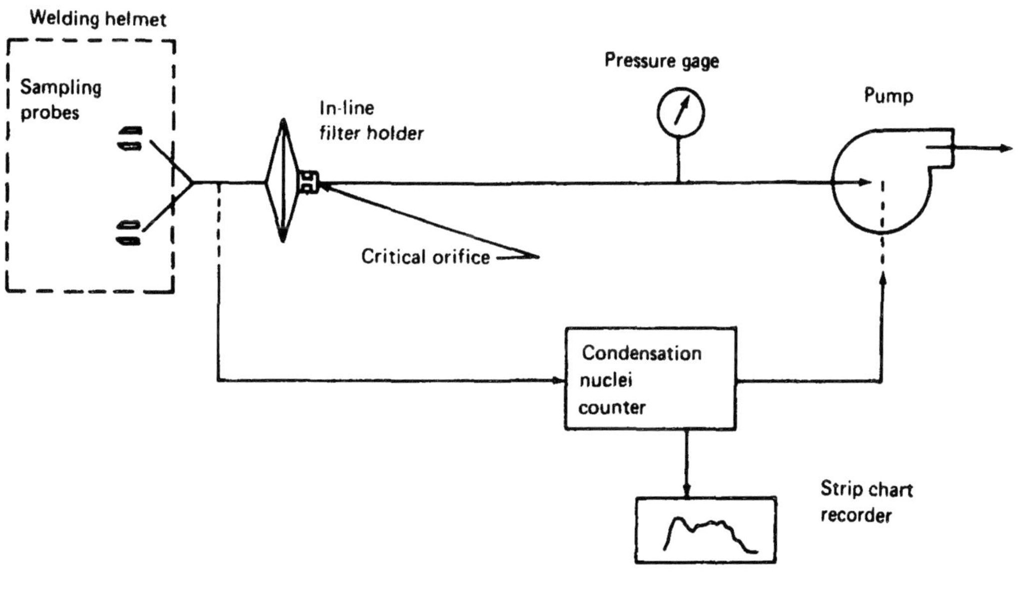

Fig. 1.2–Schematic of fume sampling system used in the filter weighing method

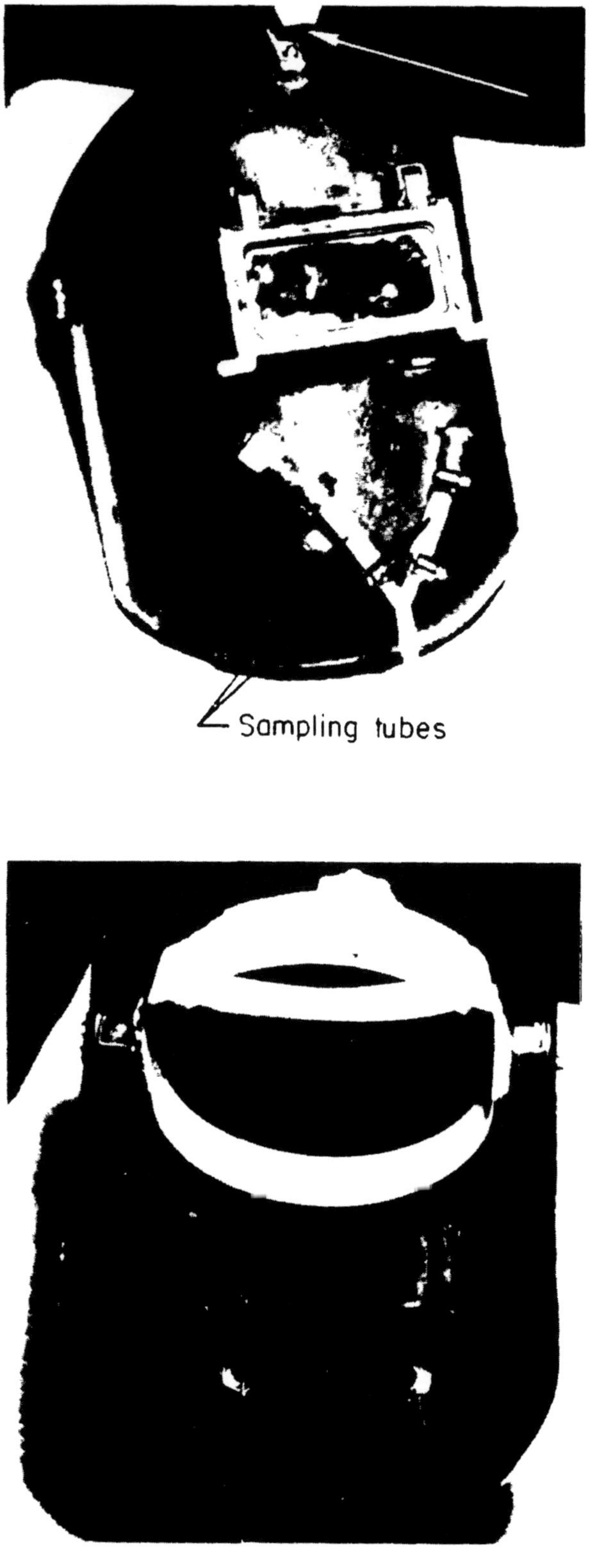

Fig. 1.3—Photograph of modified welding helmet

membrane filters (Model HA, Millipore Corporation [Ref. 1.2]) were used to collect the fume samples. The HA Millipore filter has a nominal pore size of 0.45 μm which meets the AWS specification. In addition, a previous study (Ref. 1.3) showed that the filtration efficiency of a comparable filter (Model FH, pore size 0.5 μm) was essentially 100 precent for all particle sizes. An analytical balance (Model M5, Mettler Instrument) with a nominal sensitivity of 0.001 mg was used to weigh the filters before and after use. A simple repeatability check showed that the actual accuracy was about 0.005 mg. In order to avoid any moisture absorption or chemical reaction effects on the filter and sample weight, the filters were kept in a constant temperature, constant humidity room for at least one hour prior to weighing. When a blank filter was kept under this environment and the weight was measured, the data were repeatable within 0.015 mg over 24 hours. A similar experiment using a filter-plus-fume sample showed that the measurement was repeatable within 0.020 mg. Since the collected sample weight was always higher than 5 mg, the above variations were considered acceptable.

In addition to the filter weighing method, a particle counter was also used to determine fume particle concentrations in the breathing zone. The particular instrument used was a condensation nuclei counter, Model E-1033A-001G1, manufactured by the Environment One Corporation, designed for real time measurements of airborne particle number concentrations. In this instrument the particulate-laden air sample is saturated with water vapor and then adiabatically expanded to a low pressure, causing the water vapor to condense on the particles (which, in this case, serve as condensation nuclei) and form relatively large droplets. The concentration of these large droplets is then measured optically by light transmission. The instrument can measure fume particles in the size range of 0.002 μm and larger in diameter. The standard operating flow rate for the instrument is 3 liters per minute, which was only 10 percent higher than the filter sampling rate, as will be shown. The advantage of using this instrument was that it gave real time measurements while filter measurements only provide time average values. The counter was used because it provided information on changes in particle concentration as a function of time. Of particular interest was the rate at which the fume concentration behind the helmet varied.

Air Current Generator. The air current generator was used to produce air currents or local ventilation across and around the table where the welding experiments were performed. Details of the air current generator plenum are shown in Fig. 1.4. The plenum was a plywood box, 4 ft x 4 ft x 1.5 ft (1.2 m x 1.2 m x 0.45 m), containing a thick, packed bed. A blower was mounted on the back side to force air through the assembly. The bed was filled with a foamed plastic packing material held in place with perforated plates. Measurements showed that the generator produced a uniform velocity

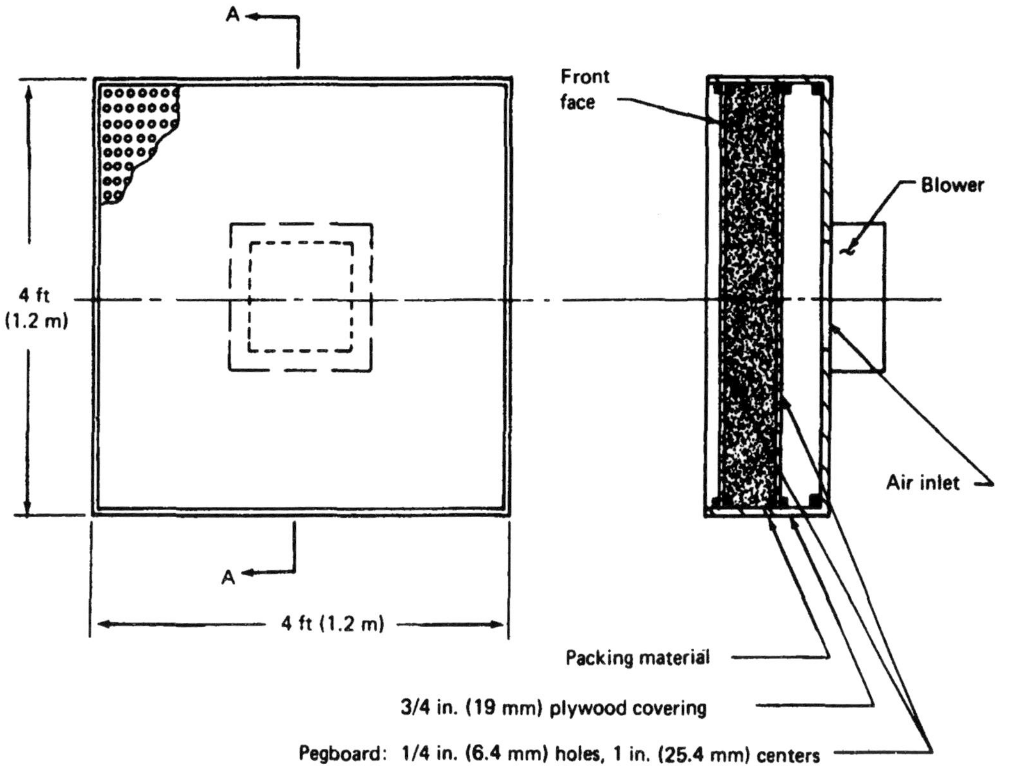

Fig. 1.4—Details of the air current generator

profile at the outlet face and over the welding area.

The air velocity in the welding zone was measured during the calibration tests using a hot wire anemometer (Thermo-System, Inc., Model No. 1050). The variation in the velocity profile at a distance of 42 in. (1.07 m) from the outlet face of the generator was ± 15 percent of the average velocity at 50 ft/min (15.24 m/min) and ± 5 percent at 200 ft/min (61.0 m/min). The air velocity was varied by varying the blower speed, which made it possible to obtain any velocity between 30 and 250 ft/min (9.14 and 76.20 m/min). Flow direction relative to the welder's normal welding position was obtained by rotating the working table relative to the generator.

Welding Equipment. Welding equipment used during these experiments included a dc power generator, a shunt bar to measure arc voltage and current, and a welding table. The electrical output of the power supply was monitored using a shunt bar arrangement which made it possible to record electrical signals (mV) that were proportional to arc voltage and current. The data signals were recorded on a Honeywell Electronic 19 Strip Chart Recorder. Welding was performed using 5/32 in. (4 mm) diameter E6013 shielded metal arc electrodes.

A working table surface of 24 in. x 24 in. (0.61 m x 0.61 m) at a height of 28 in. (0.71 m) was used in the studies. In order to maintain a consistent welder head position, a horizontal bar was set up 14 in. (0.36 m) above the table and a guide hanger was attached to the welder's helmet 4 in. (0.10 m) above the viewing window, as shown in Fig. 1.5. With this apparatus, the welder could move his head smoothly along the horizontal bar as the weld bead was made, but he could not move up and down or in and out. When the welder was in a welding position and his helmet was attached to the horizontal bar, his ear position was 13 in. (0.33 m) above the arc and 10 in. (0.25 m) away from the arc (horizontal distance).

Local Exhaust Apparatus

The test apparatus used to conduct the local exhaust experiments included: (a) local exhaust configurations, (b) fume simulator, (c) exhaust vacuum supply system, (d) welding setup, and (e) instrumentation. Components of the apparatus used in the local ventilation experiments that were also used in the local exhaust experiments are not discussed in this section; these include the air generator and components associated with the welding setup.

Local Exhaust Configurations. A number of local exhaust configurations were examined, including flanged ducts (3, 6, and 8 in. [76, 152, and 203 mm] diameters),

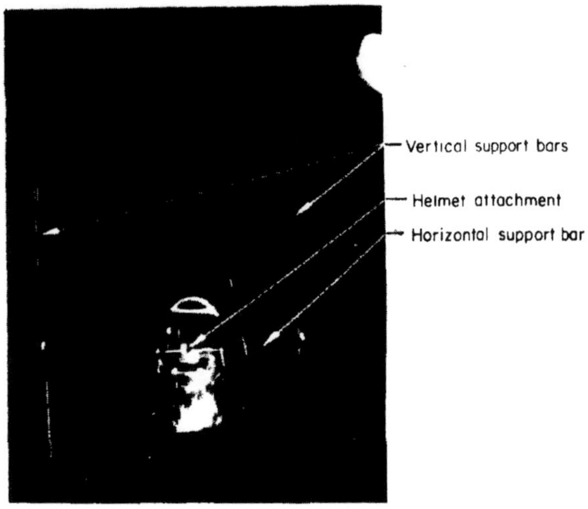

Fig. 1.5–Photograph of the helmet support mechanism

slots (single and double configurations), and two overhead hood configurations. The inlet configurations of the flanged duct exhaust systems are shown in Fig. 1.6 for the 3, 6, and 8 in. (76, 152, and 203 mm) diameter ducts. The flange size that was used is a recommended configuration set forth in Ref. 1.4. A section of sheet metal ducting was used for the exhaust inlet; the sheet metal section was connected to a flexible tube that was, in turn, connected to the vacuum supply plenum. The flanges were attached to the sheet metal ducts and sealed to eliminate air leaks between the flange and the metal duct.

Single- and double-slot exhaust configurations were examined during this study. Fig. 1.7 is a schematic of the slot configuration. Included in the drawings are the controlling dimensions for the two configurations examined (slot dimensions were obtained from Ref. 1.4). The single-slot size was 1.20 x 24 in. (30.5 x 610 mm), which is approximately equal in area to a 6 in. (152 mm) diameter circular duct. The double-slots were each 0.6 x 24 in. (15.2 x 610 mm), having a total area equivalent to the single-slot and to a 6 in. (152 mm) diameter circular duct. The flow area for the single- and double-slot exhaust configurations was made the same as the 6 in. (152 mm) diameter duct so that the face velocity at the exhaust inlet plane would be the same as that for a 6 in. (152 mm) duct at the same exhaust flow rate.

Two overhead exhaust systems, overhead shelf and slanted overhead, were examined during this study. Fig. 1.8 is a schematic of an overhead exhaust system. The basic dimensions for the overhead and slanted overhead exhaust systems were identical, the only difference being in the location of the exhaust inlet. The overhead exhaust systems were set up by attaching an overhead shelf to the vacuum plenum section.

The exhaust inlet size for both of these systems is shown in Fig. 1.8. When these configurations were tested, panels were installed on either side of the welding work area to duplicate a welding booth arrangement. The side panels helped in controlling the flow of air from a preferred direction.

Fume Simulation Apparatus. A qualitative analysis was made of the characteristics of hot, buoyant plumes containing particulate matter of very small size. From this analysis it was concluded that, relative to the objec-

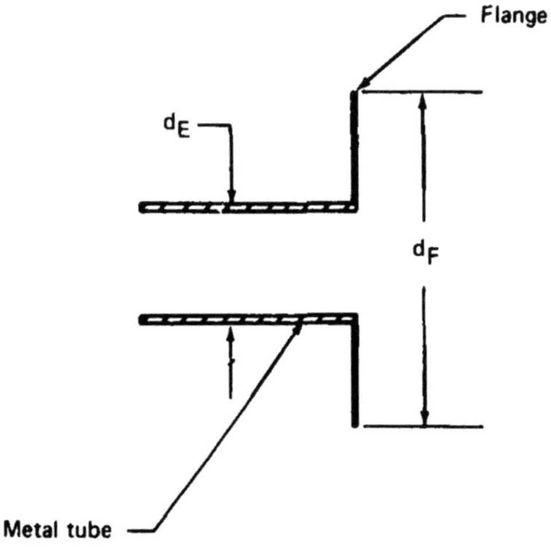

Note: Flange diameter (d_F) given below

d_E		d_F	
in.	mm	in.	mm
3	(76)	8	(204)
6	(153)	14	(356)
8	(204)	18-1/2	(472)

Dimensions in inches (millimeters)

Fig. 1.6–Details of the flanged duct exhaust inlet configurations

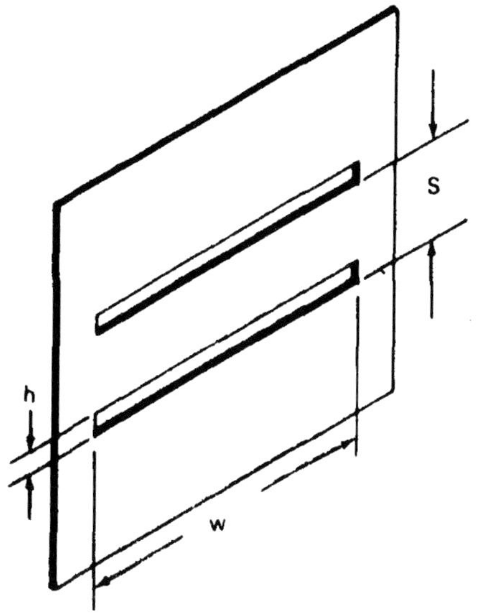

	h	w	S
Single	1.18 (30)	24 (610)	———
Double	0.60 (15)	24 (610)	4.0 (102)

Dimensions in inches (millimeters)

Fig. 1.7–Details of the slotted exhaust inlet configurations (single- and double-slot)

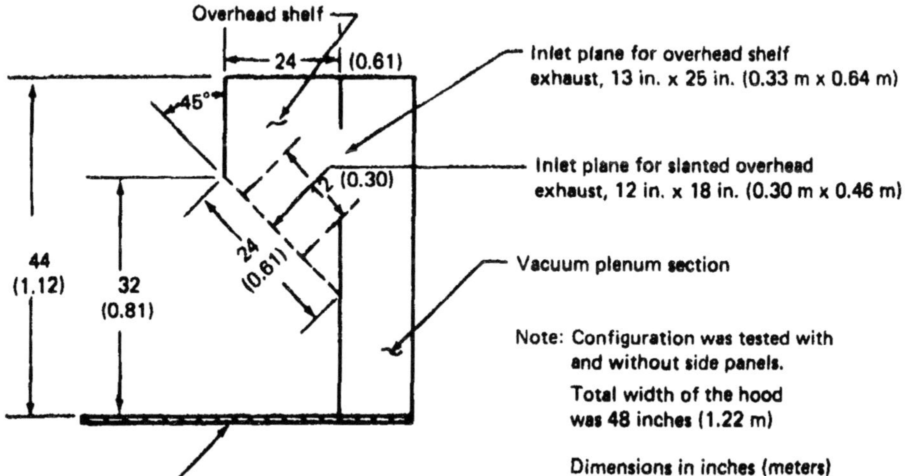

Fig. 1.8–Details of the overhead hood exhaust system

tives of this study, the required simulation of actual welding plumes would exist if the temperatures within the two plumes were equal and the response of the two plumes to a given air current was identical. Equal temperature profiles would result in equal buoyancy and, therefore, equal upward velocities. Identical response to an air current would result when the mass flow rates were equal.

The simulator was a device used to produce plumes representative of a broad range of those generated during actual welding. The reasons for using a simulator instead of actual welding plumes were twofold: it provided a means of obtaining a continuous, long-lasting plume that could be accurately reproduced and quickly changed to simulate different welding operations; and it permitted the use of a tracer gas that could be accurately measured on a real time basis. The underlying consideration was to obtain the maximum possible amount of information and design data within restraints placed on the program.

The apparatus designed, fabricated, and used to duplicate the flow of welding fumes in the vicinity of the welder included a gas heater unit, a gas supply system, and associated monitoring and control systems.

Fig. 1.9 is a schematic of the apparatus used to duplicate the flow of welding fumes. The gas supply system included a two-bottle bank of nitrogen, a two-bottle bank of carbon monoxide, and associated regulators and valves to control and select the desired gas. Nitrogen was used as an intermediate gas to bring the system up to operating condition and when CO was not being used (periods of time between test runs, etc.). The gas flow rate to the heater was controlled with a needle valve installed in the gas supply lines and was measured using a rotometer.

The gas (CO or N_2) was electrically heated by a heater designed and fabricated at BCL. A schematic of the heater assembly is shown in Fig. 1.10. Gas was introduced into the bottom of the heater and passed over electrically heated elements as it flowed up through the heater. The gas temperature was monitored at the top of the heater in the stagnation chamber using a chromel-alumel thermocouple. The thermocouple output was recorded and monitored on a strip chart recorder. At the heater outlet an orifice plate was used to obtain the desired initial plume size with various orifice plates used to duplicate different welding plumes. The heater shell was insulated to reduce heat losses. The heater operating temperature range was 100°F to 1200°F (38°C to 649°C) and was controlled using a 110 volt (ac) variac controller.

Vacuum Exhaust System. A system was set up to determine the effectiveness of various local exhaust systems. It included a blower to move the desired amount of air, an orifice plate to measure the amount of air

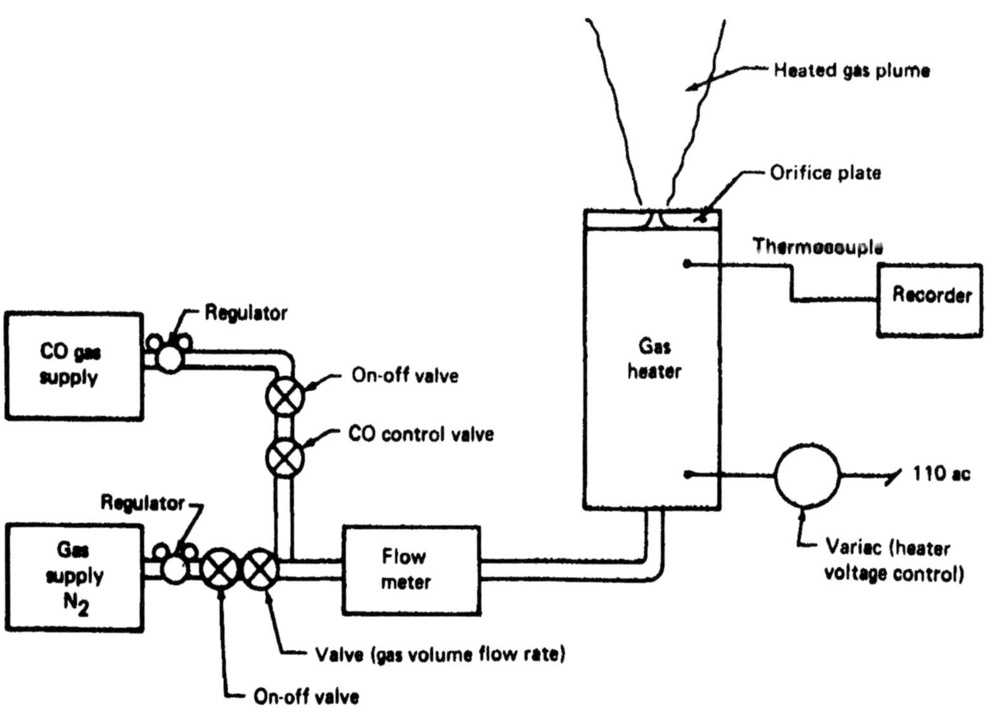

Fig. 1.9–Schematic of fume simulator apparatus

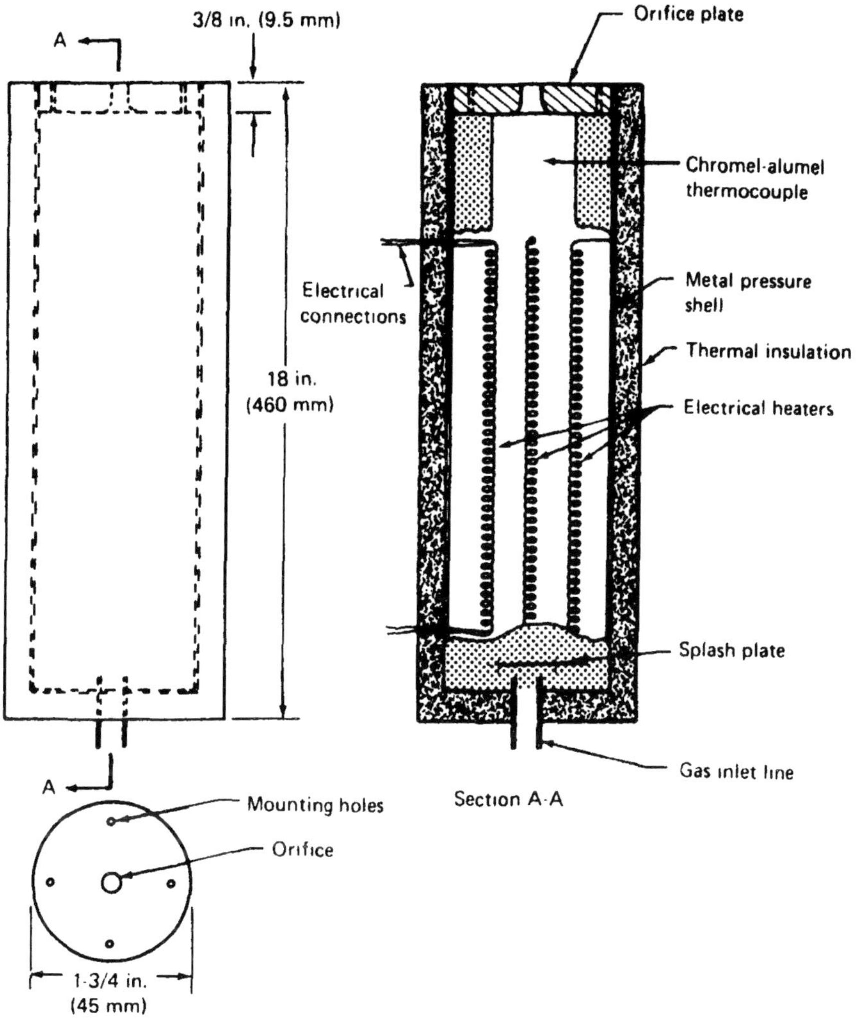

Fig. 1.10–Details of the gas heater

moving through the exhaust system, a CO sampling probe to measure the concentration level of CO in the exhaust duct, and a vacuum plenum to which the various local exhaust systems were connected.

Fig. 1.11 shows the arrangement of hardware and instrumentation used in conducting the local exhaust system experiments. The simulator apparatus and local exhaust inlet configurations shown in Fig. 1.9 were discussed in previous sections.

The duct type exhaust inlet configurations were attached to the vacuum plenum with flexible tubing so that the inlet could be positioned at various locations relative to the welding arc, or arc simulator location, or both (noted as welding area in Fig. 1.11). The vacuum exhaust plenum was connected to the blower with 12 in. (0.30 m) diameter metal ducting. A flow-measuring orifice plate was installed approximately 12 ft (3.66 m) downstream of the vacuum exhaust plenum and 10 ft (3.05 m) upstream of a 90 degree elbow in the ducting to the blower as shown in Fig. 1.11. The CO sampling probe was located 5 ft (1.5 m) downstream of the orifice plate. The blower outlet was exhausted to the atmosphere outside the laboratory.

The orifice plate pressure differential and upstream pressure were measured using inclined and vertical water manometers. CO concentration levels were measured using a Beckman CO Analyzer, Model No. 315B.

The vacuum exhaust plenum was fabricated using a 3/4 in. (19 mm) thick plywood sheet, reinforced with metal angles. The plenum was designed sufficiently

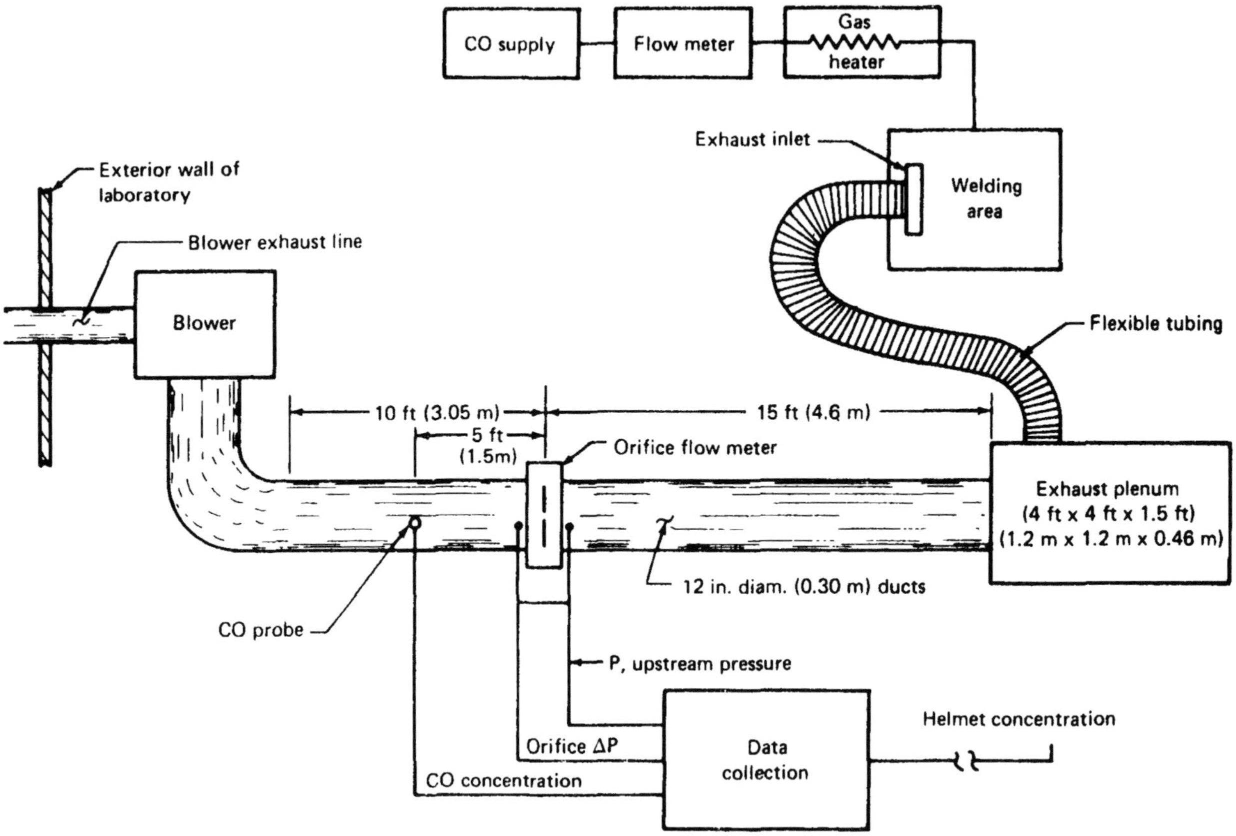

Fig. 1.11–Schematic of test apparatus used in the local exhaust experiments

large (4 x 4 x 1.5 ft [1.2 x 1.2 x 0.46 m]) to reduce non-uniformities in the velocity flow field at the entrance to the slotted and overhead hood configurations.

Instrumentation. All of the instruments used in this study were off-the-shelf items. These included recorders, thermocouples, rotometers, CO analyzer, manometers, and sharp-edge orifices. The instrumentation was checked out after installation to ensure proper operation. Two probes were designed and fabricated at BCL for this study: the CO sampling probe used in the vacuum exhaust duct (12 in. [0.3 m] diameter) to measure CO concentration level, and a multi-sensor thermocouple probe to measure temperatures within the welding plumes and the plumes generated using the simulator.

CO Probe. In order to determine the efficiency of the exhaust devices, it was necessary to measure the CO concentration level in the vacuum exhaust line. Fig. 1.12 is a schematic of the CO probe used to obtain the gas samples. The sampling holes in the probe were sized and located according to conventional practice in order to obtain an equal volume (mass) sample for equal flow areas in the 12 in. (0.3 m) diameter duct cross section. It was later verified by closing selected sampling holes that the CO concentration within the duct flow cross section was uniform.

Orifice Meter. A standard sharp-edge orifice meter was used to measure the volume flow of gas that passed through the vacuum exhaust system. A 3.6 in. (91.4 mm) and an 8 in. (203 mm) diameter orifice plate were used to cover the range of gas volume flow rates. The calibration curves for both orifices are presented in Fig. 1.13.

Temperature Probe. A temperature probe was used to measure gas temperatures in the plumes generated during arc welding and in the simulator plume. A schematic of the temperature probe is shown in Fig. 1.14. The temperature probe was positioned above the welding and simulator table so that the thermocouple array was in a horizontal plane. An array of eighteen chromel-alumel thermocouples was used to measure plume gas temperature. This large number of thermocouples was needed because of the unsteady flow of the plume. The eighteen thermocouple output signals were recorded simultaneously, with a CEC Light Beam Oscillograph, Model No. 5-124. The temperatures at various distances from the arc were easily determined from the oscillograph traces.

Photography. Pictures of the welding plume trajectories were taken as part of the characterization of the fume flow field. The pictures were taken with an Orbit 4 x 5 in. (102 x 127 mm) camera using Polaroid Type S2 film.

Welding Setup. The important aspects in setting up a welding operation in the laboratory are that field operations are represented as closely as possible and that operational variables can be repeated in a controlled manner.

A welding table having a working area of 24 x 24 in. (0.61 x 0.61 m) at a height of 28 in. (0.71 m) was set up in the laboratory and measurements were made relative to the arc location at the center of the work area. The workpiece was allowed to move opposite the weld direction in a horizontal plane in order to maintain the arc in a constant location for the plume temperature measurements and plume deflection tests.

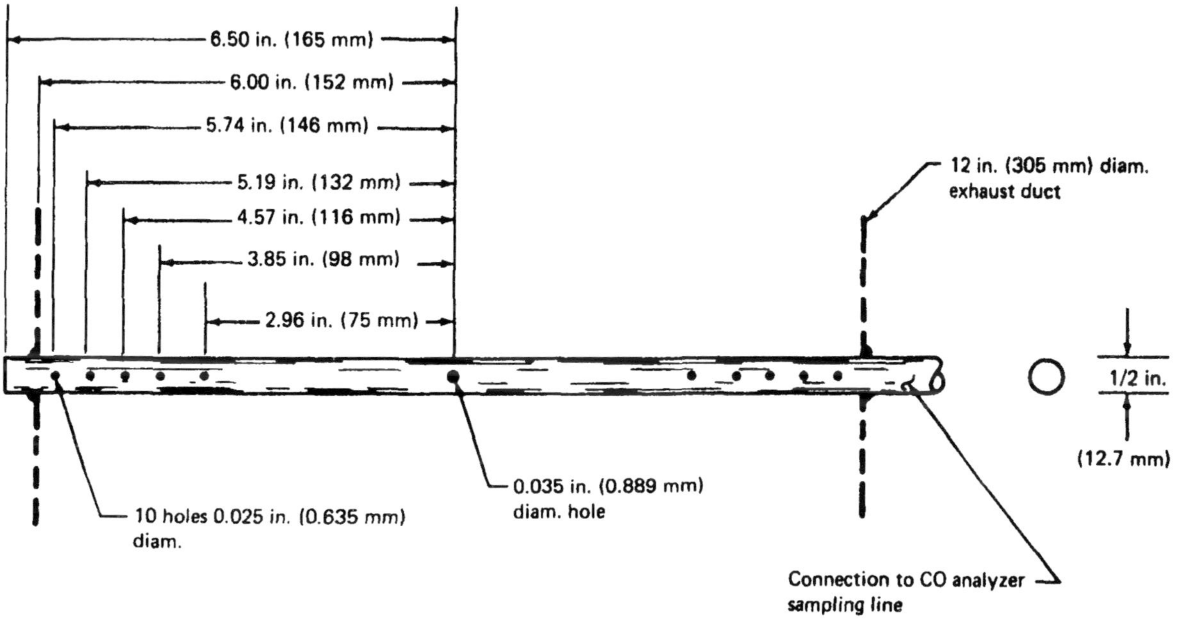

Fig. 1.12 – Details of the CO probe.

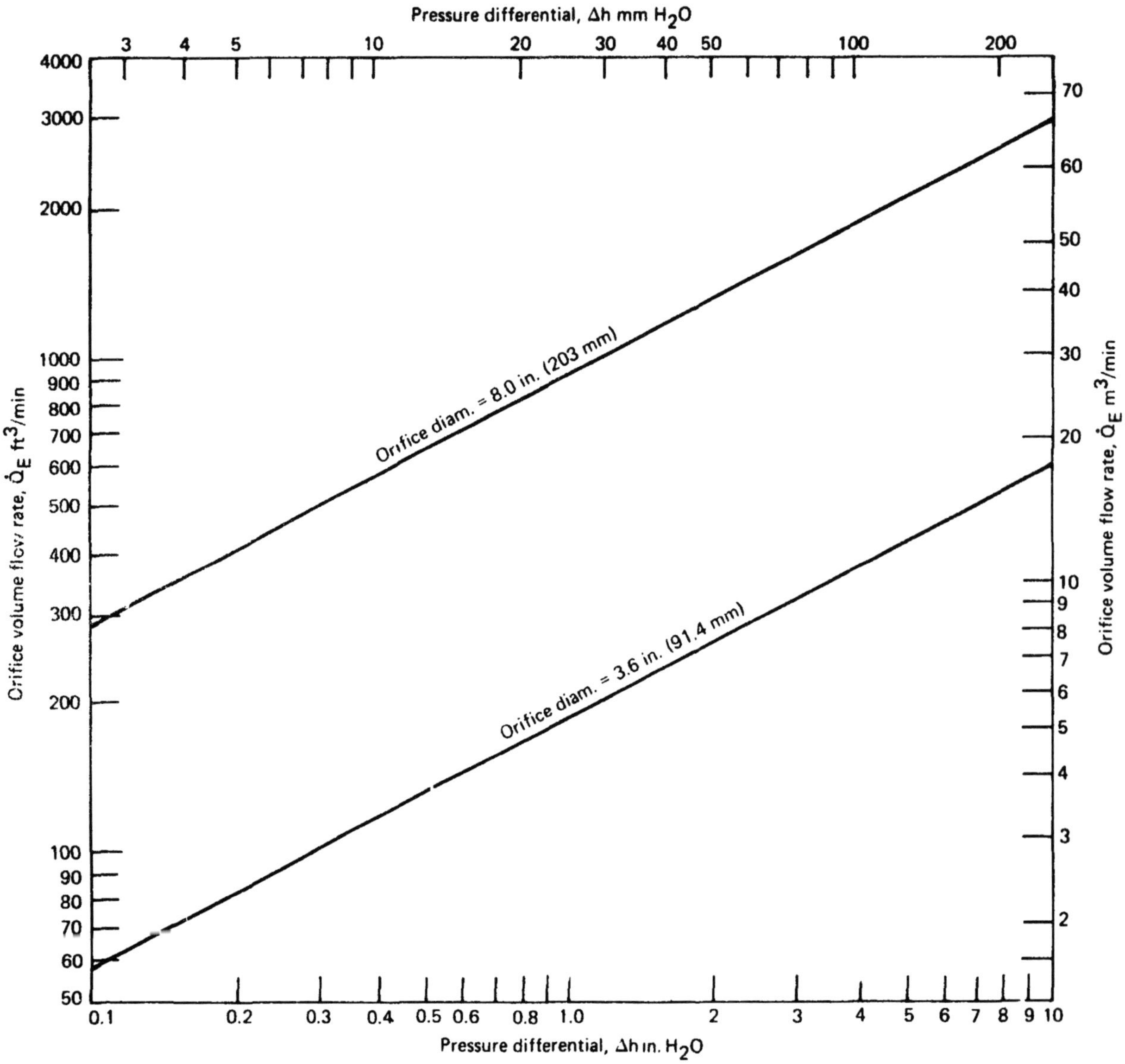

Fig. 1.13 – Calibration curves for the 3.6 and 8 in. (91.4 and 203.2 mm) diam. orifice plates

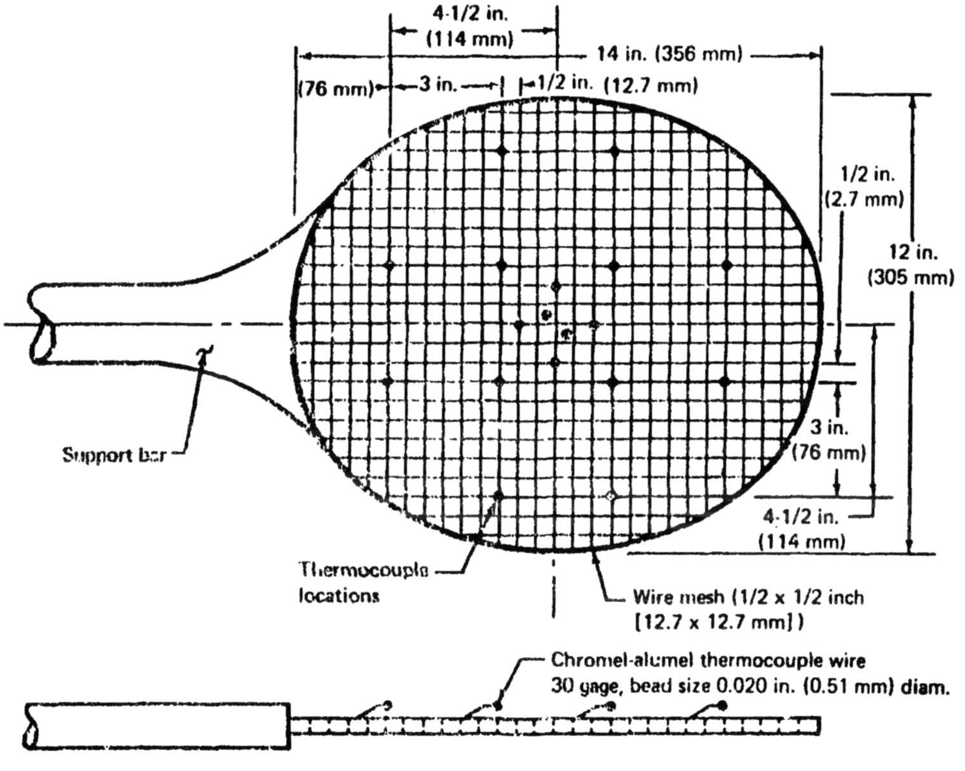

Fig. 1.14–Schematic of the temperature probe

List of Symbols Used in Sections IA, IB, and IC

A_E = Local exhaust inlet area

C = Concentration level (Equation 1)

C_0 = Contamination level in any ventilation air

C_{CO} = Carbon monoxide concentration level (Equation 5)

$C^{\circ}{}_{CO}$ = Carbon monoxide level without exhaust (Equation 7)

C_s = Steady-state concentration level (Equation 3)

C_v = Fume concentration level in welder's breathing zone

C_{v_0} = Fume concentration level in welder's breathing zone for zero ventilation velocity

d_E = Local exhaust duct diameter (Fig. 1.6)

d_F = Local exhaust flange diameter (Fig. 1.6)

d_j = Orifice diameter

$\bar{D}_p$ = Particle diameter (Equation 4)

f = Fume generation rate

h = Slot dimension (Fig. 1.7)

Δh = Orifice differential pressure

m = Particle mass concentration (Equation 4)

n = Particle number concentration

ΔP_H = Differential pressure

$\mathring{Q}_{AIR}$ = Air volume flow rate (Equation 3)

$\mathring{Q}_{CO}$ = Carbon monoxide volume flow rate (Equation 5)

$\mathring{Q}_E$ = Local exhaust volume flow rate

$\mathring{Q}_s$ = Gas flow rate through heater

$\mathring{Q}_v$ = Ventilation rate (Equation 1)

S = Slot spacing (Fig. 1.7)

T_g = Gas temperature in heater

ΔT_f = Plume temperature, fume

ΔT_g = Plume temperature, simulator gas

ΔT_{g_0} = Plume temperature above orifice

V_0 = Zero ventilation velocity

V_∞ = Local ventilation velocity

V_R = Room volume (Equation 1)

X = Horizontal distance from arc location

X_d = Plume deflection distance

X_E = Horizontal distance of local exhaust inlet from arc location (Fig. 1.40)

Z = Height above arc location

Z_E = Vertical distance of local exhaust inlet from arc location

ϵ_R = Room contamination factor (Equation 6)

ϵ_W = Welder exposure factor (Equation 7)

η_E = Exhaust collection efficiency

θ_E = Angle between exhaust device and welder (Fig. 1.40)

θ_v = Local ventilation direction (Fig. 1.23)

ρ_p = Particle density (Equation 4)

Section IA. General Room Ventilation

Fluid Dynamic Considerations

The fumes and gases generated during welding operations are influenced by the operator and interact with the ambient air currents resulting in a complex flow. The nature of the welding process and the resultant unsteady flow of the fumes and gases further complicate the flow.

In this study, the welding plume is defined as the column of gases and particulate matter resulting from the release of large amounts of energy in a very small volume. The particulate matter of importance in the control of weld fumes consists of those particles whose motion is governed by the hot welding gases and ambient air currents. Typically, the characteristic dimension of these particles (effective diameter) is approximately 1 μm or less.

To aid in understanding the physical processes of importance in the generation and control of weld fumes and gases, it is useful to consider the welding plume in terms of zones. These zones are shown schematically in Fig. 1.15 and are referenced from a horizontal plane through the arc for convenience. Three major zones can be identified: (a) arc zone, (b) welder zone, and (c) free zone. The arc zone is the region where the plume flow is predominantly controlled by the energy release and subsequent heating that occurs from the electric arc. The welder zone is defined as the region in which the presence of the welder influences the plume flow. The free zone is above the welder and is the region in which the plume is influenced solely by room air currents.

The fume and gas flows in the arc zone are strongly influenced by the arc discharge process and the intense arc heating process. The path taken by the plume in this zone is erratic (both vertical and horizontal trajectories) and the fume and gas velocities are relatively high in comparison to the corresponding velocities in the welder and free zones. Control of the fumes and gases in this zone requires external devices that can generate high, induced flow velocities in order to influence the high fume and gas velocities. However, inducing high air currents in the arc zone can have an adverse effect on the quality of the weld.

The characteristic movement of the fumes and gases in the welder zone is upward. The horizontal motion of the plume induced in the arc zone damps out quickly and is essentially zero by the time the gases and fumes reach the welder zone. The plume velocities in the welder zone are relatively low (~100 to 500 ft./min [~30 to 150 m/min]) and the gases and fumes would rise vertically if they were not influenced by ambient air currents and the presence of the welder. The major characteristic of the plume in the welder zone is that the velocities (gases and fumes) are low; therefore, the path of the plume can be influenced by air currents of small magnitude (natural room air currents can effect the plume trajectory in this zone). Another important aspect of this zone is that the welder can influence the flow pattern and resulting plume trajectory. Thus, the influence of the welder on the plume must be considered when analyzing the flow characteristics within this zone. This zone can be considered as extending from a few inches from the arc (~3.5 inches [~80-130 mm]) to approximately the elevation of the top of the welder's head. Plume velocities and temperatures decrease in the welder zone as the fumes and gases move upward. It is fairly easy to induce the necessary air currents in this zone to deflect the plume path so that the plume does not engulf the welder's helmet. In the welder zone, the plume must be controlled to protect the welder.

As the welding fumes and gases continue to rise, a well-defined plume is less apparent. In the region above the welder's head, the flow of fumes and gases is predominantly influenced by room air currents. The concentration of fumes and gases will decrease as the fumes and gases mix with the ambient room air. The mixing and diluting process involves many factors associated with the room environment. For example, under circumstances of poor room ventilation, the fumes could collect in the ceiling area. When the welding stops, the fumes would subsequently fall and contaminate the room environment. Also a portion of the fumes would be deposited on surfaces within the room (walls, floors, ceiling, etc.). The long-term room contamination level will depend upon the room ventilation rate, welding process (amount of welding contaminates generated), and the effectiveness of any exhaust system used to collect the welding fumes and gases.

Room Ventilation

The bulk of the fumes generated during welding consists of small particles that remain suspended in the air for a considerable period of time. As a result, their concentration in a closed work area builds up over a period of time, as does the concentration of any gases evolved or used in the welding process. The particles will eventually deposit on the room walls and floor, but the rate at which this occurs is extremely low compared to the rate at which they are generated by welding processes.

Since the particles tend to remain suspended in the air, their concentration can be controlled by general room ventilation. In most cases, general room ventilation is more effective in protecting personnel in the area around the welding operation than it is in protecting the welding operator.

General room ventilation may occur naturally, as when doors and windows are open, or it may require fans and blowers to force and direct the required amount of air through the building or work room. Of course, the effectiveness of general room ventilation is strongly dependent upon the design of the system, be it natural or forced ventilation. The locations at which fresh air is

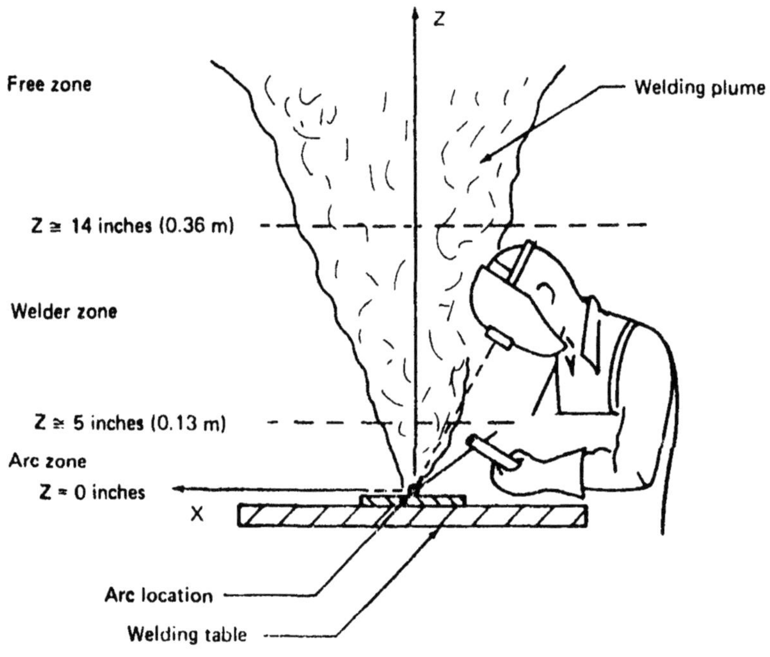

Fig. 1.15–Drawing of welding plume illustrating various zones

introduced and contaminated air is exhausted must be such that the welding fumes and gases are carried away and do not concentrate in dead zones. In some cases, it may be possible to locate the fresh air supply so the incoming air provides the required protection for the welding operator as well as for personnel in the general area.

The concentration of fumes and gases that would exist in any closed room can be accurately calculated as a function of time. The governing differential equation for the contamination level in a closed room is given as

Equation 1: $$V_R \frac{dC}{dt} = -(C - C_0)\,\dot{Q}_V + f$$

where C is the room contamination level, C_0 is the concentration level in any ventilation air, V_R is the room size, $\dot{Q}_V$ is the ventilation rate, and f is the fume generation rate.

Equation 1 applies, assuming that the contaminant is uniformly spread throughout the room, that there is no loss of fume due to deposition on various surfaces (walls, floors, etc.), and that, if the room is ventilated, the ventilation air is uniformly distributed. If it is assumed that the ventilation air is free of any fume contamination, that is, $C_0 = 0$, Equation 1 can be integrated to obtain the following expression for the room contamination level as a function of time.

Equation 2: $$C(t) = 3.53 \times 10^4 \frac{f}{\dot{Q}_V}\left(1 - e^{-\dot{Q}_V/V_R t}\right)$$

with the units for each of the variables being $f \sim$ g/min, $\dot{Q}_V \sim$ std ft^3/min, $V_R \sim$ ft^3, $t \sim$ minutes, and $C \sim$ mg/m^3

The steady-state or maximum room concentration levels can be determined from Equation 2 by assuming an infinitely large time period.[1] With this assumption, Equation 2 reduces to

Equation 3: $$C_S = 3.53 \times 10^4 f/\dot{Q}_V$$

Steady-state fume concentration levels (C_S) calculated using Equation 3 are shown in Fig. 1.16 and apply to rooms of all sizes. The concentrations presented in Fig. 1.16 can also be used to estimate room steady-state fume concentration levels if more than one welding operation is in progress. For example: assuming that there are five welding operators each generating fumes at a rate of approximately 0.6 g/min, and that room ventilation rate is 10,000 std ft^3/min (283.2 std m^3/min)[2], the room concentration level would be 10 mg/m^3 or 5 times the level for a single welding operation. Estimated fume generation rates for various welding processes are presented in

1. The time required to achieve steady-state concentration levels depends upon the room size and ventilation rate.
2. Std ft^3/min = standard cubic feet per minute, measured at 0°C and 1 atm., pressure.

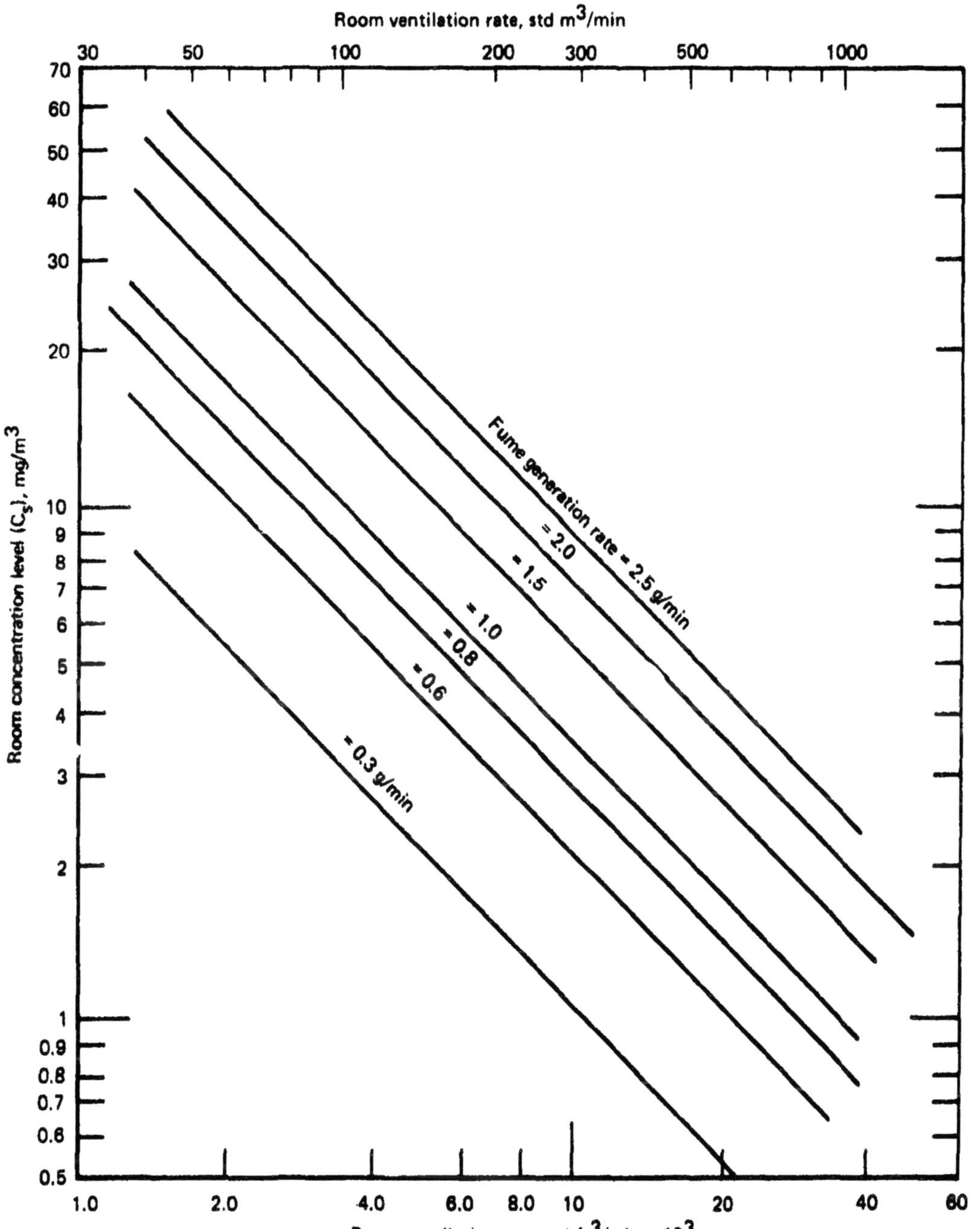

Fig. 1.16—General room ventilation requirements for various welding fume generation rates

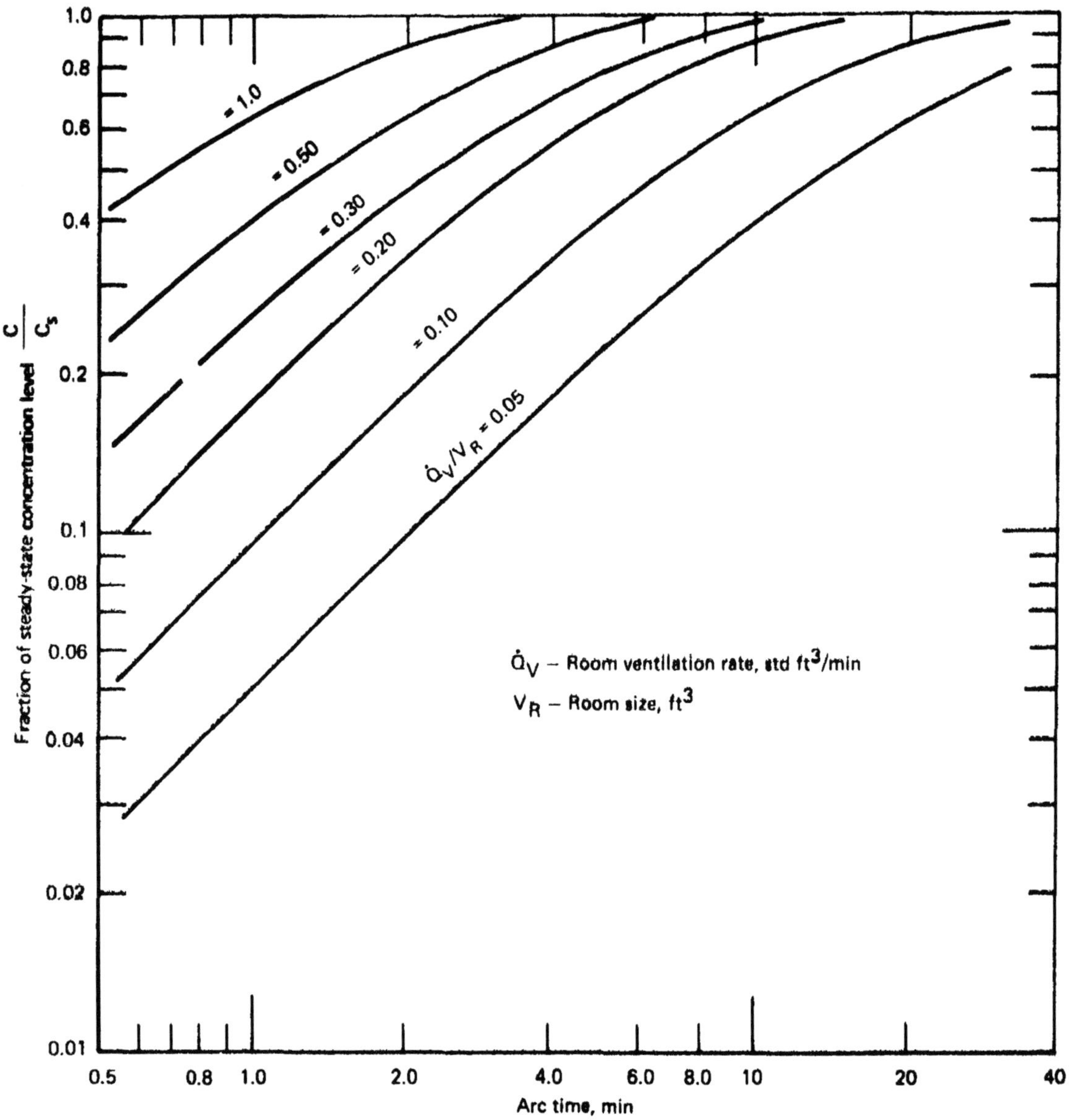

Fig. 1.17–Estimated room concentration level as a function of time for various room sizes and ventilation rates

Part II of this report, and the fume generation rates for a number of the more commonly used processes are given in Appendix A. Fig. 1.16 can also be used to determine steady-state concentration levels for arc times less than 100 percent, which is almost always the case. For arc times less than 100 percent, the time average fume generation rate is used. The time average fume generation rate is obtained by multiplying the fume generation rate, as is shown in Appendix A, by the arc time expressed as a decimal. For example: if, in a one-hour time period, 25 minutes were spent in actual welding, then the arc time would be 25/60 or 0.42.

Using Equation 2, transient concentration levels were determined for time periods less than those required to achieve steady-state levels. Fig. 1.17 shows the room concentration level as a function of time. The concentration level *(C)*, is presented as a fraction of the steady-state value (C_S), and the resulting ratio (C/C_S), is plotted as a function of arc time and $\dot{Q}_V/V_R$ (system time constant). Therefore, in order to estimate room concentration level as a function of time for a particular setup, it is necessary to use both Figs. 1.16 and 1.17.

To illustrate application of Figs. 1.16 and 1.17, the fume concentration in a room was determined assuming

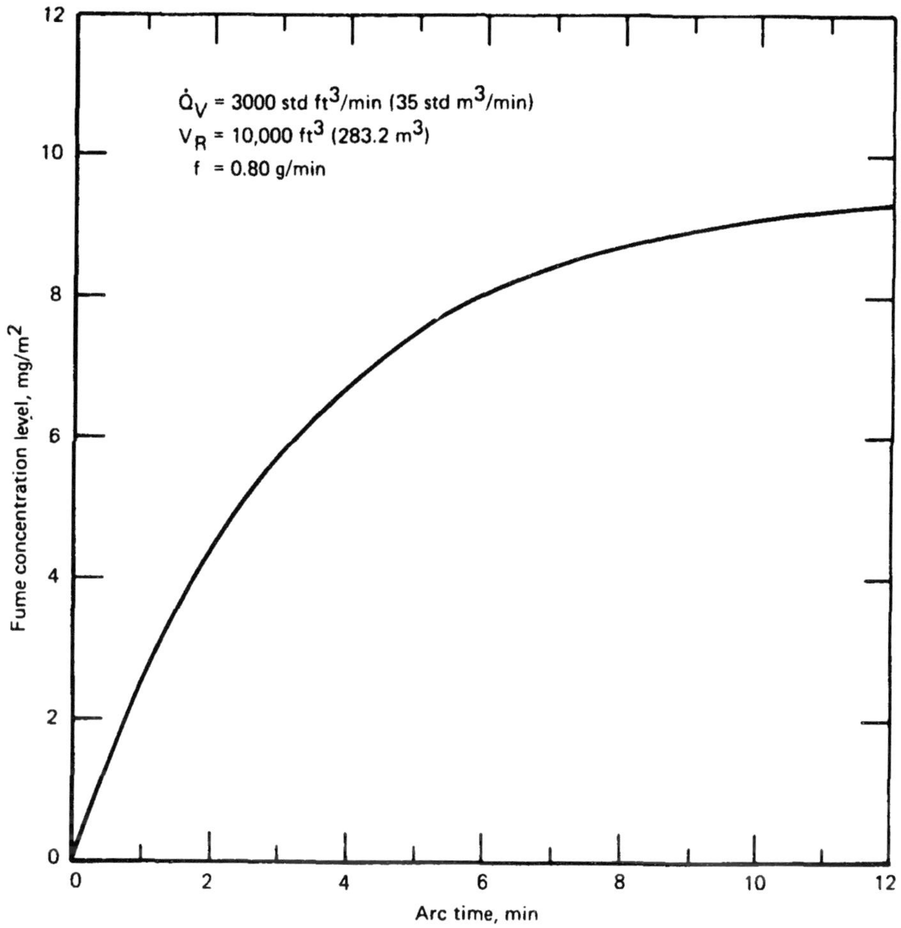

Fig. 1.18–Room concentration level as a function of time for representative room conditions

a room volume of 10,000 ft³ (283 m³), a ventilation rate of 3000 std ft³/min (85m³/min), and generation of welding fumes at a time average rate of 0.8 g/min. Fig. 1.16 shows that the steady-state concentration level for these conditions would be 9.5 mg/m.³ Referring then to Fig. 1.17, it can be seen that for the assumed conditions resulting in a system time constant of 0.3, the room concentration would be one-half of the steady-state value after about 2.5 minutes of continuous welding. Fig. 1.18 shows the room fume concentration level in mg/m³ as a function of time for this example.

Section IB. Local Ventilation

Data Needed

General room ventilation can effectively control contamination levels within a general area but, in many cases, does not provide the local control needed to protect the welding operator. Local ventilation is an effective way of providing this protection. For those cases where either room size, arc time, or natural ventilation provides adequate control of general room contamination, only local ventilation would be needed to protect the welder. Since the data needed to design for local ventilation are not available in the literature and cannot be accurately calculated, they were obtained experimentally.

Experiments were performed to determine the effect of the air currents associated with local ventilation on the fume concentration level in the welder's breathing zone. The natural tendency of welding fumes is to rise vertically, due to heating of the air and gases in the immediate vicinity of the electric arc. The welder's head is usually close to or in the plume above the arc; therefore, the welding fumes will flow around the helmet as they move vertically upward, and a portion of the fumes will also penetrate behind the helmet. The amount or concentration of fumes behind the helmet will be a function of the welder's position, the fume generation rate, and the local ventilation rate and direction. In the experimental studies, the reduction in welder exposure levels was determined over a range of air velocities and flow directions.

Calibration of Equipment

Critical Flow Orifice. The critical flow orifice was calibrated with a wet test meter. During orifice calibration, the filter holder, loaded with a filter, was connected ahead of the orifice to duplicate the actual sampling flow condition. The flow rate for the orifice used was 2.7 liters/min when choked flow existed across the orifice. To ensure that the orifice always operated with choked flow and to check for filter clogging during sampling, the pressure drop across the orifice was set and monitored with a pressure gage and installed downstream of the orifice as shown in Fig. 1.2.

Condensation Nuclei Counter. The validity of using the condensation nuclei counter for measuring fume mass concentrations was confirmed as follows. Since the background aerosol concentration ($\sim 40\ \mu g/m^3$) is negligible compared to the particle concentration caused by welding (typically $\sim 30\ 000\ \mu g/m^3$), the mean particle size should remain the same under various ventilation conditions. For these conditions, the mass concentration can be written as

Equation 4: $m = \frac{\pi}{6} n \rho_p \bar{D}^3{}_p$

where

m = particle mass concentration, mg/m^3

n = particle number concentration, number/m^3

ρ_p = particle density, $mg/\mu m^3$

$\bar{D}_p$ = particle diameter, μm

In Equation 4, it is apparent that the number concentration measured by the particle counter can be directly correlated to the mass concentration. Some measurements were performed to compare the counter output with the actual filter weighing. The comparison, shown in Fig. 1.19, demonstrates that mass measurements obtained using a filter can, indeed, be linearly correlated with the counter output as indicated by Equation 4. Therefore, the real time characteristics of this instrument were used advantageously in the subsequent tests to be described later.

Initially, tests were performed using both the particle counting method and the filter weighing method to measure fume concentration levels. Usually, it was possible to obtain the particle counter data using only one electrode while five electrodes had to be used during the filter weighing measurements to obtain a sufficient sample for weighing. The two measurements were compared and checked for agreement with the calibration curve (shown in Fig. 1.19). The calibration curve was confirmed and subsequent data were taken with the condensation nuclei counter.

Air Current Generator. The ventilation air velocity in the welding zone, 42 in. (1.07 m) from outlet face of the generator, was calibrated in terms of the pressure drop across the packed bed. The calibration curve is shown in Fig. 1.20. This pressure drop was subsequently used to set the air velocity during the experiments.

Experimental Procedure

In conducting the local ventilation experiments, several factors were maintained constant from run to run to eliminate their possible influence on the fume concentration level: (a) welder's posture and head position, (b) initial base metal temperature, and (c) surface contamination. The one important uncontrollable variable was the unsteady flow pattern of natural convection created by the welding arc. This resulted in some variation in the measured values. In order to collect samples as representative and consistent as possible, the following setup and techniques were used.

The welder's head position was controlled by the horizontal bar attached to his helmet (see Fig. 1.5) which assured that his ear was 13 in. (0.33 m) above the arc and 10 in. (0.25 m) away from the arc horizontally. When more than one electrode was needed in order to obtain a measurable fume sample, the base metal was quickly moved after each electrode so the working spot would remain in approximately the same location for each electrode. Therefore, any of the spots coated with slag from the previous electrode were not welded over. All the above procedures were followed because it was found during shakedown tests that the natural fume convection pattern and the ventilation flow pattern can be greatly influenced by these variables.

The experiments were conducted in the following manner: The welder was positioned relative to the direction of the air flow from the generator needed to obtain the desired ventilation flow angle. An arc was struck and power supply adjustments were made to obtain the arc voltage and current specified by the manufacturer for the electrode being used. After the power supply was properly set, welding was started and samples were collected. When more than one electrode was required in order to obtain a measurable sample, fume sampling was performed only during the times when welding was in progress. Therefore, the results apply to 100 percent arc time.

Fume concentration measurements were performed in the welder's breathing zone for a number of ventilation velocity conditions (magnitudes and directions). Fume measurements were also performed under conditions of zero ventilation. The results of the fume measurements follow.

Experimental Results

In order to determine the repeatability of the fume concentration measurements, eight filter measurements

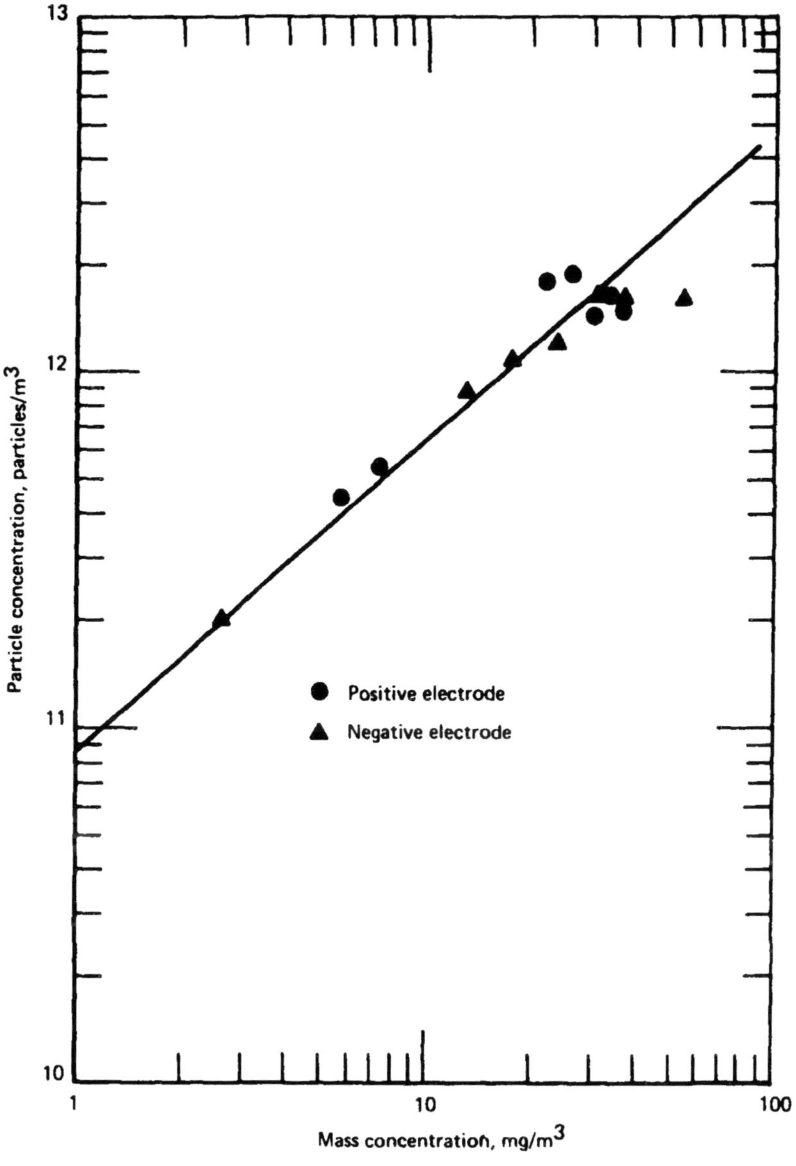

Fig. 1.19–Comparison of condensation nuclei counter and filter weighing results

were made with no ventilation. The results on the tests are shown in Fig. 1.21. The SMAW process was used during this test series using a 5/32 in. (4 mm) diameter E6013 electrode (Code 4). A welding current of 150 A and an arc voltage of 25 V were used for both polarities. The average welding time for each electrode under these conditions was approximately 70 seconds. At least five electrodes were needed to obtain measurable fume mass on the filter. The results shown in Fig. 1.21 indicate that there is a 25 percent scatter in the data.

Fig. 1.22 is a representative recorder trace of the output of the condensation nuclei counter as a function of time. The trace shows that the fume concentration in the welder's breathing zone quickly increases in the first 10 seconds after an arc is struck. The welder's breathing and the intrinsic unsteady flow pattern associated with the welding process cause the fluctuations observed in the counter signal. In reducing the data from the counter output, the area below each curve was integrated to obtain a time average value of fume concentration level.

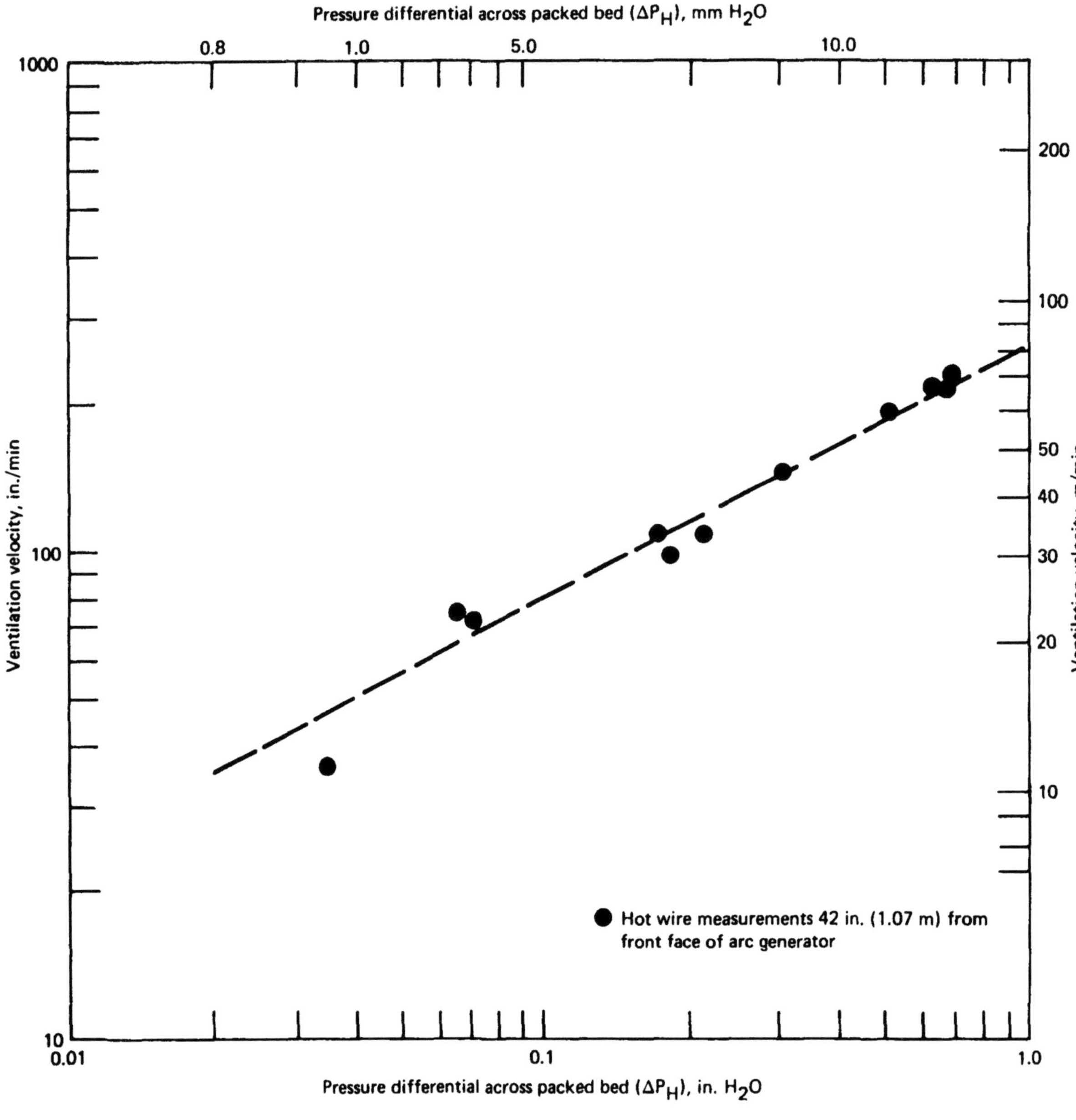

Fig. 1.20–Calibration data for the air current generator

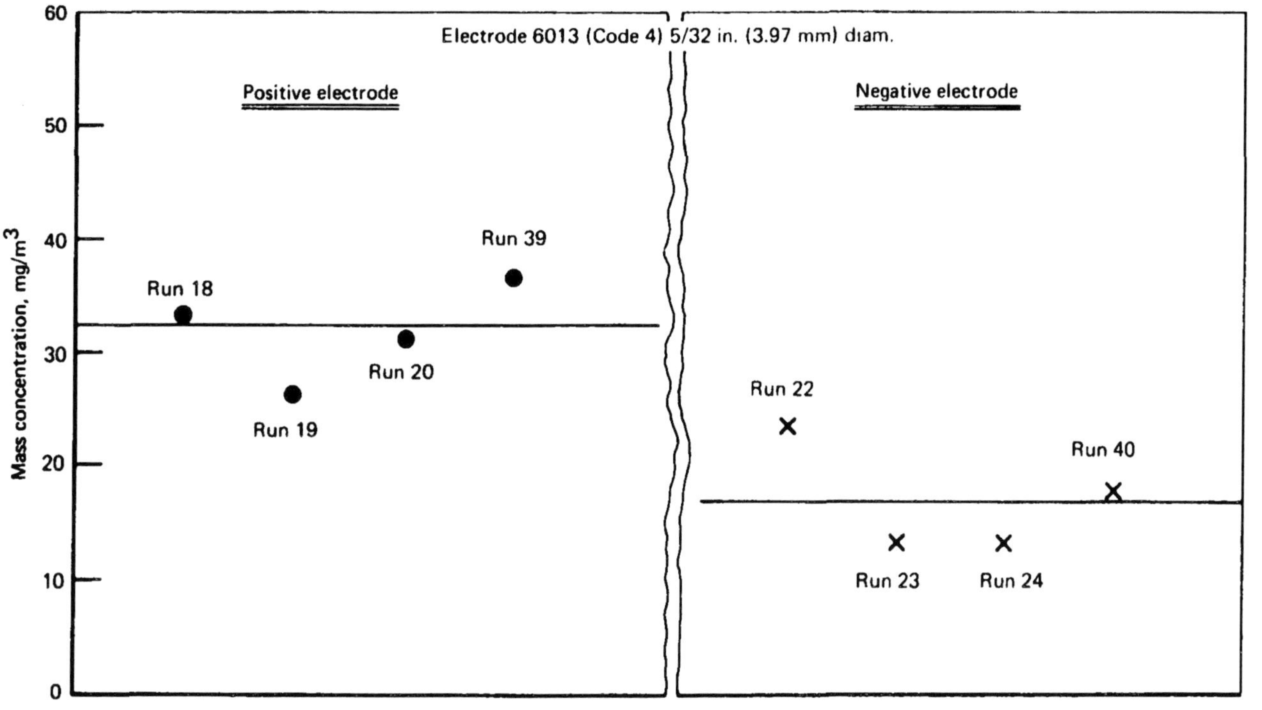

Fig. 1.21 – Results of fume concentration repeatability tests

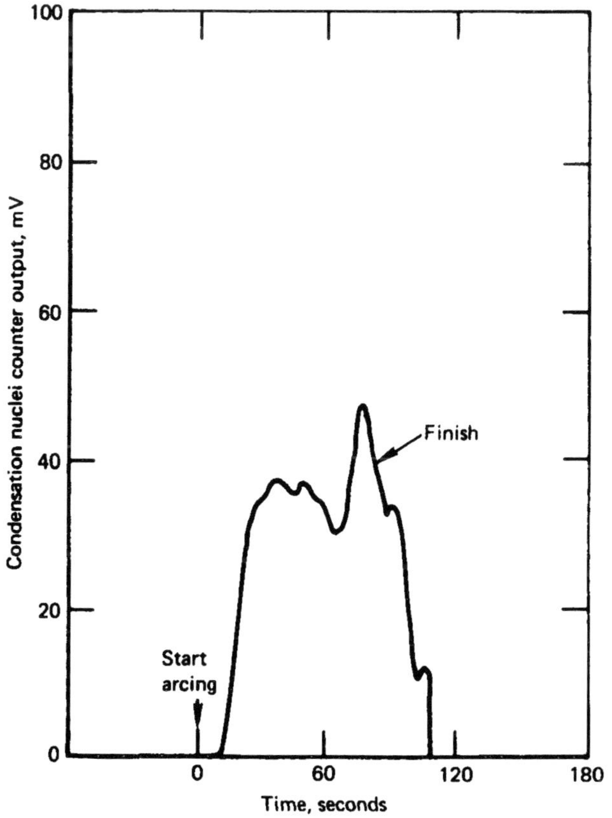

Fig. 1.22 – Representative strip chart recording of the condensation nuclei counter output signal

The reduction in fume concentration level in the welder's breathing zone is presented as a function of ventilation conditions (air speed and direction) in Figs. 1.23, 1.24, and 1.25. The fume concentration level is expressed as a percentage ratio of the fume concentration level obtained with ventilation air to that measured for zero ventilation velocity. The results from the experiments conducted with the air flow from the welder's left, θ_V = 90,° are shown plotted in Fig. 1.23 for ventilation velocities from zero to 100 ft/min (0-30 m/min). The fume concentration level drops quite rapidly from the level at zero ventilation to approximately 10 percent at about 80 ft/min (24.4 m/min). It can be assumed that the fume concentration level will decrease even further as the ventilation velocity is increased. The fume concentration level for electrode negative polarity is approximately 50 percent of the electrode positive polarity level at zero ventilation velocity. The fume concentration level for zero ventilation (C_{V_0}) was 37 mg/m^3 when operating with electrode positive polarity.

The fume concentration measurements performed with the air current generator behind the welder (θ_V=0°) and in front of the welder (θ_V=180°) are presented in Fig. 1.24. The results for θ_V=0° (ventilation from behind the welder) show that there is an increase of approximately 20 percent in the fume concentration level as the ventilation velocity is increased from zero to 40 ft/min (12.2 m/min). From 40 ft/min (12.2 m/min) to 200 ft/min to (61.0 m/min), the fume concentration level decreases linearly by approximately 80 percent and 60 percent for the elec-

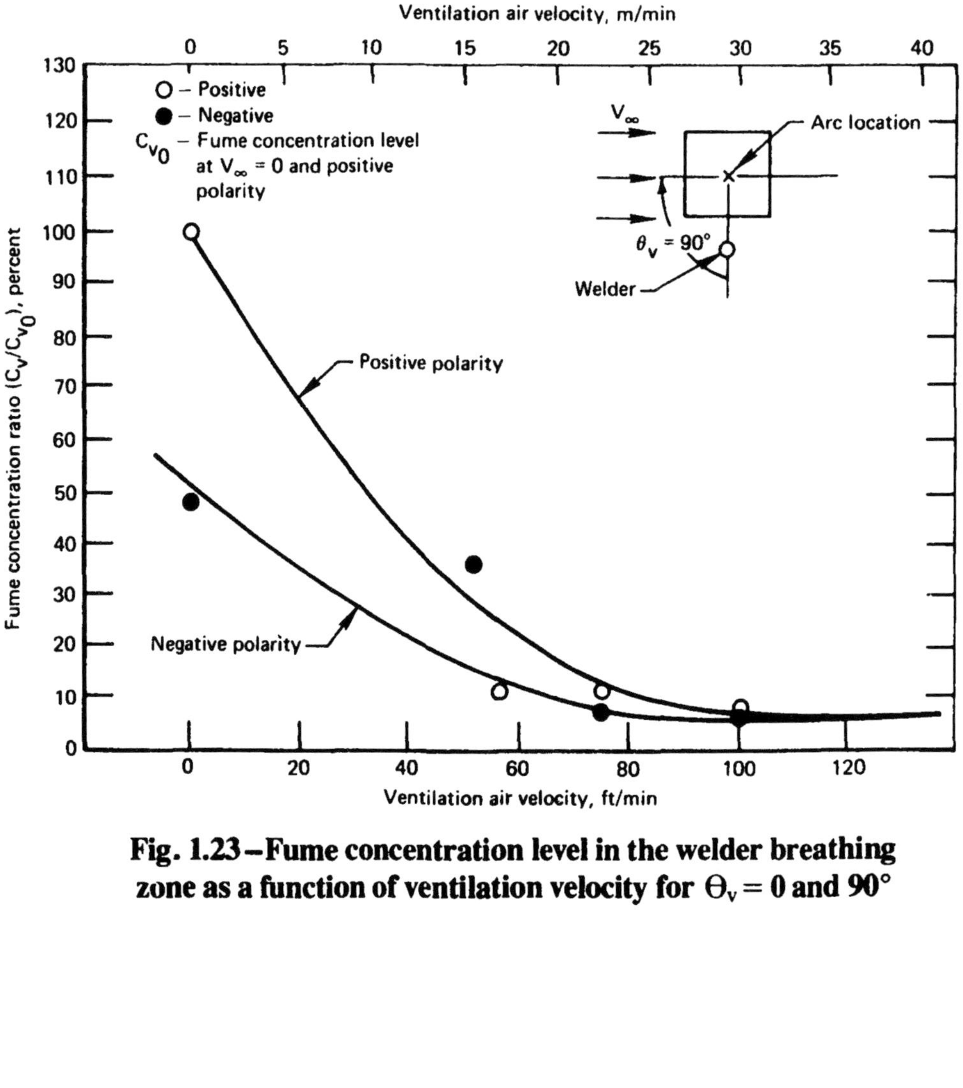

Fig. 1.23–Fume concentration level in the welder breathing zone as a function of ventilation velocity for $\Theta_v = 0$ and 90°

trode positive and negative configurations, respectively. The results obtained for $\theta_V = 180°$ (ventilation from in front of the welder) as a function of ventilation velocity showed an increase in fume concentration level of 20 to 30 percent. The fume concentration level does not decrease as the ventilation rate is increased.

The data for ventilation from directions of $\theta_V = 45°$ and 225° are presented in Fig. 1.25. It is evident from the data plotted in Fig. 1.25 that the fume concentration level is not reduced appreciably (20 percent) until the ventilation rate is increased above 110 ft/min (33.5 m/min).

In order to highlight the dependence of the fume concentration level on the direction of the ventilating air flow, the data at a velocity of 50 ft/min (15.2 m/min) are shown as a function of ventilation direction (θ_V) in Fig. 1.26. It can be observed that the fume concentration is strongly dependent upon the flow direction. A cross draft ($\theta_V = 90°$) is found to be the most effective in moving fumes from the welder's breathing zone. If this direction is not possible for a particular working condition, an angle of 45° (or 315°) between the welder and the flowing air position is the next best configuration.

Fig. 1.27 is a similar plot for a higher ventilation rate, and, generally, the same conclusions concerning the direction of air flow can be made.

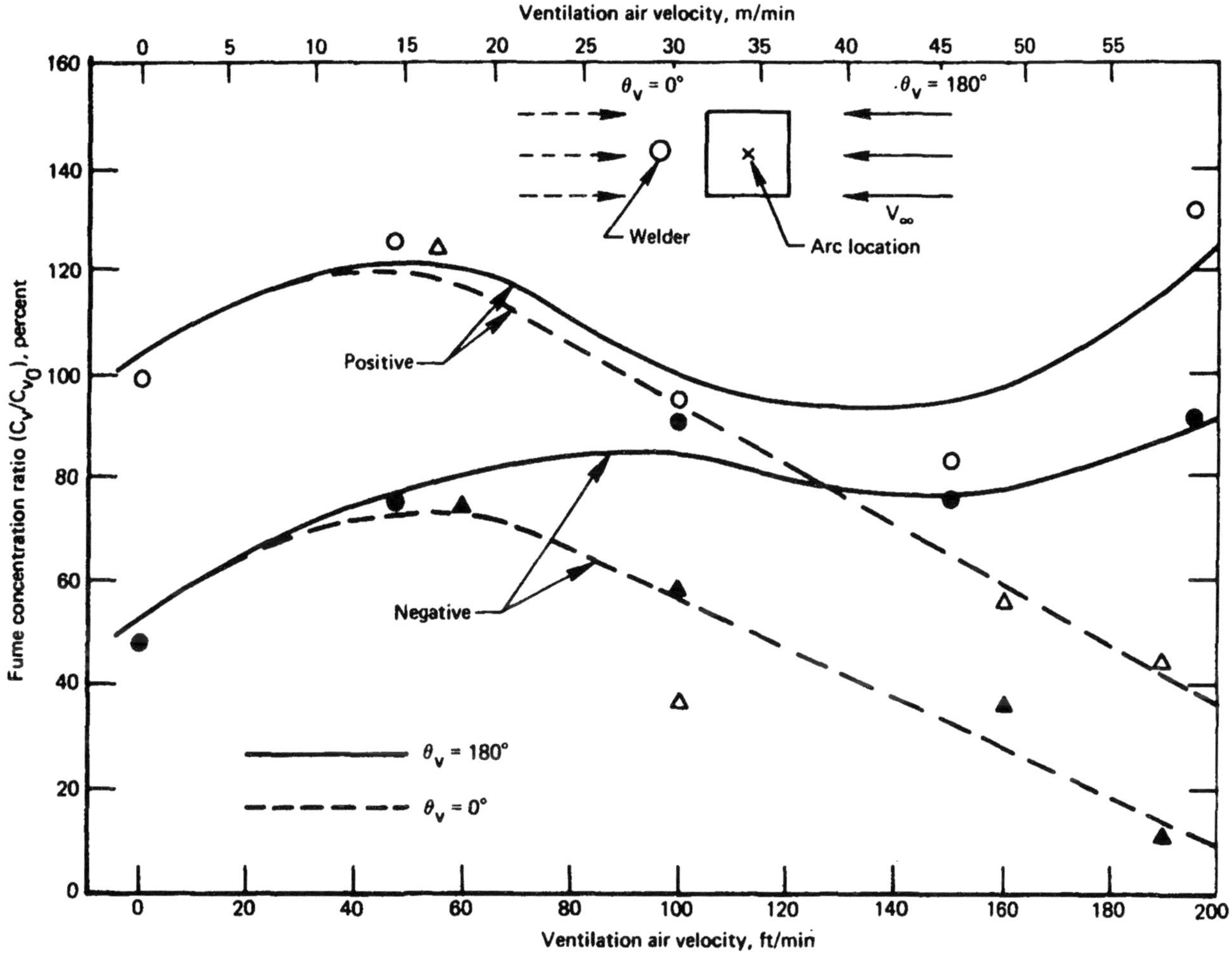

Fig. 1.24–Fume concentration as a function of ventilation air velocity for $\Theta_v = 0$ and 180°

Section IC. Local Exhaust Studies

The results presented in Section *IB*, Local Ventilation, showed how local air currents can be used to control fume concentration in the welder's breathing zone. It is evident that the environment in which welding is performed will rise in fume concentration unless general ventilation is employed. In the local exhaust experiments discussed in this section, the performances of various exhaust inlet designs are evaluated based on their effectiveness in collecting and reducing the welder's exposure to the fumes. The experimental variables were collector configuration, collector location, and exhaust flow rate.

Prior to the local exhaust experiments, two other experimental test series were carried out: (1) characterization of welding plumes for various welding processes, and (b) determination of the simulator operating conditions required to duplicate actual welding plumes.

Details of the test apparatus used, the test procedures followed, and the results obtained are presented in this section.

Calibration of Equipment

The CO probe and sampling train were checked out by introducing a metered amount of CO into the vacuum supply system and varying the air volume flow rate through the vacuum system. The CO concentration measured at the probe location should have been equal to

Equation 5: $$C_{CO} = \frac{\mathring{Q}_{CO}}{\mathring{Q}_{CO} + \mathring{Q}_{AIR}}$$

where C_{CO} is the CO concentration level and $\mathring{Q}_{CO}$ and $\mathring{Q}_{AIR}$ are the volume flow of carbon monoxide and air, respectively, if the probes were collecting a sample having an average concentration.

Figure 1.28 shows the CO concentration levels based on Equation 5, the known amount of CO introduced, and the measured CO concentrations. Agreement between the predicted and actual values is excellent, indicating that samples representative of the average concentrations were being obtained. A Beckman CO Analyzer, Model No. 315B, was used to measure the CO concentration level. This instrument was calibrated using factory calibration gases that contain precisely measured percentages of CO. Calibration curves for this instrument are given in Fig. 1.29.

Figure 1.13 shows the calibration curves for the orifice plates.

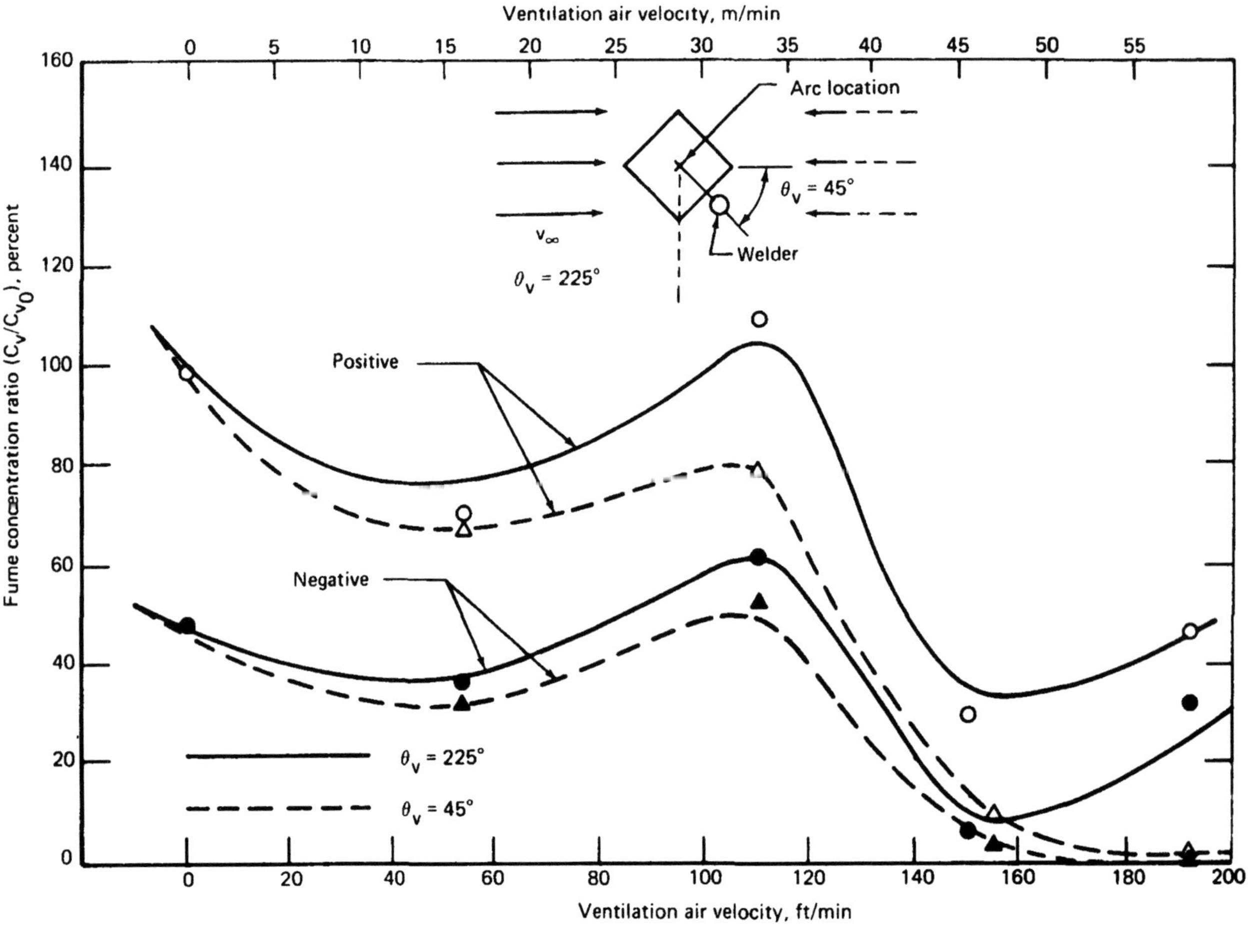

Fig. 1.25–Fume concentration as a function of ventilation air velocity for $\Theta_v = 45°$ and 225°

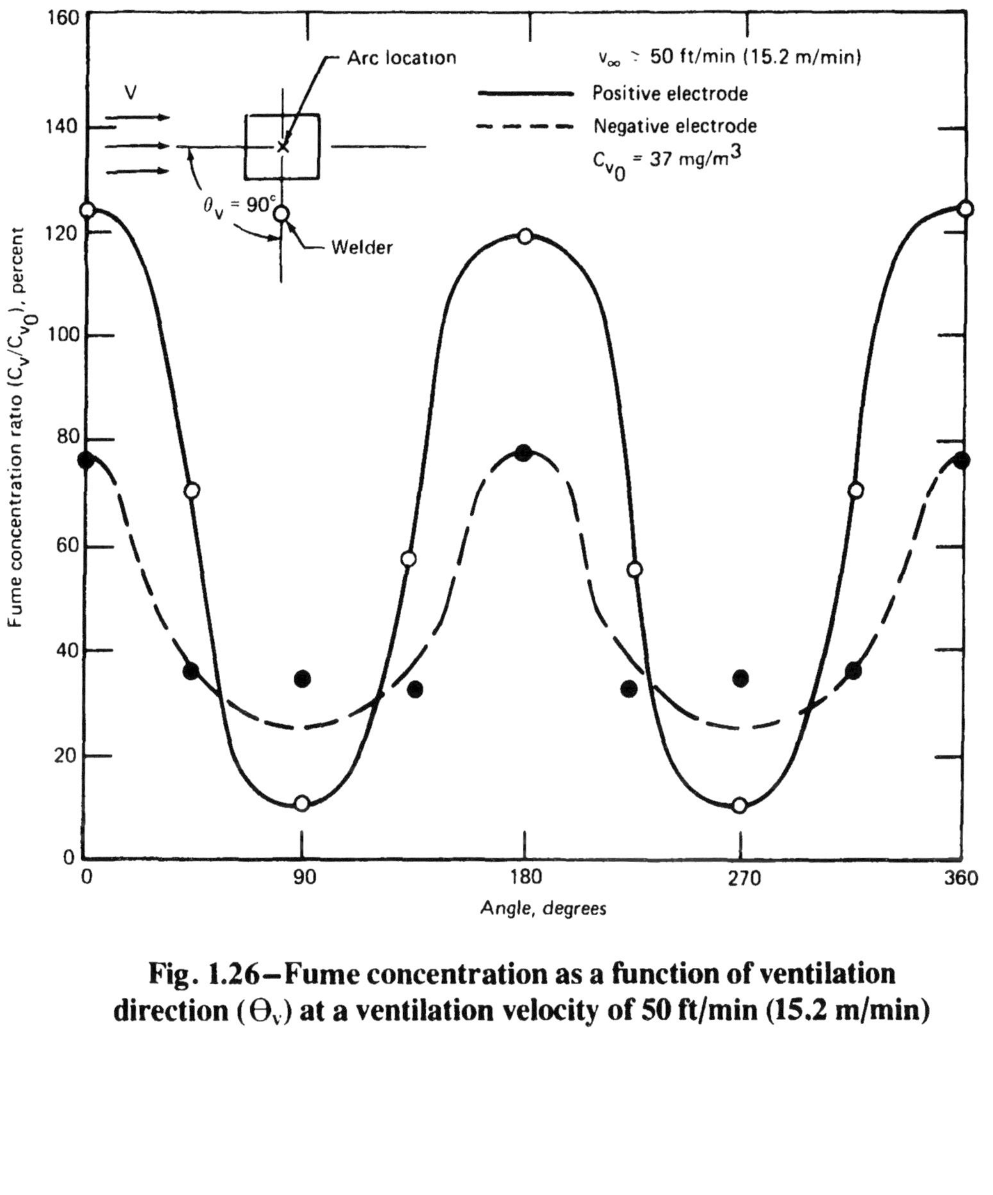

Fig. 1.26–Fume concentration as a function of ventilation direction (Θ_v) at a ventilation velocity of 50 ft/min (15.2 m/min)

Experimental Procedure

Three types of experiments were conducted as part of the local exhaust studies: (a) welding fume characterization experiments, (b) simulator characterization experiments, and (c) local exhaust system experiments. The first set of experiments was conducted in order to obtain baseline data on the welding plume temperature and deflection characteristics. The second set of experiments was conducted to establish the simulator operating conditions required to duplicate fume flow conditions measured in the first set of experiments. The third set of experiments was conducted to evaluate the performance of various inlet exhaust configurations. The procedures followed in performing these experiments are described below.

Fume Characterization. There are many welding processes in use throughout the welding community. Associated with these many processes is an even larger number of techniques that are used in welding operations. The test matrix of welding processes selected in this study as a test base included those processes most commonly used by the welding community that generate welding fumes at rates covering the spectrum associated with most welding processes.

Table 1.1 is a list of the welding processes that were used to characterize the thermodynamic and fluid dynamic behavior of welding plumes. The welding was conducted using the electrode manufacturers' recommended values of arc voltage and current. The test matrix includes shielded arc welding, flux cored arc welding (automatic), and gas metal arc welding (auto-

Table 1.1
Welding test matrix

Electrode type	Code no.	Electrode diam. in.	Electrode diam. mm	Arc current, A	Arc voltage, V	Welding speed in./min	Welding speed m/min	Wire feed rate in./min	Wire feed rate m/min	Shielding gas	Shielding gas flow rate ft^3/h	Shielding gas flow rate m^3/h	Electrode stickout in.	Electrode stickout mm
Shielded metal arc welding														
E6010	1	5/32	3.97	150	26	9	0.23	---		---	---		---	
E7018	10	5/32	3.97	170	23	9	0.23	---		---	---		---	
E7024	7	5/32	3.97	210	40	8	0.20	---		---	---		---	
E316-16	25	5/32	3.97	150	27	8	0.20	---		---	---		---	
E7018	10	3/32	2.38	65	21	6	0.15	---		---	---		---	
E7018	10	3/16	4.76	220	22	9	0.23	---		---	---		---	
Flux cored arc welding (automatic)														
E70T	40	3/32	2.38	480	30	16	0.41	205-210	5.21-5.33	CO_2	45-50	1.27-1.42	1-1/2	38.1
E70T-4	56	3/32	2.38	390	30	16	0.41	290-300	7.37-7.62	---	---		2-3/4	69.9
Gas metal arc welding (automatic)														
E70S-3	57	0.045	1.14	285	35	16	0.41	430	10.92	$AR\text{-}2O_2$	45-50	1.27-1.42	1.0	25.4
ER5356	60	0.045	1.14	165	23	16	0.41	360	9.14	AR	45-50	1.27-1.42	3/4	19.1

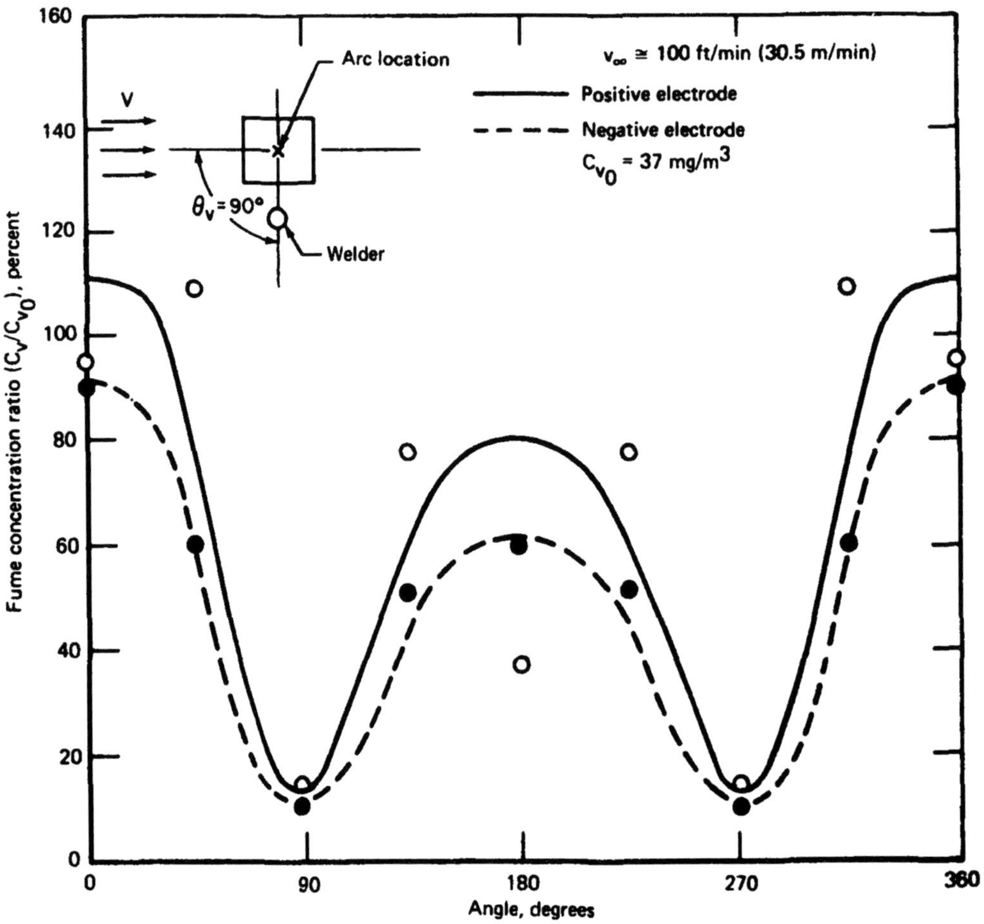

Fig. 1.27—Fume concentration as a function of ventilation direction (Θ_v) at a ventilation velocity of 100 ft/min (30.5 m/min)

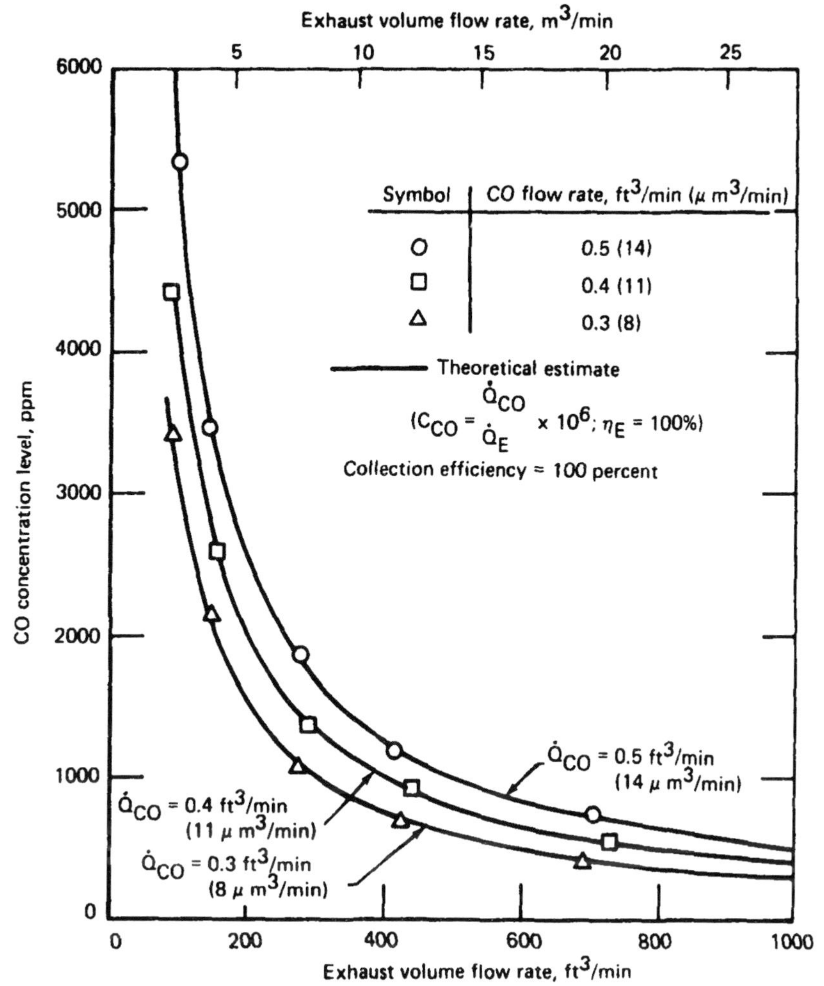

Fig. 1.28–CO probe calibration data

matic). Ten different electrodes were examined. The "code number" designations shown in the test matrix correspond to different electrode manufacturers.

Accepted criteria for establishing the similarity of plumes (such as a weld plume) require an equivalence between buoyancy forces (which are thermally induced) and momentum forces. In these tests, the buoyancy forces of the weld plume were matched with the simulator by ensuring similar plume temperature distributions with respect to height above the arc. The momentum forces of the weld plume were matched with the simulator by ensuring similar plume deflection in a given air current.

Accordingly, two properties of the plumes were measured during the characterization experiments: temperature profile and flow path in a known crosswind. Each welding process was performed alternately by using one of the six pieces of 1/2 in. (12.7 mm) thick mild steel plates. This maintained the plate temperature at a relatively constant level. Positioning the welder out of the fume flow during these experiments provided the desired unobstructed flow field that facilitated characterization of the plume temperature and deflection. It should be noted that the welder was included in the experiments to evaluate the effectiveness of the various local exhaust devices.

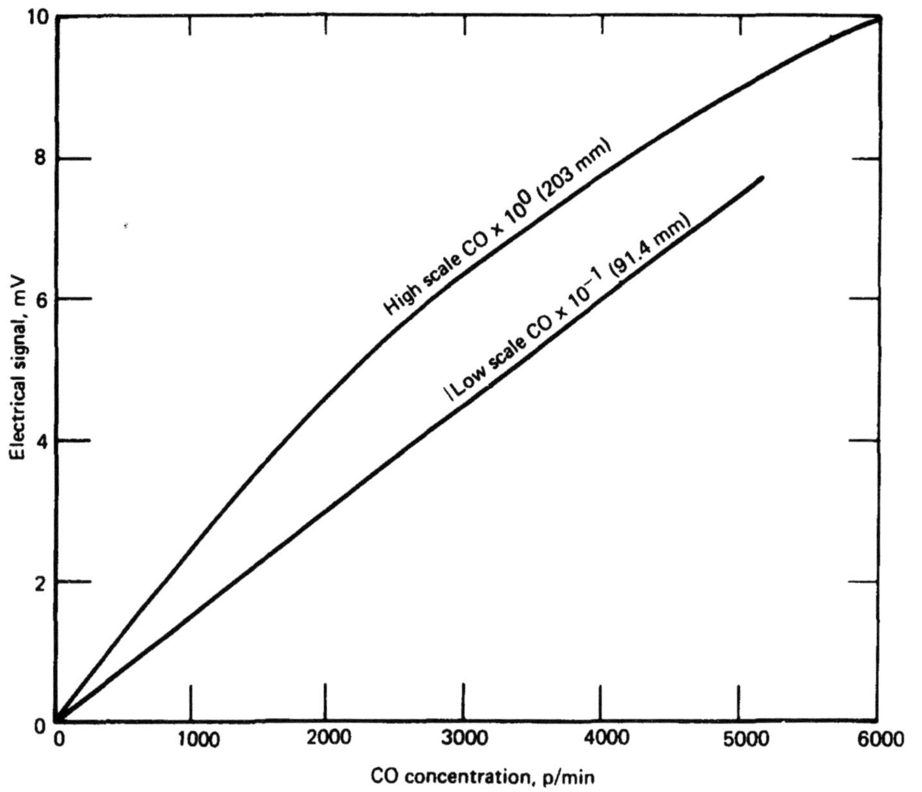

Fig. 1.29—Calibration curves for the Beckman CO analyzer

Temperature measurements were obtained using the temperature probe at various heights above the arc between 5 and 15 in. (130 and 380 mm). The maximum temperature at each elevation was determined from the oscillograph recordings of the temperature probe thermocouples. After the temperature data were obtained for a given welding process, pictures of the plume were taken to define the fume flow path (trajectory). The hot fumes and gases resulting from welding have a natural tendency to rise because of the upward buoyancy force resulting from a difference in density between the hot welding gases and the air. If no external forces acted on the rising fumes and gases, they could continue to rise in a vertical direction with little or no sideward motion until the plume was cooled by mixing with the surrounding air. The air current generator discussed previously was used to generate a controlled, uniform air flow. An air current of 80 ft/min (24.4 m/min) was used for these tests. This air speed was found to deflect the plume out of the region of the welder's helmet in the local ventilation experiments and to provide sufficient plume deflections for accurate measurement. Photographs of the plumes generated from each of the welding processes were obtained and analyzed to determine the plume deflection at a height 25 in. (635 mm) above the arc. Figure 1.30 is a representative plume photograph.

Simulator. Simulator-generated plumes were studied for a wide range of simulator operation conditions. These studies showed that the simulator could be used to produce plumes that duplicated those of hot gases and fumes generated during various welding processes.

The simulator-generated plume experiments were conducted in the following manner: A desired gas flow to the heater was set, and the heater gas temperature was allowed to reach a steady-state value. Plume temperatures were measured at various heights above the heater outlet orifice. These measurements were performed using the same instrumentation (temperature probe and recorder) and procedure discussed in "Fume Characterization." After the temperature data were obtained, a quantity of smoke was introduced into the heater plenum. The smoke made it possible to see the plume so that photographs could be taken. The air current generator was then used to generate a current that deflected the plume. The plume deflection experiments were conducted in the same manner as those conducted during the fume characterization experiments.

Figure 1.31 is a representative photograph showing plume deflection for a cross air current of 80 ft/min (24.4 m/min).

This procedure was repeated until sufficient data were obtained to indicate that the envelope of conditions characteristic of actual welding fume flows could be duplicated and to define the set points for later use. A comparison of the results obtained from the weld fume experiments and simulator experiments is included in the subsequent "Simulator Results."

Local Exhaust System Tests. The experimental procedure discussed herein was followed during studies evaluating the performance of various local exhaust inlet configurations. It should be noted that the study was limited to evaluation of inlet configurations and did not include evaluation of total exhaust system performance, which includes such things as filters, ducting, and blower designs. The present study, with respect to experiments discussed here, pertains to the flow characteristics that exist in the region between the local exhaust inlet and the area near the welder. The specific objective of these experiments was to obtain the data needed to define the effectiveness of various exhaust inlet configurations in protecting the welder from welding fumes and reducing environmental (room) fume concentration levels.

Two basic measurements were required for defining the effectiveness of a specific exhaust inlet design: CO concentration measurements in the exhaust system duct and CO concentration measurements in the welder's breathing zone behind the helmet. The simulator was run using 100 percent CO for these tests. The concentration levels in the duct were used to calculate the room contamination factors for the various local inlet exhaust configurations that were examined. Under steady-state

Fig. 1.30–Typical photograph of welding plume deflection caused by an 80 ft/min (24.4 m/min) air current

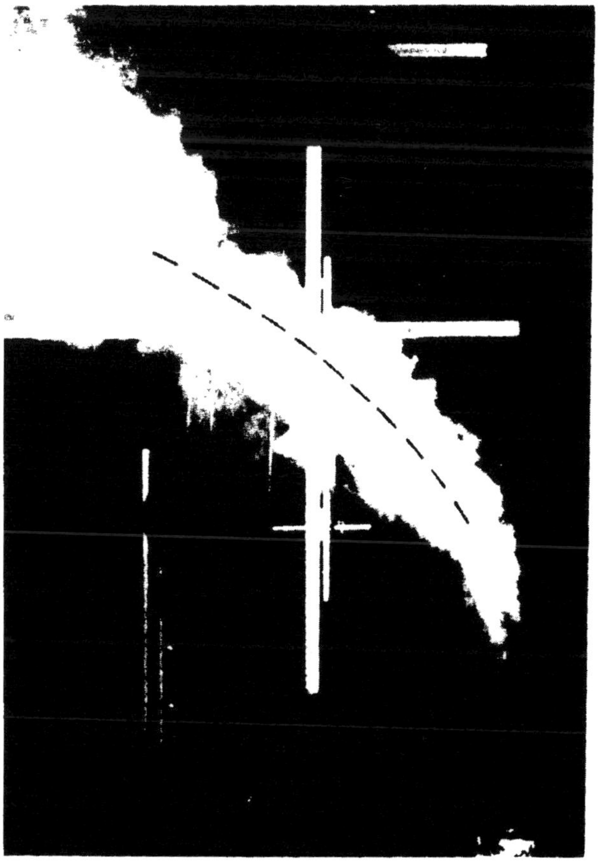

Fig. 1.31–Photograph of simulator plume deflection caused by 80 ft/min (24.4 m/min) air current

conditions, the room contamination factor can be expressed as

$$\text{Equation 6: } \epsilon_R = 1 - \left(\frac{C_{CO}\mathring{Q}_E}{\mathring{Q}_{CO}}\right) \times 10^{-6}$$

where C_{CO} is the carbon monoxide concentration level in ppm, $\mathring{Q}_{CO}$ and $\mathring{Q}_E$ are the volume flow rates of CO and gas through the exhaust in ft³/min (m³/min), respectively. In Equation 6, the term within the parentheses is the collection efficiency. Under actual welding conditions, ϵ_R is a measure of the fraction of the total fume being generated that escapes into the room. When all of the fumes and gases are collected, ϵ_R is zero.

The concentration levels measured behind the helmet were used to calculate a welder exposure factor defined as

$$\text{Equation 7: } \epsilon_W = \frac{C_{CO}}{C^{\circ}_{CO}}$$

where C_{CO} and C°_{CO} are the carbon monoxide concentration levels (in ppm) measured behind the helmet with and without the exhaust system running, respectively. The welder is fully protected from fumes when ϵ_W is zero in Equation 7. The ideal situation is to minimize both ϵ_R and ϵ_W.

For each plume collection device study, the simulator was operated at conditions that duplicated a range of welding processes. The local exhaust inlet was moved to various locations relative to the arc and the two basic CO concentration measurements discussed above were made. At the beginning of each test sequence (specific inlet configuration exhaust volume flow rate and simulator flow), two measurements were always performed: (a) prior to turning on the exhaust system, the CO concentration behind the welding helmet was measured; and (b) after the vacuum exhaust system was turned on, the exhaust inlet was moved very close to the heater outlet so that 100 percent collection was ensured.

These tests were conducted for two reasons: to obtain welder CO exposure without an exhaust system and to check that the system (CO measurement, volume flow, and CO flow rate) was operating satisfactorily. If any of the system components were not set properly, the proper CO concentration level for 100 percent collection would not have been obtained. This test procedure was followed for evaluation of the performance of each of the exhaust inlet configurations.

An outline of the test matrix of conditions used for each inlet configuration is presented in Table 1.2. The ranges of test parameter values and total number of tests for each configuration are given in Table 1.2. It should be noted that a complete test matrix was not run for all the inlet configurations. A sufficient number of tests was performed for each configuration so that the desired characteristics could be obtained.

The variables tested were:

(1) Flanged duct diameter
 3 in., at exhaust angles of 53°, 90°, and 180° (40 tests)
 6 in., at same three exhaust angles (68 tests)
 8 in., at 90° exhaust angle (12 tests)
(2) Single-slot orientation
 horizontal at 90° and 180° (25 tests)
 vertical at 180° (6 tests)
(3) Double-slot oriented horizontally at 90° and 180° (8 tests)
(4) Hood location
 overhead, oriented 180° (8 tests)
 slanted overhead at 180° (3 tests)

For each condition, the exhaust volume flow rate was varied over at least a two-fold range: the horizontal distance from the exhaust inlet to the simulated arc was varied over at least a ×1.7 range (except for the slanted hood); and the vertical distance to the arc was varied over at least a ×1.5 range (except for the vertical slot and the overhead hoods).

Table 1.2
Outline of local exhaust test matrix

Local exhaust inlet configurations	Exhaust volume flow rate, Q_E		Horizontal distance from simulated arc location, X_E		Vertical distance from simulated arc location, Z_E		Angles between exhaust device and welder, Θ_E	Total number[1] of tests
	std ft³/min	std m³/min	in.	mm	in.	mm	degrees	
3 in. (76 mm) diam. flanged duct	95-175	2.7-5.0	4-1/2-18	114-457	5-1/4-12	133-305	53, 90, 180	40
6 in. (152 mm) diam. flanged duct	95-175	2.7-5.0	6-23	152-584	7-1/2-15-1/2	191-394	53, 90, 180	68
8 in. (203 mm) diam. flanged duct	95-175	2.7-5.0	10-17	254-432	9-14	229-356	90	12
Single-slot (horizontal)	95-500	2.7-14.2	6-16	152-406	4-10	102-254	90, 180	25
Single-slot (vertical)	175-500	5.0-14.2	9-16	229-406	...		180	6
Double-slot (horizontal)	175-500	5.0-14.2	9-16	229-406	Slot heights 6, 10	152, 254	90, 180	8
Overhead hood	300-750	8.5-21.2	13-36	330-914[2]	...		180	8
Slanted overhead hood	300-750	8.5-21.2	30	762[2]	...		180	3

1. Calibration runs and 100 percent collection runs to check CO probe function, and runs to obtain welder exposure for zero exhaust flow conditions are not included in "Total number of tests."

2. Distance between front edge of hood and simulated arc location.

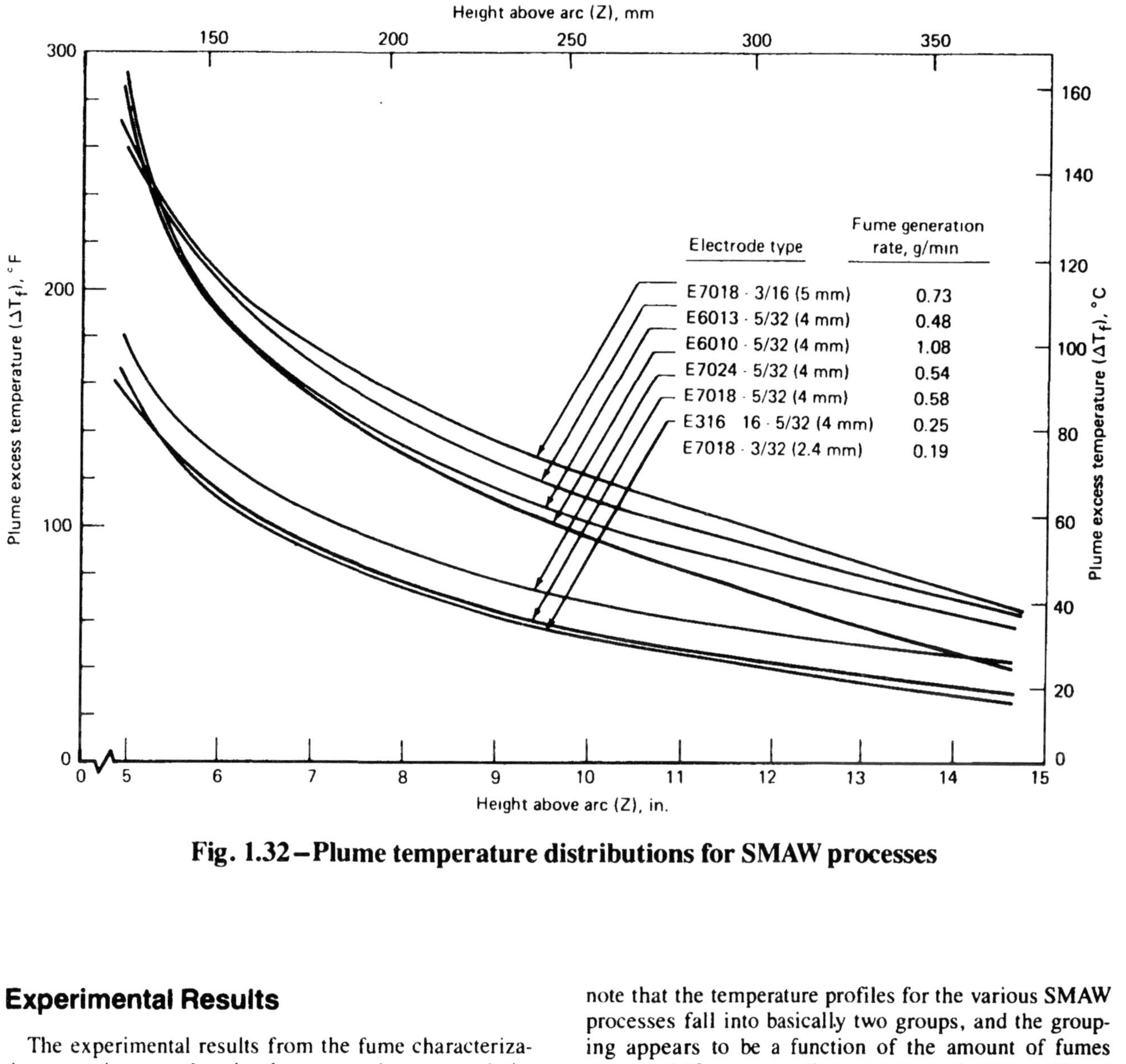

Fig. 1.32—Plume temperature distributions for SMAW processes

Experimental Results

The experimental results from the fume characterization experiments, the simulator experiments, and the local exhaust inlet evaluation experiments are presented in a form that can be used easily to estimate the performance of various local exhaust configurations or to define required exhaust operating conditions to achieve a specific performance level.

Fume Characterization Results. Experiments were performed for various arc welding processes to measure the fluid dynamic characteristics of welding plumes. Figures 1.32 and 1.33 are plots of the increase in plume temperature above ambient for SMAW, GMAW, and FCAW processes. The increase in plume temperature (plume excess temperature) is defined as the local plume temperature minus the environment temperature (in this case, laboratory room temperature). The welding conditions (voltage, current, wire stickout, etc.) for each process were presented in Table 1.1. It is interesting to note that the temperature profiles for the various SMAW processes fall into basically two groups, and the grouping appears to be a function of the amount of fumes generated (fume generation rate shown in Fig. 1.32 and next to electrode identification in Fig. 1.33). In general, the plume temperatures at various heights above the arc for electrodes with relatively high fume generation rates are greater than the plume temperatures for those electrodes (5/32 in. [4 mm] E316-16 and 3/32 in. [2.4 mm] E7018) that have a low fume generation rate. The temperature profile for the larger E7018 electrode was the only one that fell outside the grouping based on fume generation rate. The reason for this observed deviation is not known. The temperature profiles for the GMAW and FCAW processes shown in Fig. 1.33 are similar to the results obtained for the SMAW processes, except that the initial temperatures at a height of 5 in. (127 mm) are lower than the corresponding initial temperatures for SMAW. It is possible that the lower initial temperatures associated with automatic welding are due to gas shield-

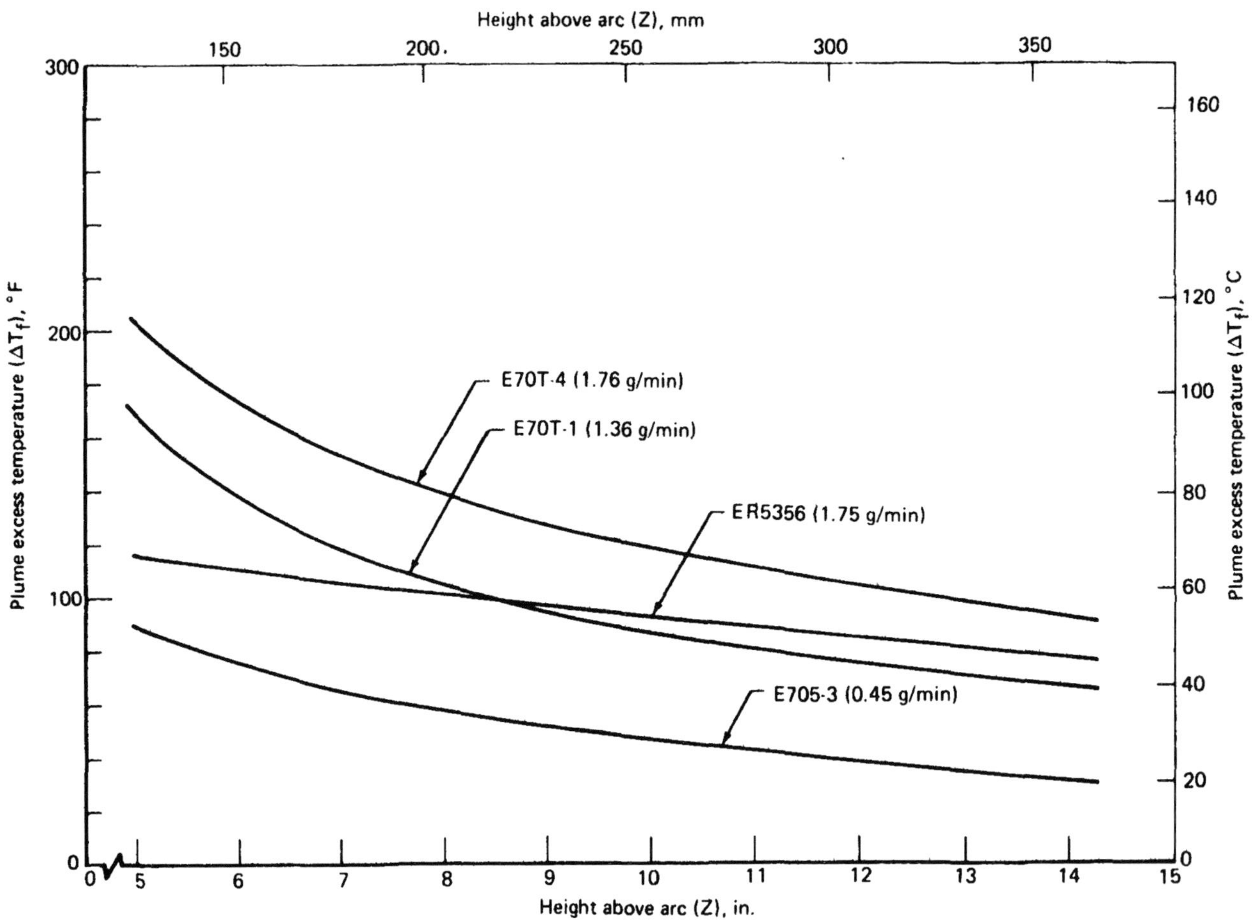

Fig. 1.33—Plume temperature distributions for automatic welding processes

ing. It should also be noted that there is a less rapid temperature decay for the semiautomatic welding process than for the SMAW process.

Additional information on the dynamic nature of heated welding plumes was obtained by observing the fume flow under the influence of a known cross current of air. Measurements were performed to determine the plume deflection distance resulting when a uniform flow of air at 80 ft/min (24.4/min) interacted with the plume. This method of characterizing the fluid dynamic nature of heated welding plumes was used because the technique involves flow interactions (plume and air cross current) similar to those existing in the exhaust evaluation experiments. Plume deflection measurements were performed for all the arc welding processes presented in Table 1.1, except for the E70S-3 electrode. The fume flow for the E70S-3 electrode was not visible.

The results of the plume deflection measurements are shown in Fig. 1.34. The plume deflection distance (X_d) was measured at a height of 25 in. (635 mm) above the arc, in the downstream direction (see Fig. 1.30 for representative photograph). Because of the unsteady nature of the plume, the plume deflection varied with time. This is reflected in the results presented in Fig. 1.34 by the length of the bar corresponding to plume deflection for each of the welding processes. The results indicate that there is a tendency for further plume deflection to decrease with increasing plume temperature. This is expected because the greater buoyancy forces characteristic of higher plume temperature result in larger vertical accelerations.

These results on plume temperature and deflection and the simulator results were used to determine the simulator operating conditions necessary to duplicate the selected welding plumes.

Simulator Results. A comparison of the results from the simulator and welding plume experiments was made to establish the simulator operating conditions for use in evaluating the performance of the local exhaust systems.

Plume temperature data as a function of vertical distance above the heater orifice which represents the arc location are shown in Figs. 1.35 and 1.36 for orifice diameters of 11/32 in. (8.73 mm) and 1.0 in. (25.4 mm), respectively. Each curve in the figures is labeled with a test number that is also used in Table 1.3, where the simulator operating conditions for each test run are tabulated. Included in Table 1.3 for each test is orifice diameter (d_j), heater voltage and current, gas flow rate through the heater ($\mathring{Q}_S$), gas temperature in the heater (T_g), and plume temperature 5 in. (127 mm) above the orifice (ΔT_g). The gas temperature heater was varied by adjusting the voltage to the heating element; plume temperatures from approximately 80° to 305°F (27° to 152°C) were achieved. The gas flow rate was varied from 0.3 ft³/min to 2.0 ft³/min (10 to 60 μm³/min). A comparison of the plume excess temperature profiles obtained with the simulator and those measured during actual welding operations showed that the required temperature profiles and characteristic temperature decays could be achieved with the simulator.

The simulator plume deflection data (X_d) are shown in Fig. 1.37 for the 11/32 in. (8.73 mm) and 1.0 in. (25.4 mm) diameter orifices. An air current of 80 ft/min (24.4 m/min) was used during the plume deflection experiments (the same as was used during the welding plume experiments). The plume deflection data appeared to correlate quite well with the term $\mathring{Q}_S{}^2\ \Delta T_{g_0}$.[3] For a given orifice size, as simulator plume volume flow rate and temperature were increased, the plume deflection distance approached a limit. This limit was approximately 14 and 6.5 in. (356 and 165 mm) for the 1.0 in. (25.4 mm) and 11/32 in. (8.73 mm) diameter orifices, respectively. By comparing the data on plume deflection obtained from the welding and simulator experiments, it is apparent that the required range of welding plume deflections could be duplicated with the simulator.

3. The term $\mathring{Q}_S{}^2\ \Delta T_{g_0}$ represents an indication of the buoyancy and momentum forces.

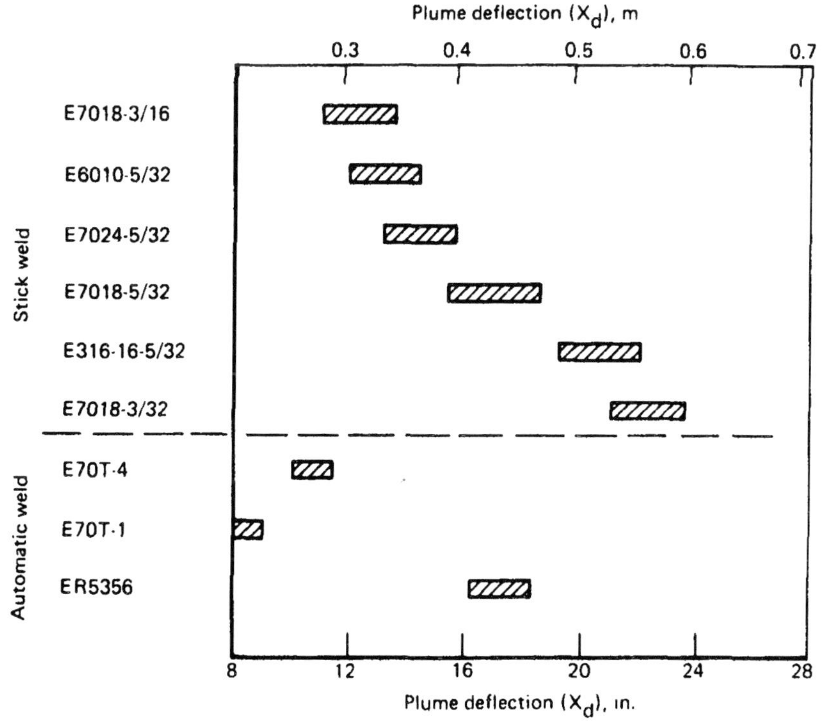

Fig. 1.34—Measured plume deflections for SMAW and automatic welding processes for 80 ft/min (24.4 m/min) cross current

Table 1.3
Simulator operating conditions

Test no.	Orifice diameter		Heater voltage	Heater current	Volume flow rate		Heater stagnation temp.		Plume temp., 5 in. (127 mm) T_{g0}	
	in.	mm	V	A	std ft³/min	std μm³/min	°F	°C	°F	°C
1	11/32	8.73	75	46	.45	13	950	510	272	133
2	11/32	8 73	89	52	30	08	1010	543	296	147
3	11/32	8.73	99	60	.45	13	1075	579	308	153
4	11/32	8.73	110	67	.78	22	1050	566	250	121
5	11/32	8.73	90	53	.78	22	840	449	210	99
6	11/32	8.73	50	30	45	13	600	316	175	79
7	11/32	8.73	48	30	.30	08	520	271	168	76
8	11/32	8.73	40	24	45	13	340	171	82	28
9	1.0	25.40	30	2	.30	08	200	93	---	
10	1.0	25 40	29	2	.78	22	200	93	109	43
11	1.0	25.40	25	2	2.0	57	150	66	67	19

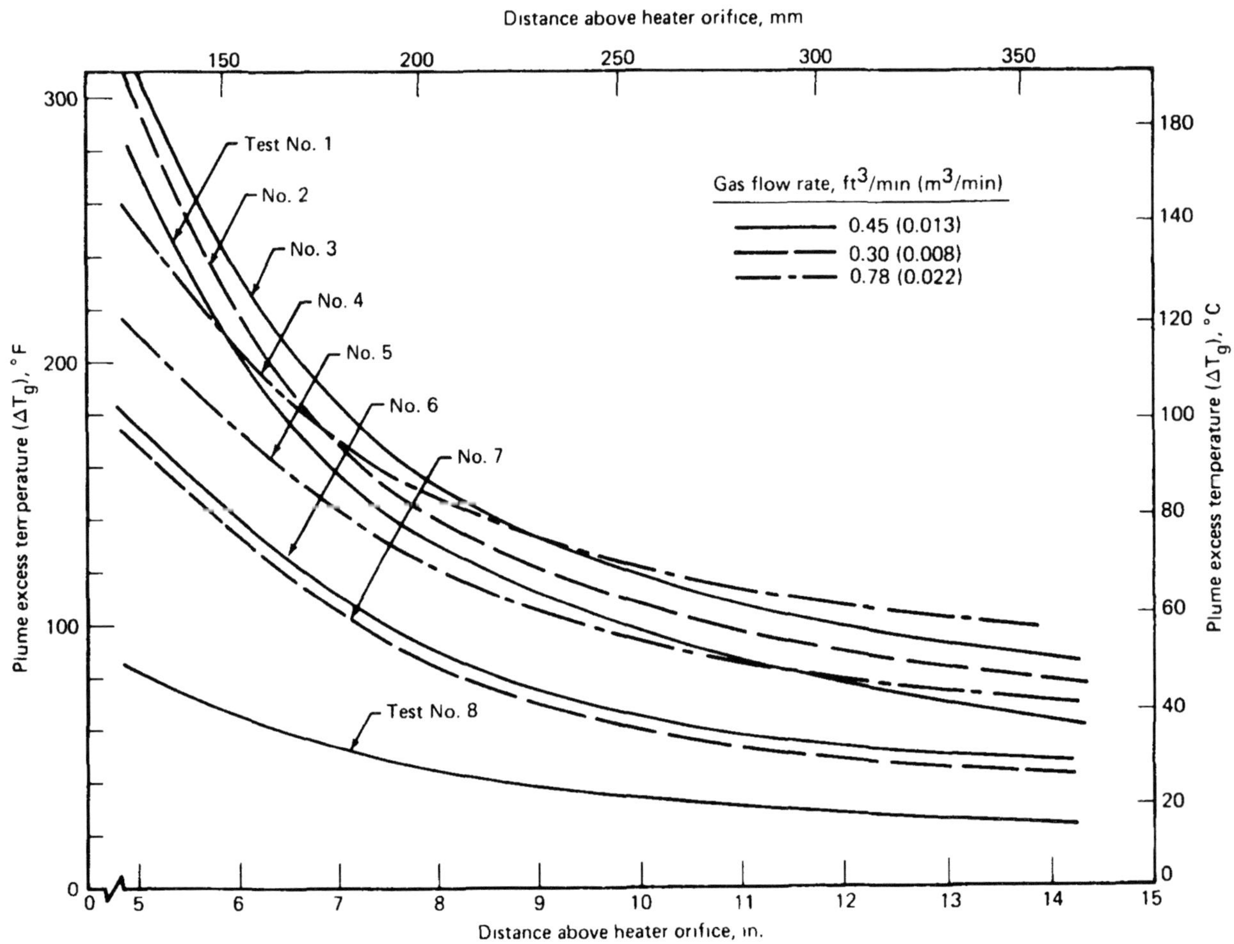

Fig. 1.35–Simulator plume temperature distributions for various simulator operating conditions using an 11/32 in. (8.75 mm) diameter

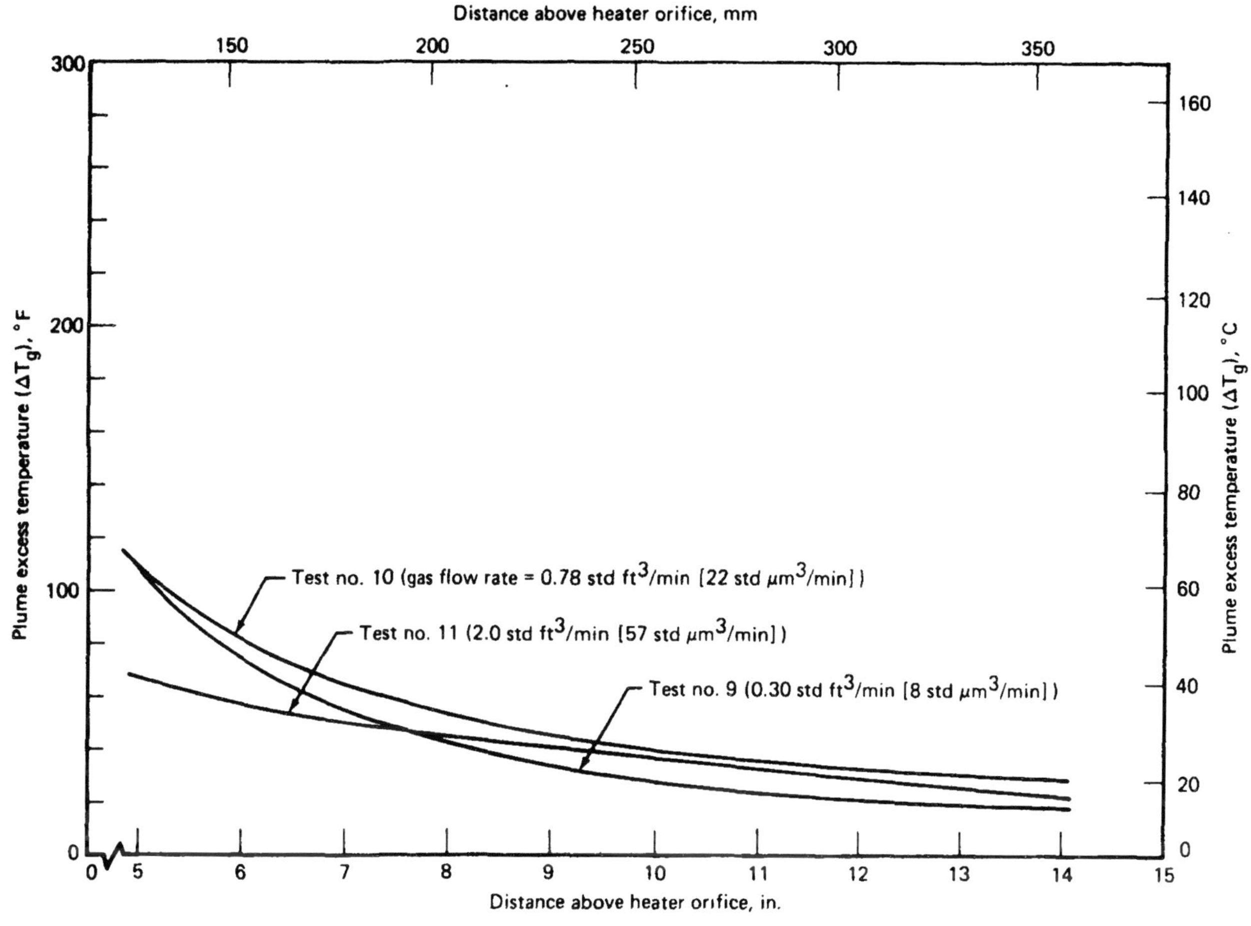

Fig. 1.36–Simulator plume temperature distribution for various simulator operating conditions using a 1 in. (25.4 mm) diameter orifice

An analysis was conducted to determine which simulator operating conditions should be used during the local exhaust experiments. The simulator was found to be capable of independently matching welding plume temperature profiles and deflections, and it was necessary to match these simultaneously to properly duplicate a specific welding plume. Based on the results from the welding plume experiments, the worst plume conditions existed for the E70T-1, the E70T-4, and a number of SMAW processes that are characterized by the results obtained from the E7018 (3/16 in. [4.8 mm]) welding processes. The plumes generated during these welding processes were collectively judged to be the worst case in comparison to other welding plumes studied because of their relatively higher gas temperatures and smaller deflections (plumes reflected less by air currents).

The simulator results were reviewed and compared to the plume results for the E7018-3/16, E70T-1, and E70T-4 welding processes. It was concluded from this evaluation that the plume characteristics obtained in simulator test No. 3 closely duplicated the E7018-3/16 plume characteristics. Figure 1.38 shows a comparison of the simulator and actual plume excess temperatures. The plume deflection for the simulator was 8 in. (0.2 mm) in the 80 ft/min (24 m/mm) air current. Although this deflection is less than the 11 to 14 in. (0.3 m) for the E7018-3/16 in. (5 mm) welding plume, this simulator condition was selected for evaluating the performance of the local exhaust devices because of the unsteadiness of the spatial location of the welding plume.

The simulator plume result labeled condition B in Fig. 1.38 represents welding plumes that exhibited lower temperatures and higher deflection distances.

Ventilation tests were conducted using a 5/16 in. (8 mm) diameter E7024 electrode and the simulator to duplicate the E7024 plume conditions. Those tests were performed using a 6 in. (153 mm) diameter exhaust duct operated at a constant volume flow rate. The duct inlet

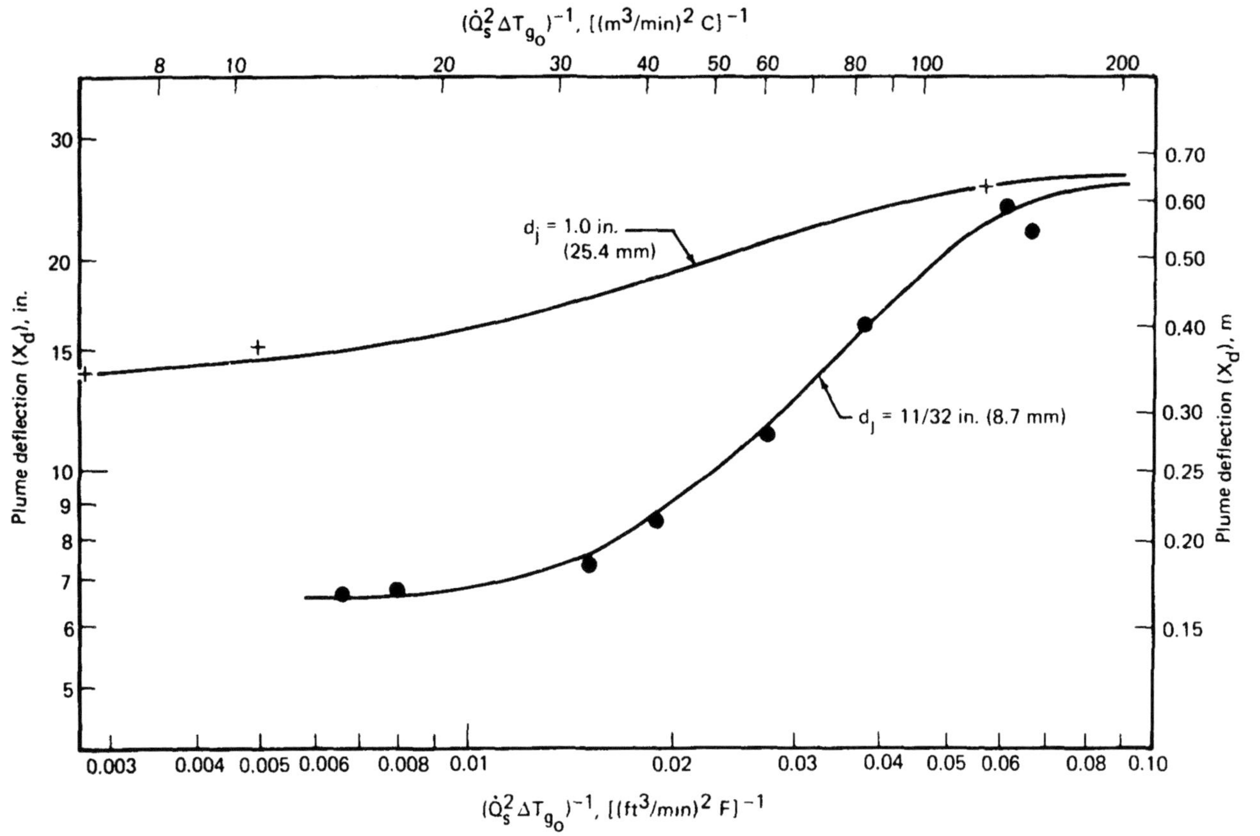

Fig. 1.37—Simulator plume deflection results for the 11/32 (8.7) and 1 in. (25.4 mm) diameter orifices

was positioned at various distances from the arc. The fumes are collected on a filter located 12 in. (0.37 m) from the inlet of the exhaust duct. The procedures followed to collect the fume samples and to weigh them were basically the same as discussed in "Local Ventilation." The data from the tests are shown in Fig. 1.39 in terms of the room contamination factor (ϵ_R) as a function of exhaust inlet distance from the arc. The room contamination factor is equal to zero when all of the fumes are collected (100 percent collection efficiency). Simulator data were obtained using the same exhaust system arrangements, and CO concentration measurements were performed in the exhaust duct to determine the collection efficiency. The room contamination factor was then calculated from the exhaust efficiency data (shown in Fig. 1.39). Comparing the results from the actual welding tests and the simulator tests indicates that the simulator can be used to duplicate the gross dynamic behavior of welding plumes.

Local exhaust experiments were conducted using both simulator conditions. It was determined from preliminary tests using a 6 in. (153 mm) diameter duct that the local exhaust volume flow rate required to capture the plume gases for simulator conditions A and B was approximately the same. Therefore, subsequent local exhaust experiments were conducted using simulator plume condition A.

Local Exhaust Configuration Results. Herein are the results of the experiments conducted to determine the performance of eight local exhaust inlet configurations that were tested at various exhaust volume flows at various distances from the simulated arc location (heater orifice outlet). Two simulated plume conditions representing worse and moderate welding plume conditions were used for these tests.

The results of the room contamination studies for the flanged, circular duct inlet configurations are presented in Figs. 1.40, 1.41, and 1.42. These figures show the room contamination factor in terms of the nondimensional horizontal distance between the exhaust inlet and the arc and the flow parameter $\mathring{Q}_E/A_E$. These correlations were based on tests conducted with the three duct sizes: 3, 6, and 8 in. (76, 152, and 203 mm) diameters. The room contamination factor was defined in Equation 6 as one minus the fraction of total fume being generated that is collected by the exhaust system. The location of the exhaust inlet is specified by three coordinates defined as: X_E, the horizontal distance between the inlet plane and a vertical line through the arc location; and θ_E, an angular measurement defined in Fig. 1.40. The room contamination data were correlated using the parameter $\mathring{Q}_E/A_E$ where $\mathring{Q}$ is the exhaust air volume flow rate and A_E is the duct area. Baseline data at Z_E/d_E equal to 1.5 were obtained for exhaust angular positions of 53, 90, and 180 degrees (Figs. 1.40, 1.41, and 1.42, respectively).

The effectiveness of the local exhaust device will vary from those levels shown in Figs. 1.40 through 1.42 depending upon the distance between the exhaust inlet and the arc. As the exhaust inlet height is increased above a value of Z_E/d_E = 1.5, from approximately 5 in. (127 mm) above the arc to the bottom of the welder's helmet, the collection efficiency will increase. Accordingly, the volume flow rate can be reduced in order to maintain a constant collection efficiency. Alternatively, the horizontal location of the exhaust inlet can be changed (exhaust inlet moved further from the arc) at constant volume flow rate while maintaining a constant collection efficiency. The effect of such a change, i.e., exhaust inlet heights greater than 1.5 duct diameters (Z_E/d_E = 1.5), can be determined from the room contamination factors of Figs. 1.40 through 1.42 and the displacement correlations given in Fig. 1.43. For example: assuming that a local exhaust device is operated at a $\mathring{Q}_E/A_E$ = 10 and the exhaust inlet is located at θ_E = 90° at a height above the arc of Z_E/d_E = 2.0, then from Fig. 1.41, at a volume flow rate of $\mathring{Q}_E/A_E$ = 10, the inlet of the device must be located at an X_E/d_E = 2.75 in order to achieve

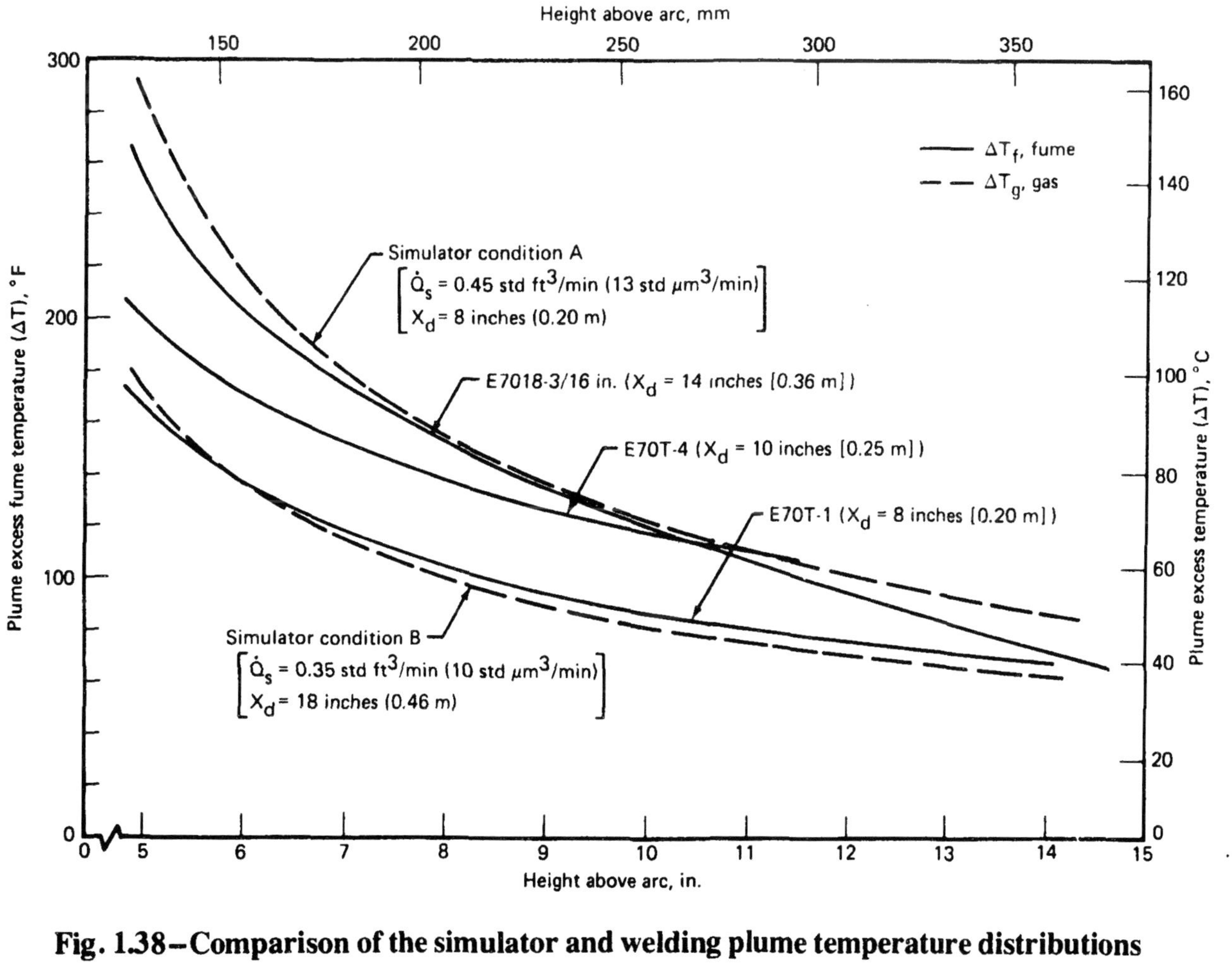

Fig. 1.38–Comparison of the simulator and welding plume temperature distributions

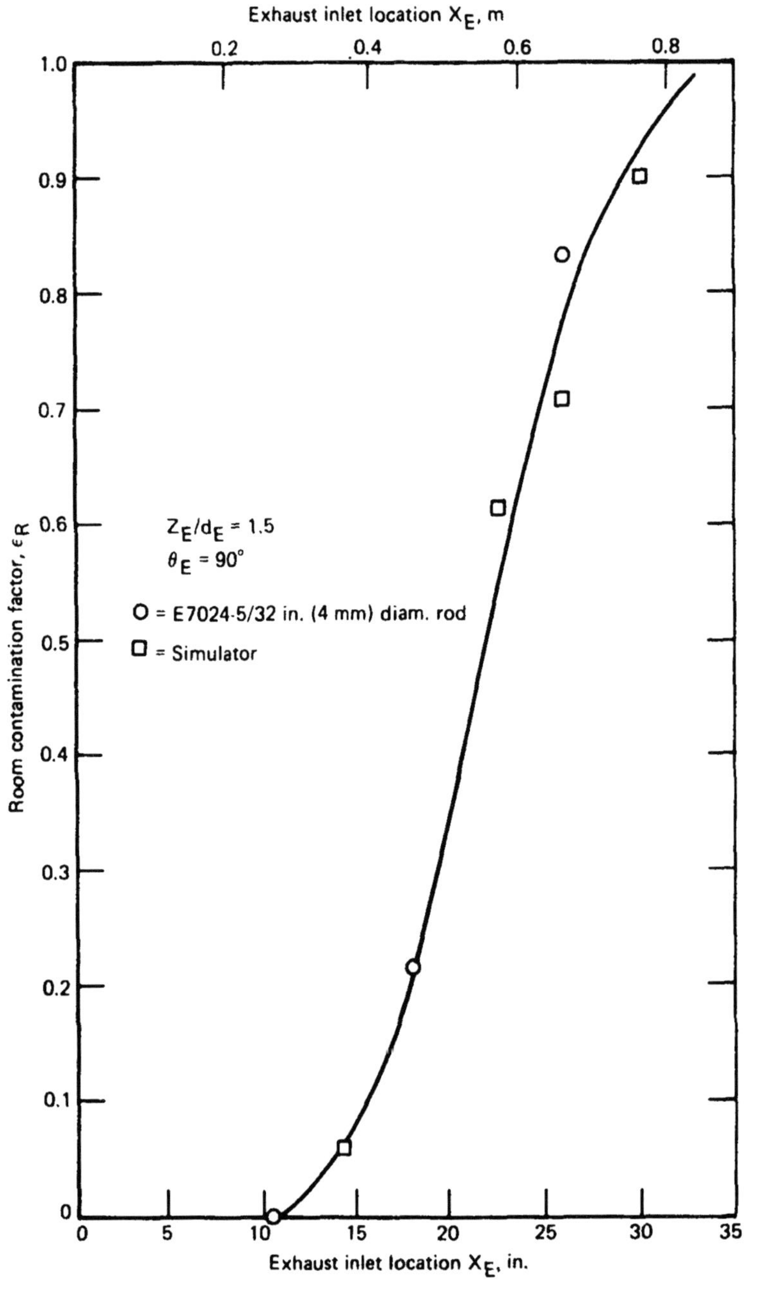

Fig. 1.39–Comparison of local exhaust performance results using an actual welding process and the welding plume simulator

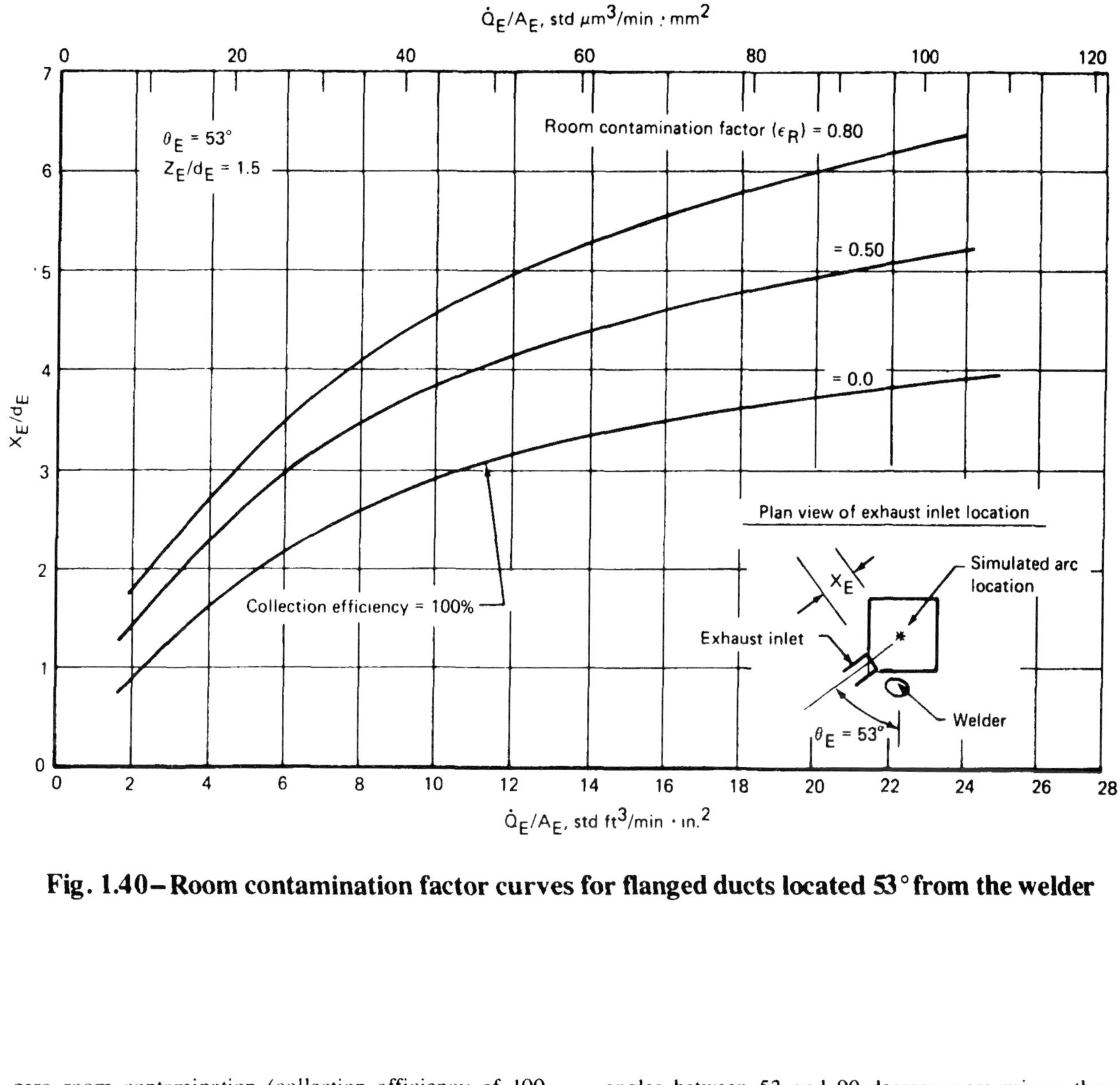

Fig. 1.40–Room contamination factor curves for flanged ducts located 53° from the welder

zero room contamination (collection efficiency of 100 percent) for a height of Z_E/d_E = 1.5. If the inlet of the exhaust device were raised to Z_E/d_E = 2.0, from Fig. 1.43 at the new height, $\Delta X_E/d_E$ = 0.5. Thus, the inlet of the exhaust device can be moved to an X_E/d_E = 3.25 while maintaining a zero room contamination factor (collection efficiency of 100 percent).

The effect of the local exhaust systems (flanged duct inlets) on the contamination in the welder's breathing zone is shown in Figs. 1.44 and 1.45. In these figures, the welder exposure factor (ϵ_W) is shown as a function of exhaust location (X_E/d_E) and exhaust flow rate ($\dot{Q}_E/A_E$). An exposure factor (ϵ_W) of zero means that the local exhaust system is reducing the fume concentration level to zero in the welder's breathing zone. Likewise, a value of ϵ_W = 0.8 indicates a reduction of 20 percent. Differences in the welder exposure factor for angles between 53 and 90 degrees were minor; therefore, the data for these two wind flow directions were correlated and only one set of curves is presented (Fig. 1.44). On the other hand, there were major differences in the exposure factor for the exhaust location corresponding to θ_E = 180 degrees and the other two directions. Results for this wind direction are shown separately in Fig. 1.45. In applying the data for θ_E between 90 and 180 degrees, the assumption of a linear variation of exposure factor with angle will yield conservative results.

Studies using a single-slot (1.2 x 24 in. [32 x 610 mm]) inlet configuration were performed with the inlet face located at 90 and 180 degrees. The 90 degrees case is similar to a cross-draft welding table arrangement and the 180 degrees case is similar to a back ventilation table. The flow area of the exhaust slot was designed to be

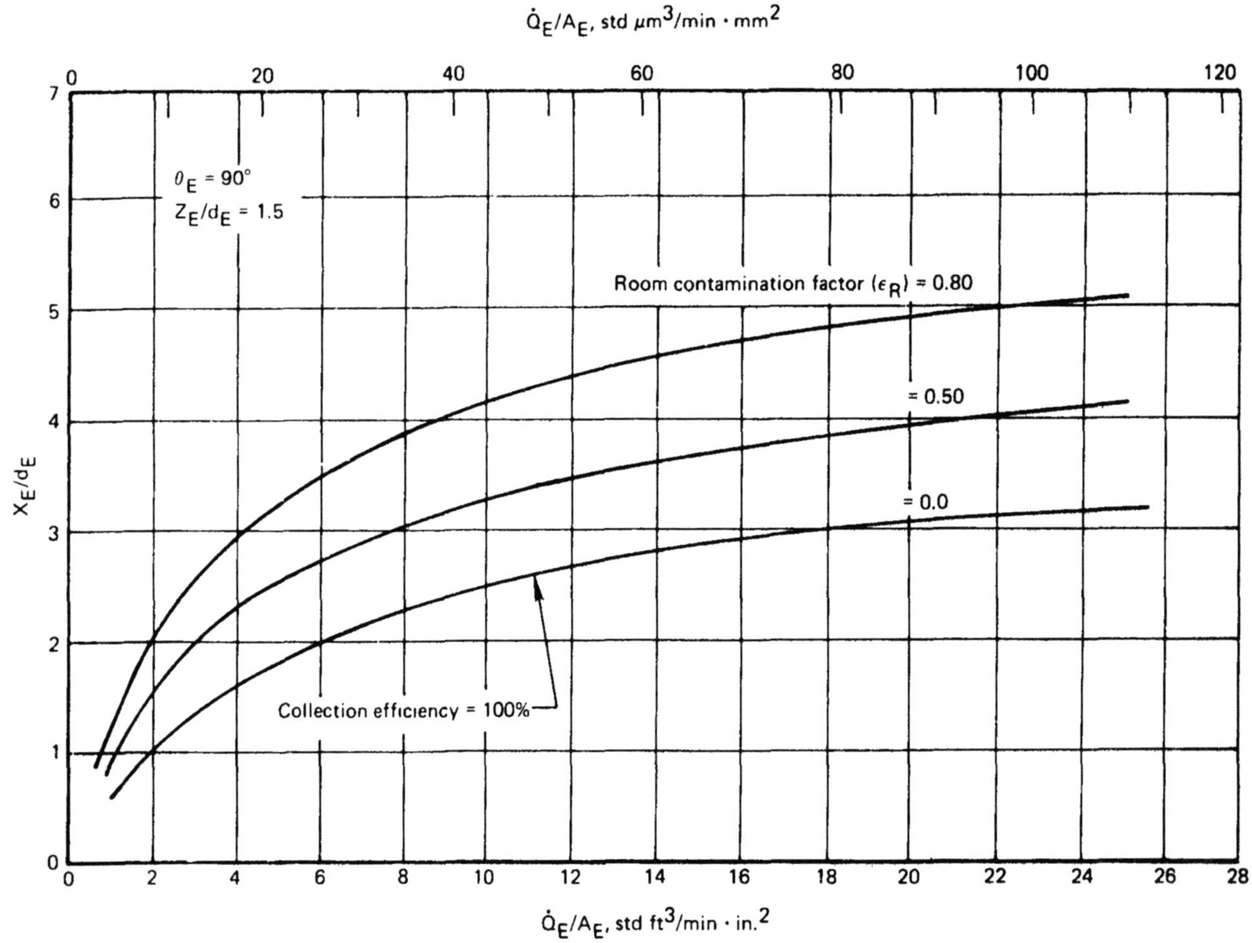

Fig. 1.41–Room contamination factor curves for flanged ducts located 90° from the welder

equal to the flow area of the 6 in. (152 mm) diameter circular exhaust duct. The slot aspect ratio (height/width ratio of open flow area) was selected from data presented in Ref. 5. Single-slot room contamination factors (ϵ_R) are presented in Figs. 1.46 and 1.47 for exhaust inlet angular locations of 90 and 180 degrees, respectively. Correlations of exhaust performance for horizontal distances from the arc (X_E) of 6 and 16 in. (152 and 406 mm) are presented as a function of exhaust volume flow. As shown in Fig. 1.46, in order to achieve 0 room contamination for the exhaust inlet at 6 in. (152 mm), an exhaust volume flow of 300 std ft³/min (8.5 std m³/min) is required.

Welder protection factor data for the single-slot inlet exhaust configuration are presented in Fig. 1.48. To reduce the fume concentration level to approximately 0 in the welder's breathing zone (ϵ_W=0), it can be seen from Fig. 1.48 that an exhaust flow rate of greater than 300 std ft³/min (8.5 std m³/min) is required. The single-slot configurations were also set at various vertical heights above and below Z_E = 7-1/2 in. (191 mm). Only minor performance decreases were seen in the room contamination factor and basically no increase in the welder protection factor. Therefore, the data presented in Figs. 1.46, 1.47, and 1.48 can be used for exhaust inlet height positions (Z_E) from 4 to 10 in. (102 to 254 mm) without modification.

The single-slot exhaust configuration was operated with the slot oriented with the 24 in. (610 mm) dimension in the vertical elevation. The slot inlet plane was located 16 in. (406 mm) from the arc at θ_E = 90 degrees. The results from these tests were approximately the same as those obtained for the horizontal single slot at θ_E = 90 degrees, and Z_E = 7-1/2 in. (191 mm).

A double-slot exhaust inlet configuration was tested and the resulting room contamination and welder protection factors are presented in Figs. 1.49 and 1.50, respectively. When the θ_E = 180 degrees case was run, panels were used on either side of the welding table to simulate a booth. Each slot was designed with equal flow area (0.60 x 24 in. [15.2 x 610 mm]) and a total flow area equal to the single-slot configuration. The vertical height above the arc of each of the slots' centerlines was 6 and 10 in. (152 and 254 mm).

Two overhead hood type exhaust systems were investigated; their performance characteristics are shown in Figs. 1.51 and 1.52. As shown in Fig. 1.51, these configurations are quite efficient and will collect a large portion of the fumes generated at low exhaust volume flow, resulting in low room contamination factors. These systems are efficient because they take advantage of the natural rising tendency of the hot fumes.

Reviewing the results presented in Fig. 1.52 for welder exposure, it is seen that ϵ_W remains high at conditions where all of the fumes are being collected, and the room contamination factor is 0 (volume flows greater than 500 std ft^3/min [14.2 std m^3/min]). This result is consistent with visual observations of the plume wherein it was observed that a portion of the plume was still flowing past the welder's helmet at conditions of 100 percent exhaust collection.

Application of Results of Part I

To show how the results given in this report can be used in the design of an effective ventilation system, the following example is presented.

Assume for illustrative purposes only:

(1) Three welders (at most)

(2) SMAW, E7018 5/32 in. (4 mm) diameter electrodes

(3) Shop size 50 x 30 x 20 ft. (15.2 x 9.1 x 6.1 m)

In this example, the results of this study are used to answer the following questions:

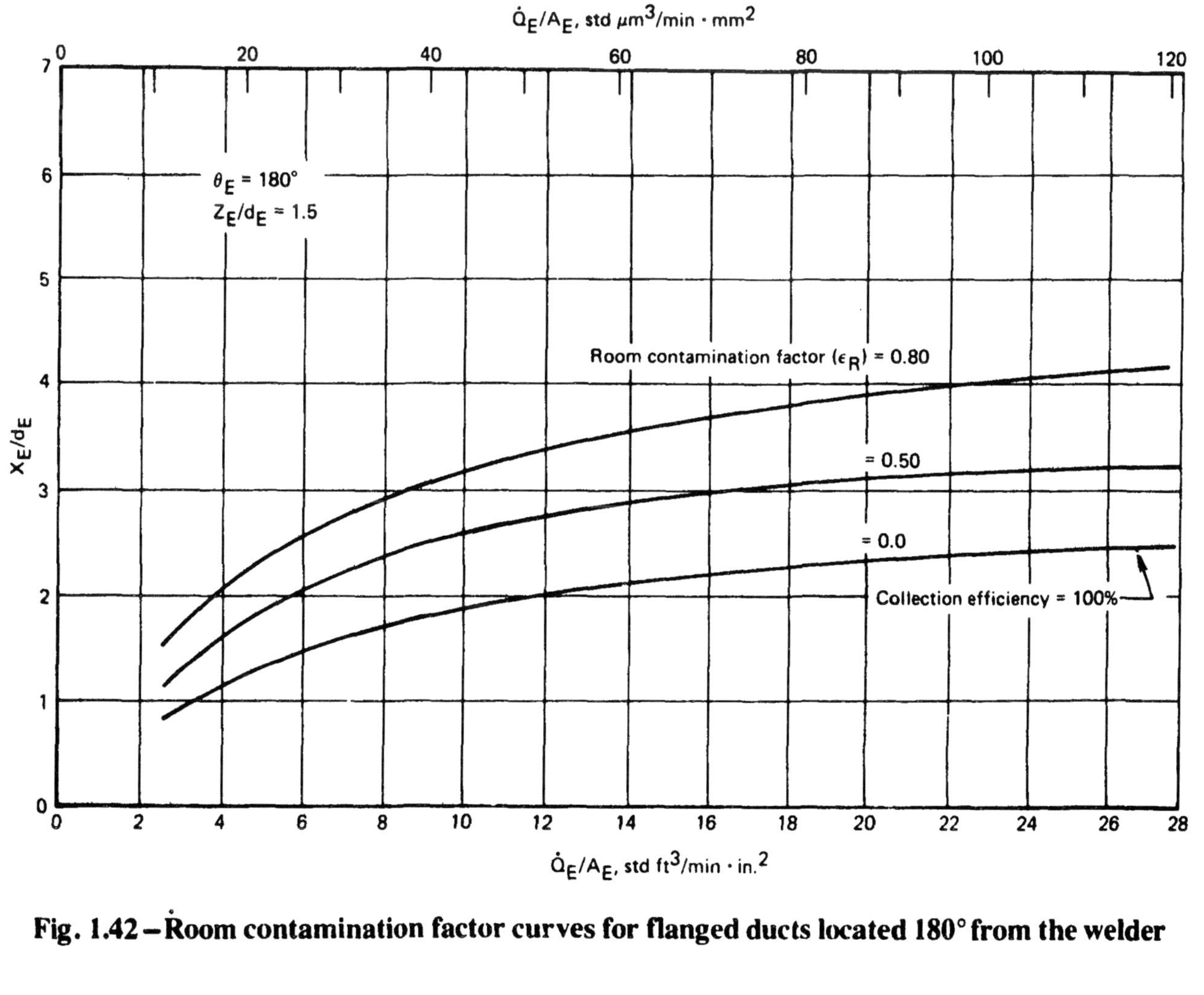

Fig. 1.42 – Room contamination factor curves for flanged ducts located 180° from the welder

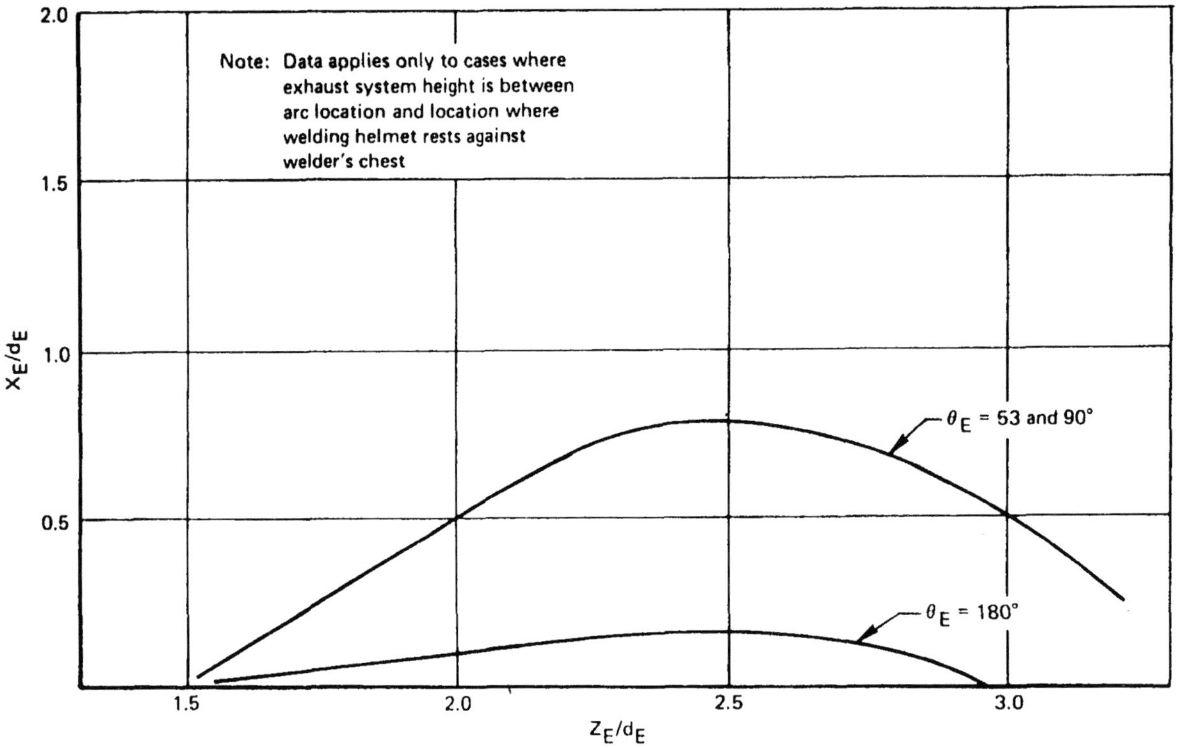

Fig. 1.43 – Increment of X associated with exhaust weight variation for constant collection efficiency

(1) Under worst condition (all three welders operating), how much room air is required to dilute the fumes to various concentration levels?
(2) What local ventilation conditions are needed to protect the welder from welding fumes?
(3) For a flanged duct inlet design, what operating condition would be required to collect 50 and 100 percent of the fumes?

It is assumed that 100 percent of the fumes generated are used in specifying dilution air requirements and local ventilation requirements. If it is known or decided by some other input requirements that only 50 percent of the fumes being generated have to be handled, then instead of using 100 percent of the fume generation rate, 50 percent would be used and the solution would be obtained as though the 50 percent rate were a total generation rate.

Question 1

The amount of air required to dilute the fumes to an acceptable level will depend upon the level considered acceptable. For the E7018 electrode, approximately 0.6 g/min of fumes will be generated per welding operation (see Fig. 1.53). The volume flow rate of air required to dilute the fumes to various levels is given in Fig. 1.2, and presented in Table 1.4, which also presents required ventilation rates for one and three operating welders. The ventilation air flow rates shown are needed if the arc time is 100 percent.

If the arc times for the three welders are such that, on the average, only one welding operation is in progress continuously, then the one welding operation results presented in Table 1.4 should be used.

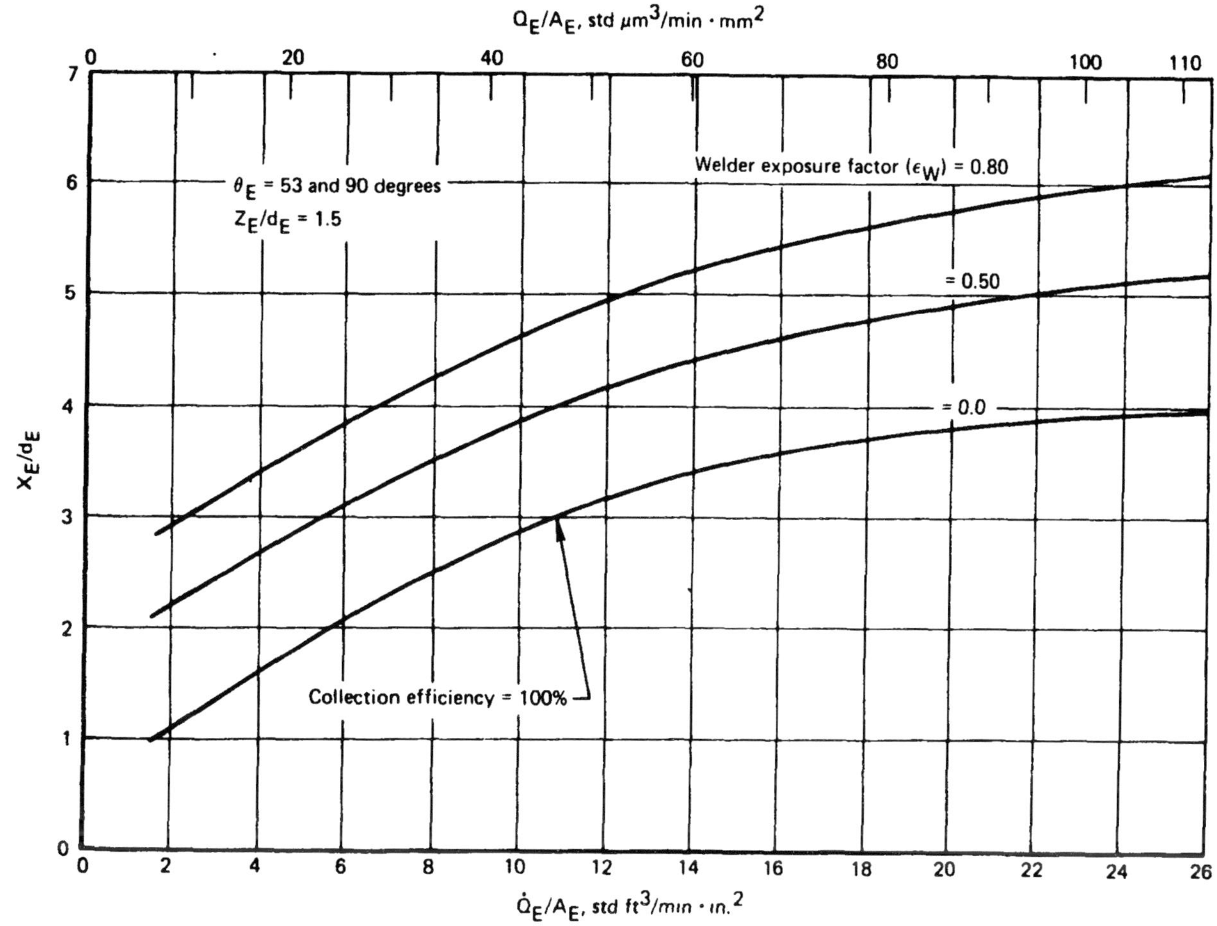

Fig. 1.44 – Welder exposure factor data for flanged ducts located at 53° and 90° from welder

Table 1.4
Dilution air required to achieve various room contamination levels

Room contamination level, mg/m^3	One welding operation		Three welding operations	
	std ft^3/min	std m^3/min	std ft^3/min	std m^3/min
2.0	10,500	297	21,500	692
5.0	4,300	122	12,900	365
10.0	2,000	57	6,000	170
20.0	1,000	28	3,000	85

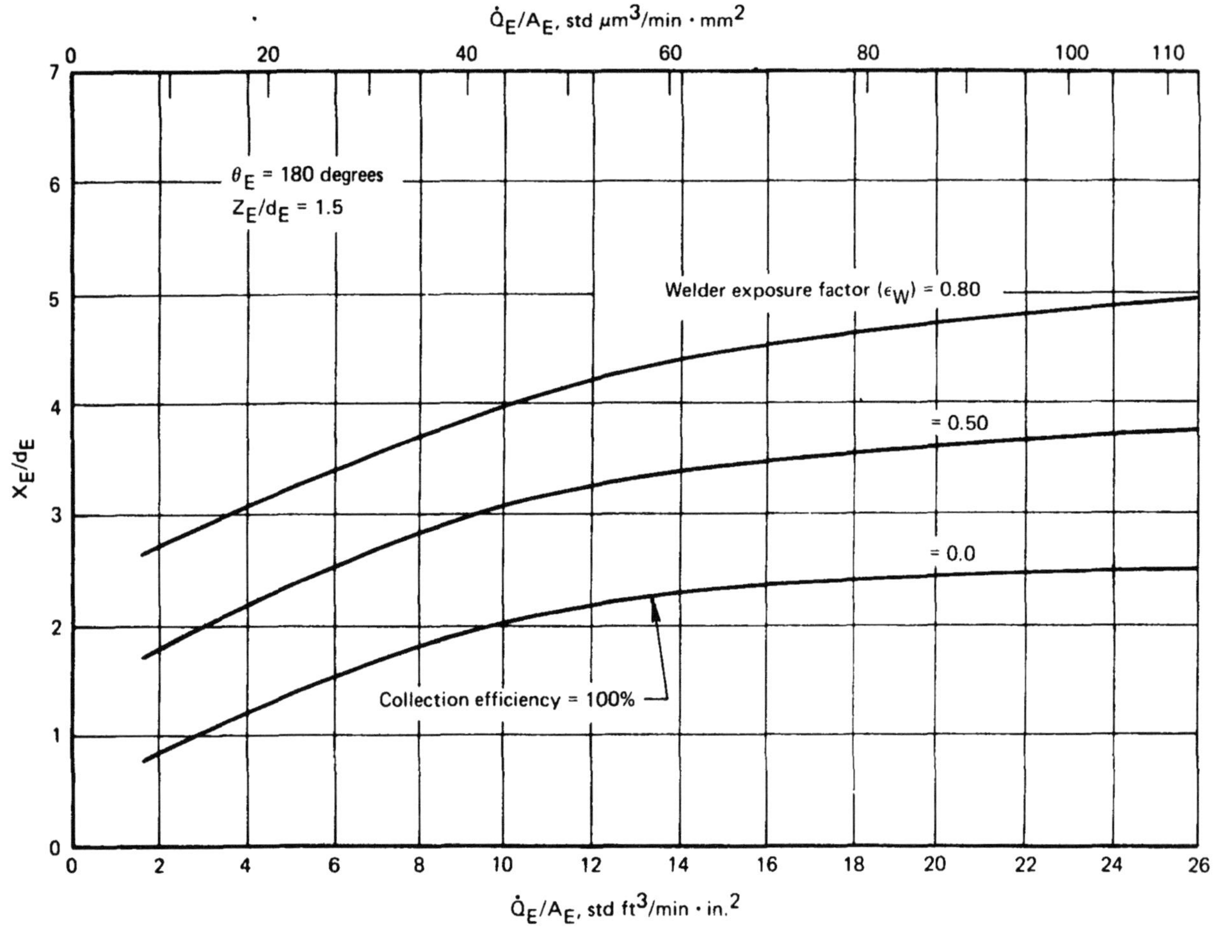

Fig. 1.45—Welder exposure factor data for flanged ducts located at 180° from welder

Question 2

The local ventilation conditions required to protect the welder can usually be treated on an individual basis, unless their working conditions require that they be close together in a general working area that must be treated as a unit. In this example, it is assumed that each welder has sufficient area surrounding him so that it is possible to treat the problem on an individual basis.

The welder's estimated exposure condition during welding is given in Fig. 1.53. At a fume generation rate of 0.6 g/min, the estimated fume concentration level in the welder's breathing zone (C_{V_0}) is 40 mg/m³ for zero ventilation.

Assuming that a cross draft is generated in the welder's area using a fan or blower, the ventilation air velocity required to reduce the welder's exposure is given in Fig. 1.23. The reduction in welder's exposure (C_V) can be determined for various values of ventilation velocity using the 100 percent C_V/C_{V_0} curve (referenced as electrode positive polarity) in Fig. 1.23. The results obtained from Fig. 1.23 for C_{V_0} mg/m³ are presented in the following tabulation.

Ventilation velocity		Welder exposure, C_V
ft/min	m/min	mg/m³
20	6.1	27
40	12.2	16
60	18.3	8.4
100	30.5	4

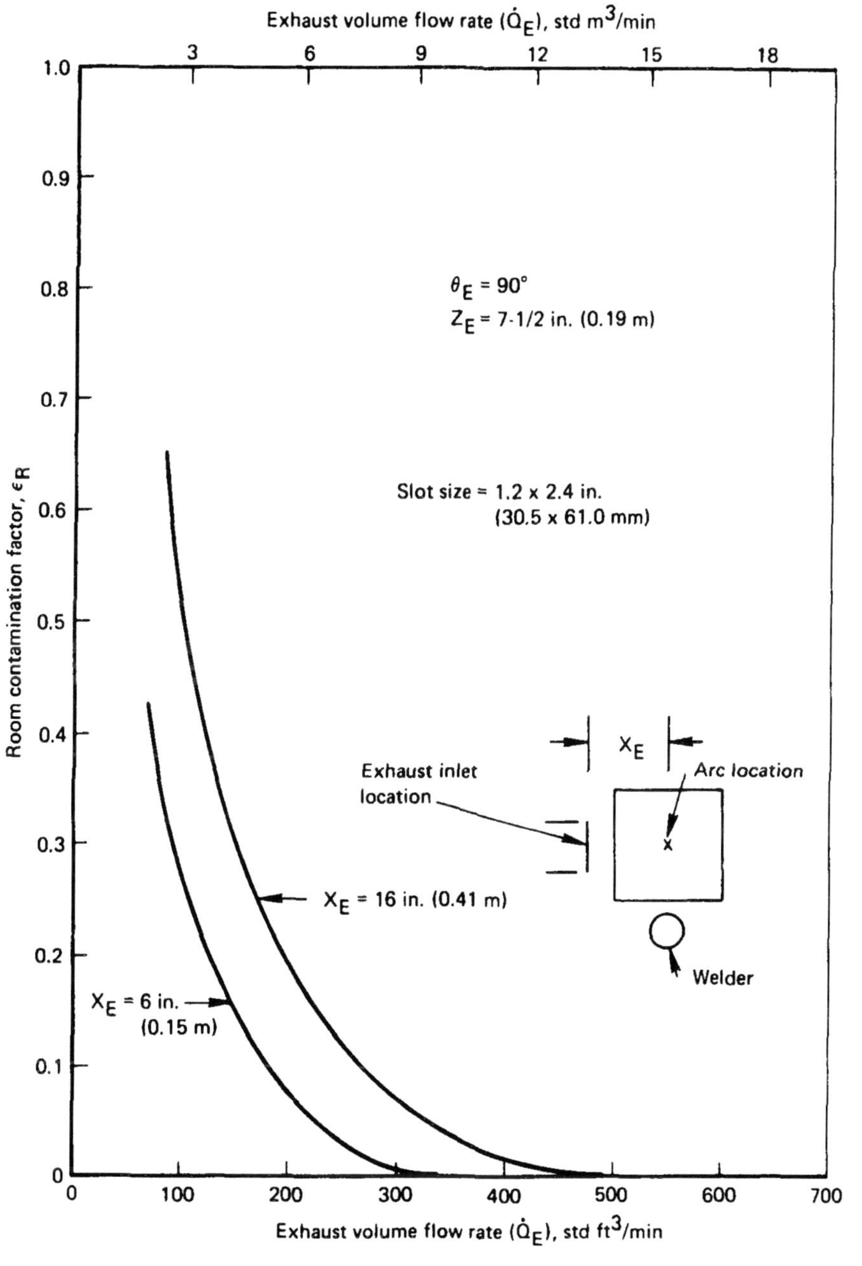

Fig. 1.46–Single-slot room contamination factor for $\Theta_E = 90°$

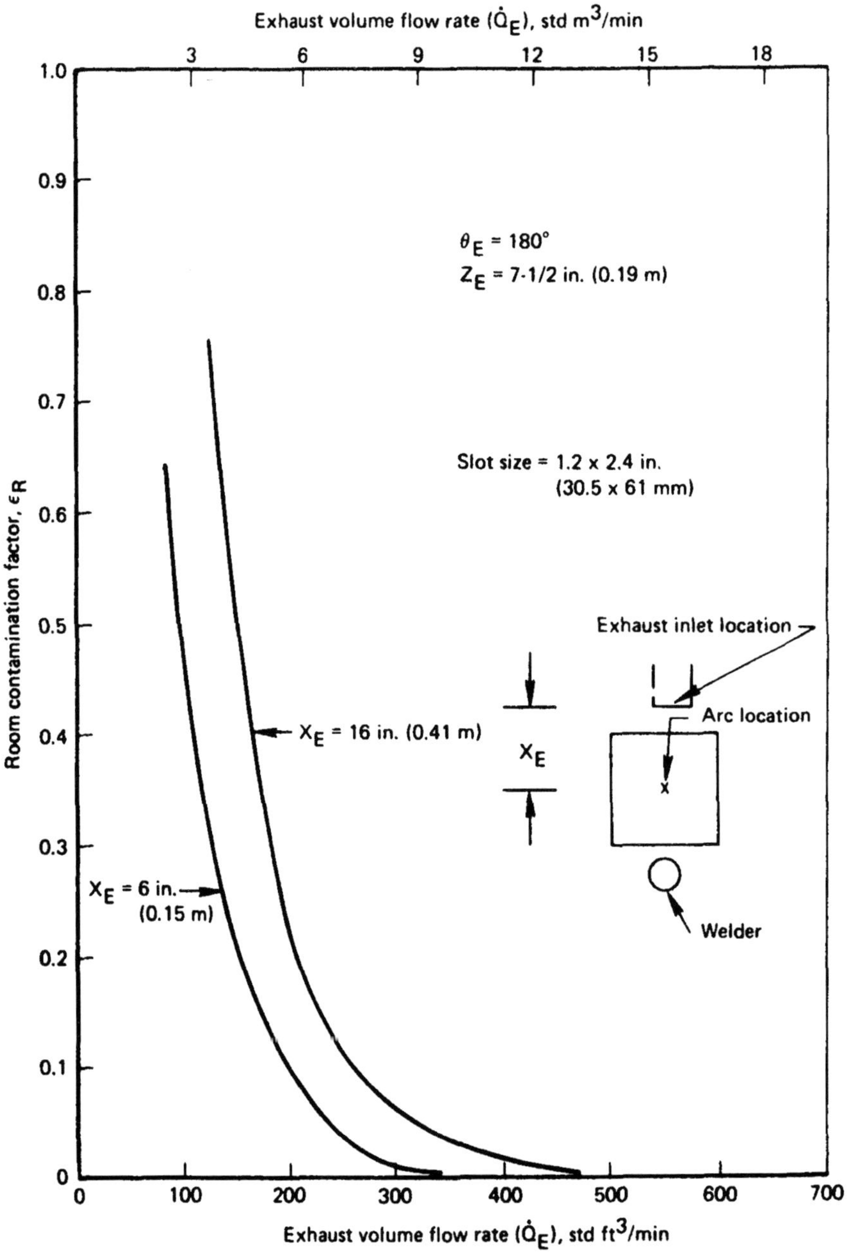

Fig. 1.47–Single-slot contamination factor for $\Theta_E = 180°$

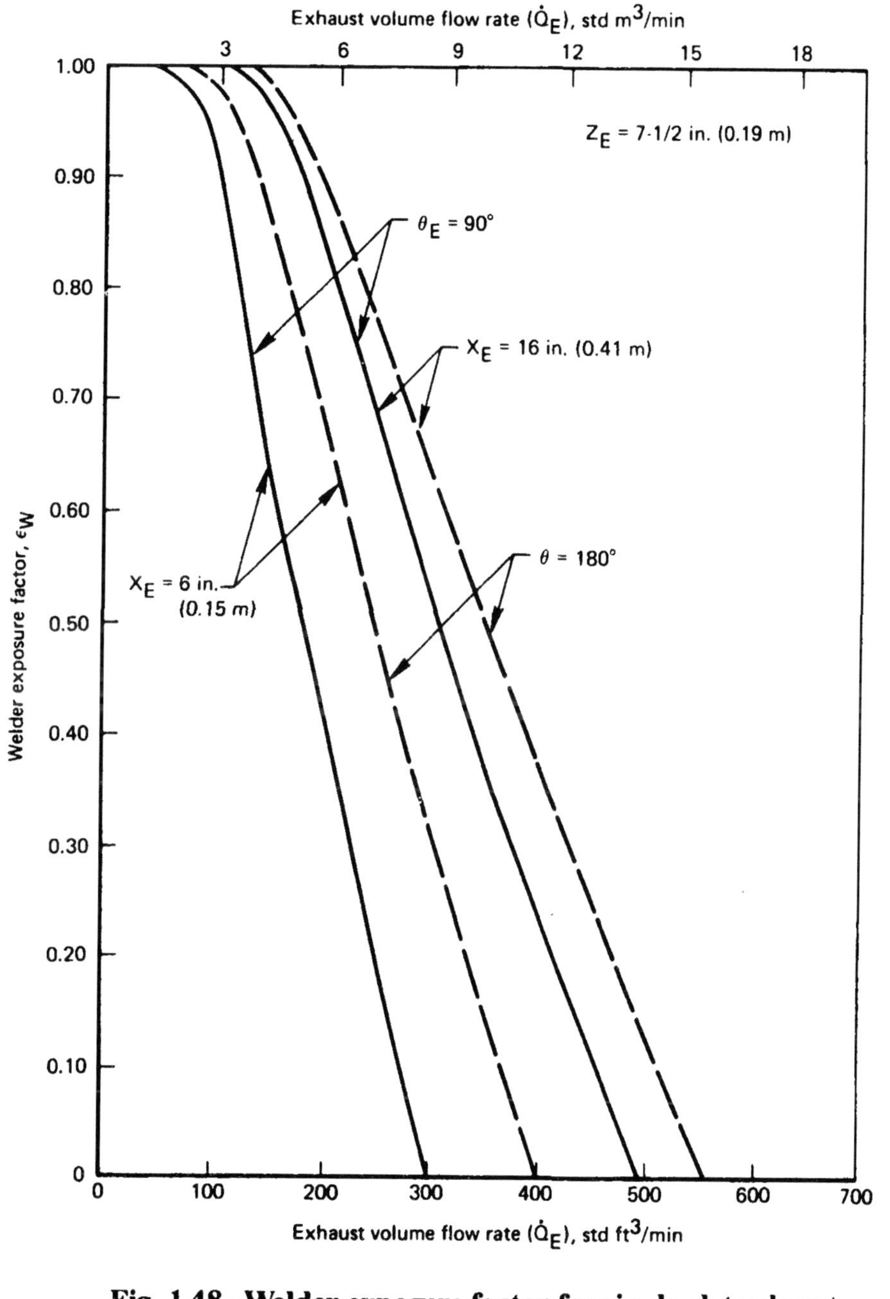

Fig. 1.48–Welder exposure factor for single-slot exhaust configuration

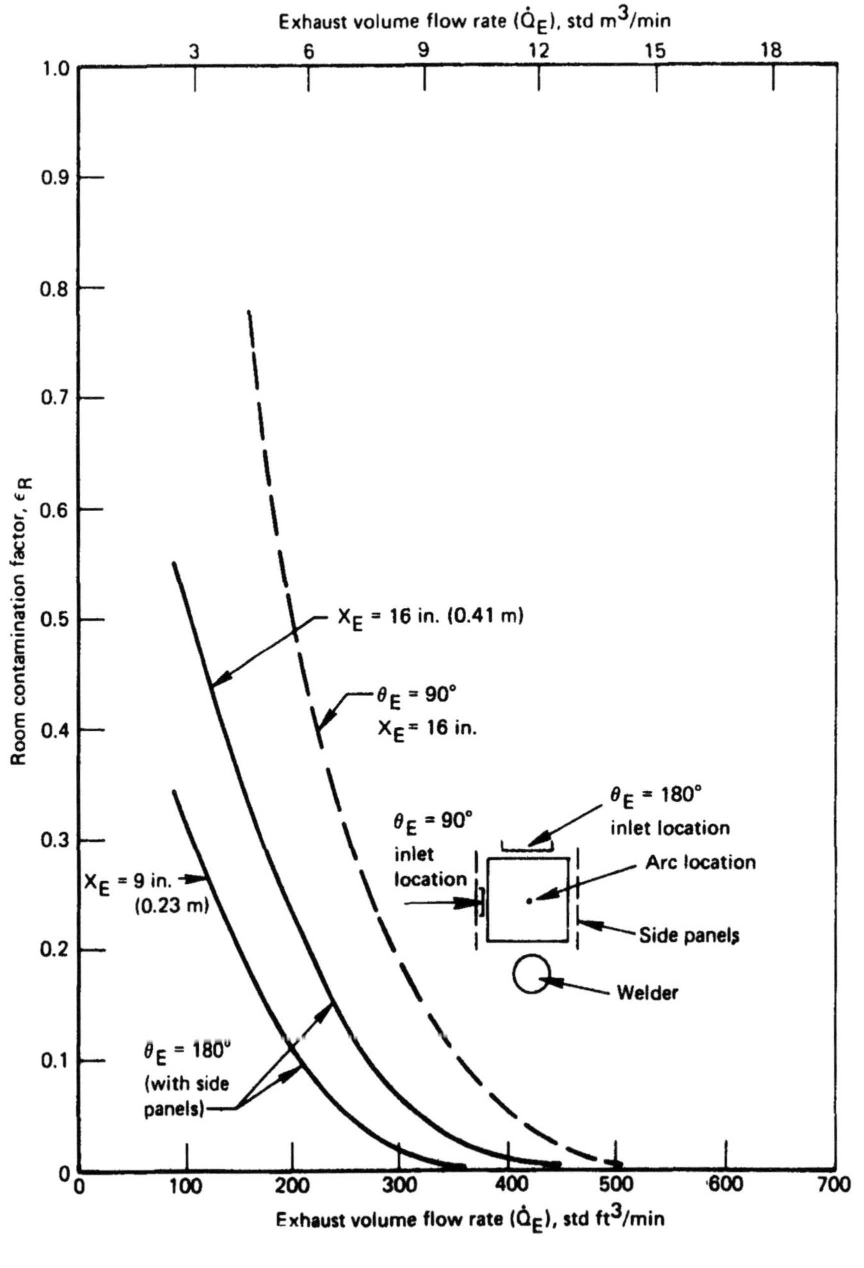

Fig. 1.49–Room contamination factors for double-slot inlet exhaust

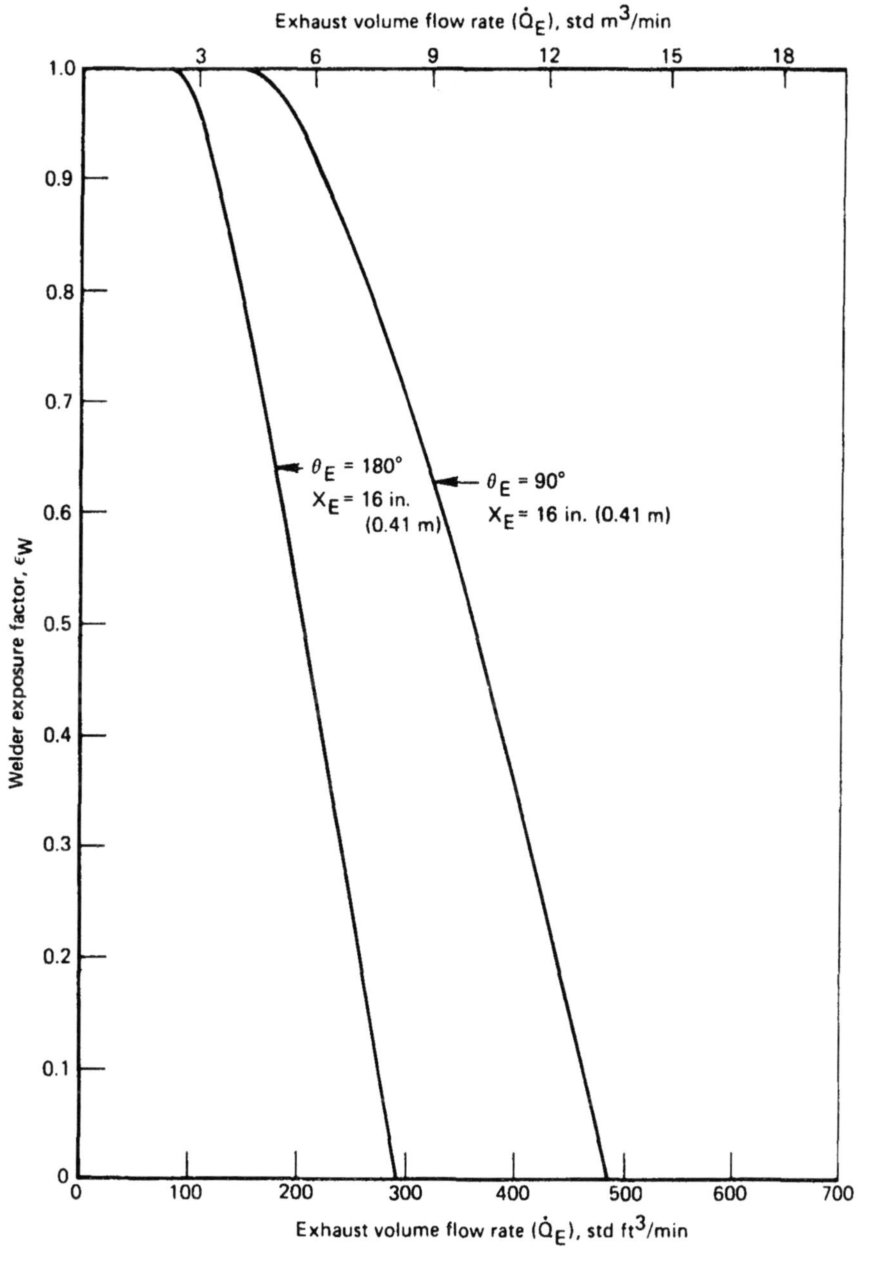

Fig. 1.50–Welder exposure factors for double-slot inlet exhaust

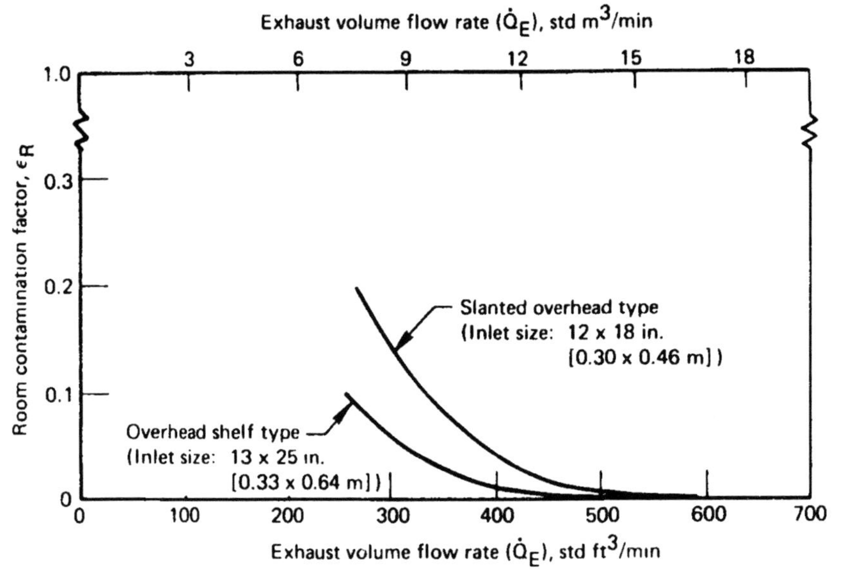

Fig. 1.51 – Room contamination factors for overhead shelf and slanted overhead type exhaust hood

If it is not possible to set up a fan, blower, or other system at θ_V = 90° (cross-draft configuration), then Fig. 1.24 or Fig. 1.25 would be used instead of Fig. 1.23. As indicated in the report, ventilation from behind and in front of the welder is not effective and should not be used (welder is better off without ventilation from these directions). It is possible to generate air currents of the level indicated above using either floor or table fans.

Question 3

Local ventilation usually provides protection only for the welder. If environmental protection is also required, a local exhaust device could be used. For this example, it will be assumed that a flanged 4 in. (102 mm) diameter duct will be used to collect the fumes. Figure 1.44 can be used to specify where the inlet should be located and to determine the volume flow rate required to achieve various collection efficiencies, or room contamination factors, or both. For example, the exhaust flow rates needed for a collection efficiency of 50 percent and no room contamination are given in Table 1.5 for various exhaust inlet locations (X_E) at θ_E = 90.°

Due to the unsteadiness of the welding plume with respect to its spatial location, the exhaust volume flow rate should be increased by at least 30 percent or higher.

Table 1.5
Exhaust volume flow rates (std ft³/min)

	Room contamination (ϵ_R)				Welder exposure condition
		Flow rate			
	std ft³/min	std m³/min	std ft³/min	std m³/min	C_V, mg/m³
Inches	50 percent		100 percent		50 percent
6.0	25	0.71	46	1.30	8
8.0	38	1.08	77	2.18	12
10.0	60	1.70	125	3.54	16
12.0	96	2.72	225	6.37	24

It should be noted that the welder's exposure to fumes may be reduced to zero at less than 100 percent exhaust collection or zero room contamination. For example, at X_E = 12 in. (305 mm), an exhaust volume flow rate of 225 std ft³/min (6.37 std m³/min) is required to achieve zero room contamination, whereas a flow rate of only 138 std ft³/min (3.91 std m³/min) is required to fully protect the welder (ϵ_W = 0). This shows that either performance criterion can dictate the design of the system and that both must be checked.

An alternate method of solving the problem is to determine the exhaust conditions required to fully protect the welder (ϵ_W = 0). An estimate of the exhaust volume flow rate for various exhaust inlet locations (X_E) can be determined from Fig. 1.44 (assuming an exhaust duct diameter of d_E = 4 in. [102 mm] for ϵ_W = 0). The

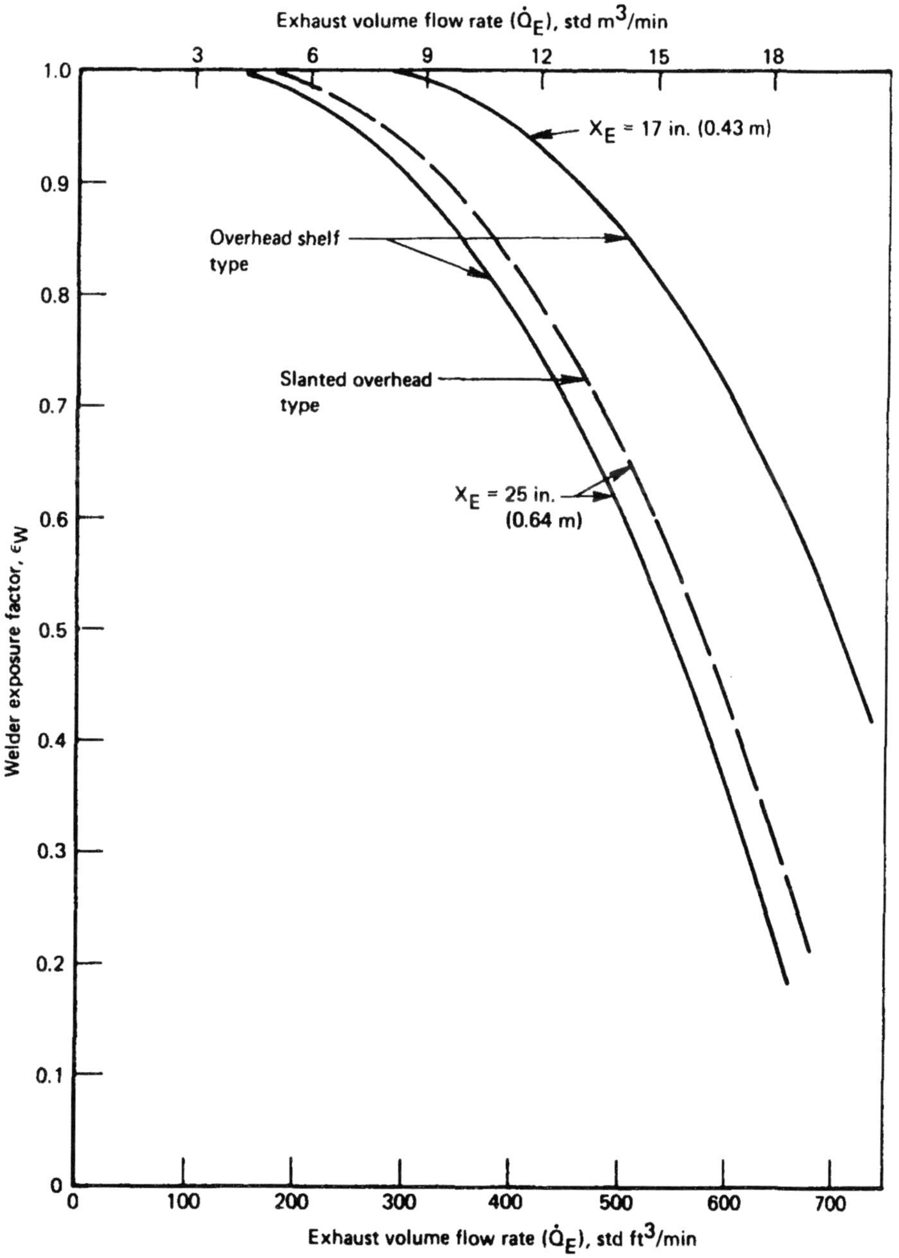

Fig. 1.52–Welder exposure factors for overhead shelf and slanted overhead hood exhaust configurations

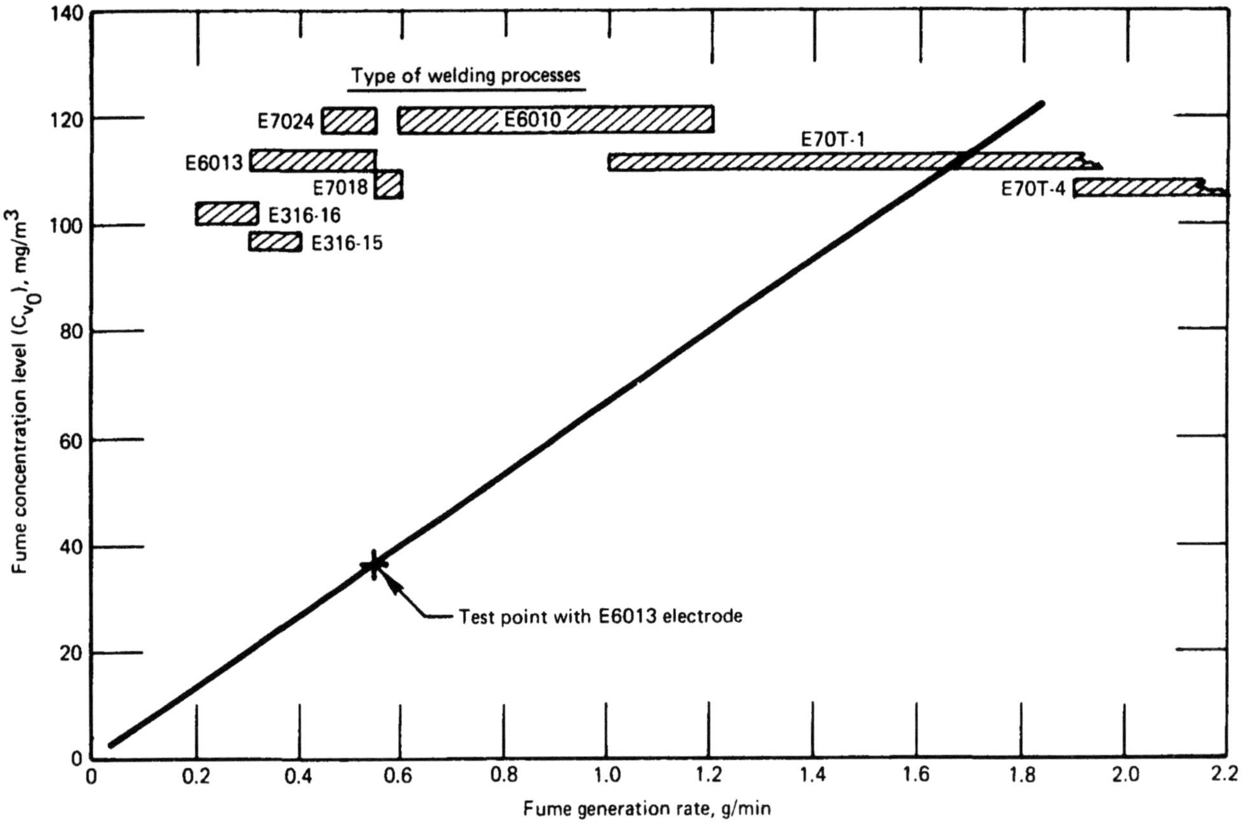

Fig. 1.53 – Estimated levels of welder fume exposure for zero ventilation and 100 percent arc time

volume flow rates are presented below.

Horizontal distance, X_E		Flow rate, $\mathring{Q}_E$	
in.	mm	std ft³/min	std m³/min
6.0	152.4	46	1.30
8.0	203.2	73	2.07
10.0	254.0	102	2.89
12.0	304.8	138	3.91

The welder exposure conditions (ϵ_W) for the exhaust configurations defined above can be determined from Fig. 1.41. For example, assuming X_E = 8.0 and ϵ_R = 0.5: therefore, X_E/d_E = 2.0 and $\mathring{Q}_E/A_E$ = 3.0 for d_E = 4 in. (102 mm) and $\mathring{Q}_E$ = 38 std ft³/min (1.08 std m³/min). From Fig. 1.44, ϵ_W = 0.3, and the welder exposure would be C_V = 12 mg/m,³ assuming a zero ventilation level of C_{V_0} = 40 mg/m³ (refer to Question 2 for C_{V_0} determination). The welder exposure conditions (C_V) for the 50 percent room contamination factor were given in Table 1.5. For zero room contamination, the welder exposure to the fumes would be zero.

Summary of Part I

The objective of this study was to generate data that can be used in the design of systems used to control the flow of fumes and gases generated during various welding processes. Design data for control methods were developed from experimental and theoretical studies in terms of (a) the level of exposure to which personnel within an area or room would be subjected as a result of welding operations and (b) the level of exposure that the welder would experience. These data were generated by investigating general room and local ventilation and local exhaust devices.

The effectiveness of local exhaust (collection) devices in reducing welding fume concentration levels in the welder's breathing zone and in reducing room concentration levels by collection of the welding fumes was defined experimentally. For circular cross-section flanged exhaust duct devices, the experimental results are presented in terms of room contamination factors for such devices as related to the required exhaust flow rate and physical conditions of the welding setup (e.g., location of the inlet of the exhaust device relative to the arc). Design guidelines are given for a large range of physical conditions and exhaust flow rates to achieve desired room and welder exposure levels. This information can, in part, be used to determine blower capacity, to determine the associated energy requirements for operation, and to perform systems analyses to evaluate various design options.

A room contamination factor and a welder exposure factor are defined for use in determining the effectiveness of the local exhaust devices. Experimental results are given which show that an exhaust device that reduces the room contamination of fumes (reduces room contamination factor) may not reduce the fume contamination level (welder exposure factor) in the welder's breathing zone. It is shown that both factors should be considered for comprehensive fume control.

Correlation of data for single- and double-slot local exhaust devices was obtained. The correlation is given in terms of the room contamination and welder exposure factors. The correlations are expressed in terms of weld process physical variables and exhaust device collection flow rate.

Correlation of data for overhead shelf and slanted overhead types of local exhaust devices was obtained. This relationship is also given in terms of room contamination and welder exposure factors. The correlations are expressed in terms of weld process physical variables and exhaust device collection flow rate.

The effect of local ventilation in reducing the fume concentration level in the welder's breathing zone was experimentally determined. Reductions in fume concentration level in the welder's breathing zone are presented as a function of ventilating conditions (air speed and direction). The effect of the ventilating air flow on reducing fume concentration levels was found to be strongly dependent upon the ventilation direction. A cross draft was found to be the most effective in removing fumes from the welder's breathing zone, whereas there may be an increase in the fume concentration level when the ventilating source is behind the welder.

Theoretical studies were conducted to determine the rate of buildup of contaminants within any confined space resulting from welding fume generation rate. The room concentration level is expressed in terms of the fume generating rate and average ventilation rate.

Part II. ARC WELDING FUMES AND GASES

Introduction

Welding fume particles are exceedingly small. Virtually all such particles are smaller than 1.0 μm (micrometers). In previous studies conducted for the American Welding Society, the observers found that a significant fraction of the fumes produced by covered, solid, and tubular or flux cored electrodes contained particles that were smaller than 0.25 μm in diameter.

Such particle size ranges are in substantial agreement with the observations of other investigators. Particle size is important since it is an indication of (a) the depth to which particles will penetrate into the respiratory system and (b) the number of particles that will be retained therein. Welding fume particles are in the size range of particles that penetrate deep into the respiratory system, but they are smaller than particles that are retained with maximum efficiency assuming agglomeration does not occur.

In common with other gases, the gases associated with welding follow the normal laws of diffusion and mix freely with the general atmosphere. They are affected by air movements and by gravity. They pass through the respiratory system along with the other gaseous constituents of the atmosphere.

Origin of Fumes and Gases

Fume generation processes involved during welding are complex and not completely understood. It is generally agreed that the principal mechanism is elemental and oxide vaporization, followed by extremely rapid condensation to form particles. In gas metal arc welding (GMAW), Heile and Hill have suggested that the Mn component comes strictly from elemental vaporization, the Si component from SiO formation, and the Fe component from elemental plus FeO vaporization (Ref. 2.3). Regardless of the precise mechanism, vaporization occurs mainly at or very near the tip of the electrode where the temperature is highest.

The major source of fumes is the metallic part of the electrode and the electrode covering or flux. Only a small fraction of the total quantity of fumes is contributed by the molten weld metal pool because the temperature of the pool is much lower than the temperature at the tip of the electrode. Significant contributions from the base plate can occur, especially if the plate is protected against rusting or corrosion with a metal that has a high vapor pressure: for example, zinc or cadmium.

The constituents in welding fumes can be predicted to a degree on the basis of the composition of the electrode and base metal. If an E70S-3 solid electrode is used for welding carbon steels, iron oxide will be present as the major fume constituent along with small amounts of the oxides of manganese and silicon; copper oxide will be present if the filler wire is copper-coated. If a covered electrode is used for the same purpose, predictions regarding fume constituents are more difficult. The oxides of iron, manganese, and silicon will be present in the fumes, but constituents associated with the electrode covering will also be present. Electrode coverings are quite complex and include gas formers, deoxidizers, binders, slag formers, arc stabilizers, and alloying additions; low hydrogen electrodes contain fluoride compounds as major ingredients. Constituents in the fumes produced by flux cored electrodes are similar to those associated with covered electrodes and are just as difficult to predict.

Welding gases have several origins; depending upon the specific welding process, these include:

(1) Shielding gas
(2) Decomposition products of electrode coverings and fluxes
(3) Reactions of the arc with atmospheric constituents (and atmospheric contaminants)
(4) Ultraviolet irradiation of the atmosphere

In gas shielded arc welding operations, for example, the most likely source of gases in the welding environment is the shielding gas or its decomposition products. When carbon dioxide is used for shielding, carbon monoxide as well as carbon dioxide may be present in the welding area. Similarly, argon will be present when it is used to shield the arc. While argon is not considered toxic, it is heavier than air and can act as an asphyxiant when it displaces atmospheric oxygen in a closed or confined space.

Other gases that may be present include nitrogen dioxide (or other oxides of nitrogen) and ozone. These gases can be formed by the photochemical reaction of ultraviolet radiation from the arc on nitrogen and oxygen in the atmosphere.

Care must be exercised when welding in an area (a) where chlorinated hydrocarbons (e.g., trichloroethylene, perchlorethylene, and others) are used as solvents for degreasing purposes or (b) where vapors from such operations can be carried by air currents. A toxic gas may be formed by photochemical reactions between ultraviolet radiation from the arc and these solvent vapors. Toxic gases can also occur in small amounts as the result of thermal decomposition of such vapors as they pass over hot surfaces (Ref. 2.4). Other toxic gases may be formed by the thermal decomposition of phosphate coatings used to prevent rusting of steel surfaces.

Concerns Regarding the Welding Environment

The presence of fumes and gases in the welding environment is a matter of concern to those responsible for the well-being of welding personnel. Since constituents are a potential hazard to the health of the welder, potential problems can be anticipated by estimating the concentrations of fume and gas constituents in the welding area and comparing these data to established exposure standards.

To consider a concrete example: it will be assumed that fumes are being generated at the rate of 0.40 g/min during the semiautomatic gas metal arc welding of carbon steel with an E70S-3 electrode in an unventilated room with a volume of 283 m^3 (10 000 ft^3). If it is assumed that such fumes diffuse uniformly throughout the room (a simplistic assumption, since fume particles are subject to gravitational and other forces), the total fume concentration in the room will exceed the established threshold limit value, TLV, as defined by the American Conference of Governmental Industrial Hygienists (ACGIH) of 5 mg/m^3 [4] for not otherwise classified total fume particulates in about 3.5 minutes. Ventilation provided at the rate of about 85 m^3/min (3000 ft^3/min) should be sufficient to maintain the fume concentration below 5 mg/m^3 indefinitely.

To examine this example further, it will be assumed that the composition of the fumes is as follows:

4. Note that the value accepted by OSHA for iron oxide, the major constituent in the fumes produced during the welding of steel, is 10 mg/m^3. See General Industry Safety and Health Regulations Title 29 CFR Part 1910.1000.

Element (oxide)	Amount, weight percent
Fe (Fe_2O_3)	63 (90.0)
Mn (MnO)	5 (5.2)
Si (SiO_2)	2 (4.3)
Cu (CuO)	0.4 (0.5)

With a total fume generation rate of 0.40 g/min and uniform distribution of the fume constituents in the room, the time required to exceed the ACGIH time-weighted-average threshold limit values (TWA-TLV) for the constituents listed above is shown in the following table:

Constituent	TWA-TLV, mg/m^3	Time to exceed TWA-TLV, min.
Fe_2O_3	5	3.9
Mn	5	70.8
SiO_2	2*	32.9
Cu	0.2	70.8

*Amorphous, respirable.

The tabular data indicate that iron oxide is of most concern from the hygienic viewpoint, because the time for this compound to exceed its limiting value is the shortest.

As has been observed, the assumption that the fumes are dispersed uniformly throughout the welding area may be in error since the fume particles possess mass and are subject to settling. Fume concentrations are highest in the immediate welding area during the actual welding operation. Nonetheless, an example of this nature is useful in demonstrating possible problem areas.

Existing Literature

The technical literature of the past two decades contains numerous references to research conducted to (a) sample fumes and gases in the welding area, (b) determine individual fume and gas constituents, and (c) calculate concentrations of these constituents for comparison with recognized allowable limits. Many of these references are included in the AWS publication, *The Welding Environment* (Ref. 5). These investigations generally had limited objectives because they were undertaken to study the fumes associated with a particular welding application, to evaluate a specific electrode, to assess the performance of a ventilation system, or for some other definite purpose. The usefulness of some of this work is questionable today because supporting data on welding procedures, sampling methods, and analytical techniques are often incomplete or totally lacking. Nonetheless, these studies are important because the results form the basis for much of today's knowledge.

Until recently, little data were available that could be used to determine the rate at which fumes are produced as a function of welding process, welding conditions, consumables, and other variables. The lack of such data is due in part to the difficulties associated with determining fume generation rates accurately. This information is valuable to those engaged in the layout of welding areas and the design of ventilation systems as well as those concerned with the health, safety, and comfort of welding personnel. Some of the more significant investigations are reviewed below; the cited references should be consulted for additional details.

Kobayashi, Maki, and Ohe investigated the fumes produced by several electrodes used for shielded metal arc welding and determined fume generation rates as a function of the type of electrode covering (ilmenite, lime, titania, and iron powder) and selected process variables (current, voltage, power, electrode polarity, and welding speed) (Ref. 2.5). In the course of these studies, equations were developed to show the dependence of fume generation rate on welding current. For electrodes with ilmenite or lime coverings, the rate increased almost in proportion to the square of the welding current. For electrodes with lime-titania coverings, the rate increased almost linearly with current. Increases in arc voltage (or arc length) were accompanied by significant increases in fume generation rate.

Swedish investigators developed procedures to determine fume generation rates for various types of electrodes to obtain data for use in establishing a system for classifying electrodes according to a calculated theoretical air requirement factor (Refs. 2.6, 2.7). The magnitude of this factor depends upon the weight of each potentially hazardous fume constituent as well as on the fume generation rate.

Studies were conducted by Heile and Hill to investigate the fume generation and metal deposition characteristics of electrodes used for shielded metal arc welding, flux cored arc welding, and gas metal arc welding (Ref. 2.3). These data and those obtained from analyses of fume and thermodynamic calculations were used to support theoretical concepts for the mechanisms of fume generation. Elemental vaporization, condensation, and oxidation-enhanced vaporization processes appeared to control fume generation during welding.

Other investigations concerned with the mechanisms of fume generation during shielded metal arc welding were conducted by Kobayashi, Maki, Hashimoto, and Suga (Ref. 2.8, 2.9) Using a high speed motion picture camera, evidence was obtained that fumes originated at the tip of the electrode and were propelled away from the tip by magnetic forces. Various tracer metals (manganese, chromium, and nickel) were used to determine the contributions of the core wire, electrode covering, and base metal to the total fume production rate. Only a small fraction of the fumes originated in the molten weld metal pool.

Morita and Tanigaki measured the fume generation rates of 60 different covered electrodes used for welding carbon and low alloy steels (Ref. 2.10). Fume generation rates were highest for electrodes with cellulosic coverings; rates were lower for electrodes with low-hydrogen, mineral, and iron powder iron oxide coverings. By correlating these data with the composition of the fumes, electrode coverings, and slags, a relationship between fume generation rates and the presence of specific elements and compounds was established. This information was used in the development of several experimental electrodes with low fume generation rates.

Recent Soviet research has been directed toward determining the fume generation and metal deposition characteristics of electrodes used in various welding processes. Katrenko, Kassov, and Karpenko investigated the effect of slag-forming constituents in flux cored electrodes on the fume generation rate and the weight of fumes as a function of the weight of deposited metal (Ref. 2.11). The effect of various shielding gases on the fume generation characteristics of electrodes used in the gas metal arc welding of ferrous metals was studied by Erman and Kabasin (Ref. 2.12) and by Vorontsova and Mosolov (Ref. 2.13). The fume generation characteristics of aluminum gas metal arc welding electrodes have been determined by Psaras and Popova (Ref. 2.14). In a related investigation, Erman demonstrated that the fume generation rate for an aluminum electrode was lower during pulsed arc welding than during conventional gas metal arc welding (Ref. 2.15).

Studies of various types of electrodes have also been conducted by Pantucek to establish an electrode classification system based on fume generation rates and fume compositions (Ref. 2.16).

Of these investigations, the one conducted by Heile and Hill was most useful in the AWS program because experimental data were obtained with many of the same domestic electrodes. Data from this paper have been incorporated in this report wherever they served to supplement or assist in the interpretation of the experimental results. The results of foreign studies are useful because they indicate trends in expected fume generation rates when welding is done with electrodes similar to those produced domestically.

Data on fume generation rates are often based on the weight of a sample that represents only a fraction of the total amount of fumes produced during welding. As a result, sample weights are subject to variations caused by inconsistencies in the sampling procedures and in the behavior of the fume plume. The laboratory procedures used in this investigation were established to minimize these sources of error by collecting all the fumes produced during a measured interval of welding.

Scope

This task was designed to provide basic information on the fume generation characteristics of commonly used arc welding electrodes as a function of welding process, electrode classification, and selected operation variables. It emphasized a comprehensive laboratory investigation of arc welding fumes to determine the rate at which they are produced with selected welding conditions in order to relate the fume generation rate to various electrode characteristics. More limited studies were also conducted to (a) detect gases in the welding area, (b) determine fume compositions, and (c) characterize fume particles in terms of size, crystalline structure, etc. Such data will be useful to industry in planning industrial hygiene programs and determining ventilation requirements.

At the request of the AWS Research Committee, a short study was made to investigate the influence of humidity in the sampling area on fume sampling results. This study is reported in Section II E.

Limitations

Detailed laboratory investigation of the fume generation characteristics of representative electrodes was restricted to the following processes:

(1) Shielded metal arc welding
(2) Flux cored arc welding
(3) Gas metal arc welding
(4) Gas tungsten arc welding

Research was also conducted to investigate the fume generation characteristics of selected electrodes as a function of various process variables: current, voltage, arc length, etc.

During the course of this investigation, limited research on fume particle characterization, fume analysis, and gas detection during welding was included.

Program

Fume generation rates were determined for all electrodes under study under selected operating conditions. Relationships were also established between this rate and various electrode characteristics (melting rate, metal deposition rate, etc.) to obtain data that could be used effectively by welding engineers, industrial hygienists, ventilation specialists, and others concerned with the health and safety aspects of arc welding.

Equipment, Materials, Procedure

Fume Collection Chamber

The Battelle fume collection chamber is a sheet metal structure with a volume of about 33 ft^3 (0.93 m^3). The upper portion of the chamber tapers inward to form an area in which the filters and holder are located; the fumes are removed from the chamber by an exhaust blower attached to the exit side of the filter holder by flexible ducting. The chamber can be sealed to prevent the leakage of fumes during the sampling interval.

The chamber is equipped with a variable-speed turntable on which the workpiece is placed. A steady-rest is provided to assist the welder in maintaining a constant arc length during welding. The welding torch is manipulated by means of a glove port. Several holes are located around the base of the chamber to insure a positive flow of air through the structure during welding and sampling. Views of the chamber are shown in Figs. 2.1A and 2.1B.

Filter System. Samples of the total fumes produced during welding were collected with an assembly consisting of an absolute filter preceded by a prefilter.

Absolute Filters. Glass fiber absolute filters with a high capture efficiency for particles below one micrometer in size were used because they were well-suited for gravimetric analysis and could retain large quantities of fume particulates. Also, moisture pick-up problems were minimal with these filters, since they were non-hygroscopic. However, glass fiber filters are ill-suited for chemical analysis because it is difficult to separate the fume particles from the filter mesh. The absolute filters were 8 by 10 in. (0.20 x 0.25 m) in size.

Prefilter. Despite the ability of glass fiber absolute filters to capture and retain large amounts of fumes, they were subject to clogging or overloading. To eliminate this problem, a prefilter consisting of two layers of glass fiber insulation was placed in front of the absolute filter.

Analytical Samples. Fume samples were also collected for chemical analysis. Cellulose membrane filters with a pore size of 1.2 micrometers were used for this purpose because such filters could be ashed, dissolved, or otherwise treated to recover the fumes for analysis. Since the amount of fumes that could be retained by membrane filters without clogging was limited, these filters were also preceded by a glass fiber insulation prefilter.

The fume collection area of the chamber is shown in Fig. 2.1C; the filters are shown in Fig. 2.1D.

Other Equipment

Welding with covered electrodes was done with a conventional motor-generator power unit and a standard electrode holder.

For welding with wire electrodes, a torch with a water-cooled nozzle was mounted in the fume collection chamber on a steady-rest that permitted the operator to maintain the selected electrode stickout. Power was supplied by a constant potential source and a commercial wire feeder was used to feed wire at the desired rate.

Current and voltage were continuously monitored during welding operations. Since voltage was measured between the wire feeder and base plate, the measurement included the voltage drop in the welding torch and cables as well as the drop across the arc.

Electrodes

The fume generation characteristics of electrodes in a single classification were generally determined by evaluating three electrodes, each made by a different manufacturer. Only one or two electrodes per classification were evaluated in a few instances when the number of manufacturers producing a particular electrode was limited (some nickel-base electrodes) or when the evaluation of more than one electrode appeared to be unnecessary (solid electrodes used for gas metal arc welding).

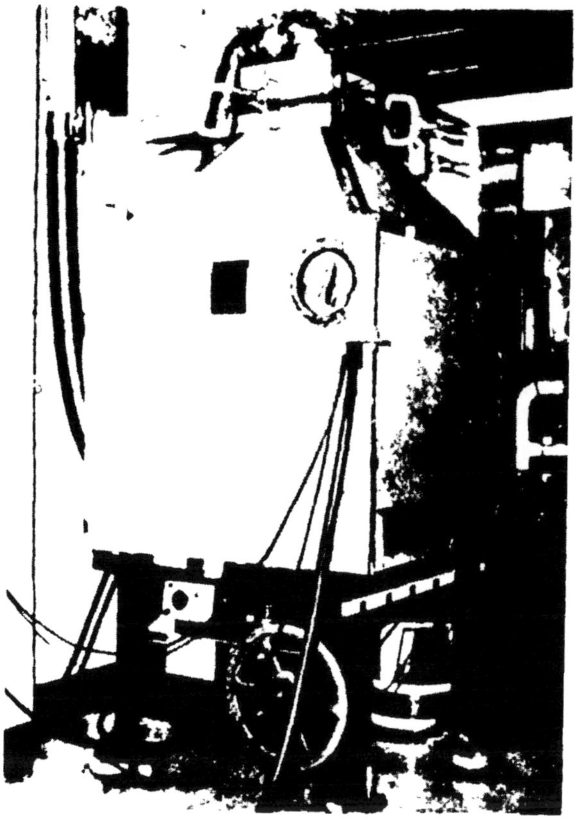

Fig. 2.1A—Fume collection chamber

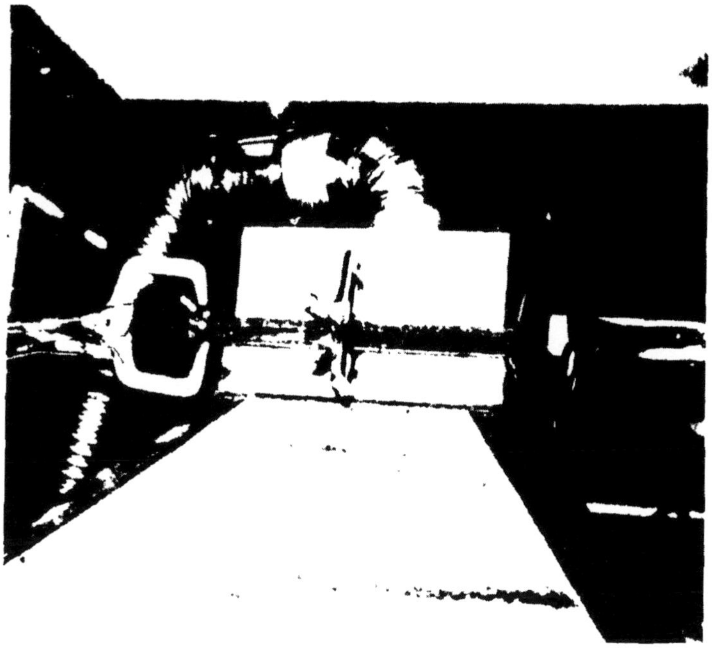

Fig. 2.1C—Fume collection section

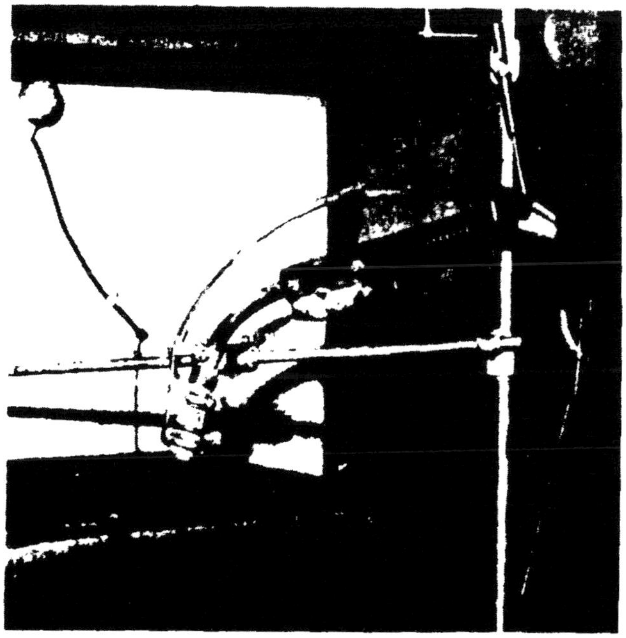

Fig. 2.1B—Torch arrangement inside chamber

Fig. 2.1D—Collection area with a used absolute filter and glass fiber prefilter

The electrodes used in this investigation were obtained from local commercial sources. Electrodes used for some specific studies (e.g., studies of the fumes produced by electrodes with different diameters) were obtained from one manufacturer.

The electrodes included in this investigation are listed in Appendix A, along with the purposes for which they were used.

Section IIA, Laboratory Test Method

Under the guidance of the AWS Research Committee, a laboratory test method was established to study the fumes associated with arc welding and related processes. Specifically, this method was designed for the laboratory collection of total fume samples in order to determine fume weight as a function of welding time, various electrode characteristics, and procedural variables.

Procedures

Experimental procedures were established to determine the fume generation characteristics of electrodes used in various arc welding processes by gravimetric analyses of fume samples. Samples of the fumes produced by specific electrodes were collected in triplicate to average the inconsistencies of sampling; in some instances, one or two extra samples were collected to obtain additional data or to verify specific results. With appropriate changes to fit the process, the sampling procedures were generally applicable to cutting and spraying as well as arc welding.

After the desired welding conditions were established, the fume samples were collected in accordance with the following procedures:

(1) The filter-prefilter assembly was weighed to 0.1 mg on an analytical balance.

(2) The filter-prefilter assembly was placed in the holder at the apex of the fume collection chamber, the sealing gaskets were installed, and the holder was clamped in place.

(3) The chamber was sealed and the blower that exhausted the fume chamber was turned on. Immediately thereafter, a bead-on-plate weld was made. The welding interval was one-half to one minute, depending on the expected fume generation rate.

(4) With the blower functioning, fumes were collected until the chamber atmosphere was clear.

(5) The filter-prefilter assembly was weighed immediately after its removal from the filter holder to determine the fume weight.

The following were recorded during sampling: (a) filter-prefilter weight before and after sampling, (b) welding conditions, and (c) electrode physical characteristics (diameter, length, and weight of electrode consumed).

To avoid problems caused by the pickup of moisture by the absolute filter, prefilter, or fume sample, welding and sampling were done in a clean room where temperature and humidity could be controlled. Also, glass fiber filters were used because they were nonhygroscopic. Nevertheless, moisture pickup was possible in view of the long periods (15 to 30 minutes) during which fume samples were collected. To examine this possibility, fume samples were collected in accordance with (a) established procedures and (b) procedures in which the filter-prefilter was baked before each weighing operating to remove absorbed moisture. Moisture did not appear to be a problem, because there was essentially no difference in the fume weights. However, the possible effects of moisture pickup on the accuracy of experimental data should be considered if welding and sampling are done in a high humidity area. Additional information on the effect of humidity is given in Part IIE of this report.

Procedures were also established to measure the metal deposition rate during welding. To obtain these data, bead-on-plate welds were deposited on small sections of plate that were weighed before and after welding to determine the amount of metal deposited during the welding interval. In the case of welds made with shielded metal arc or flux cored arc welding electrodes, slag was removed before the weighing operation.

Chamber Evaluation

Studies were conducted to evaluate the performance of the Battelle fume collection chamber and filter system. To provide a common basis for comparing the performance of this chamber with that of equipment designed and used by members of the AWS Research Committee, these studies were undertaken using specific lots of E7018 and E70T-1 electrodes set aside for this purpose. To further insure comparability, the welding conditions used by the subcommittee members for each of the respective electrodes were also used by Battelle; these conditions are shown in Table 2.1A. The data obtained by the AWS Research Committee and by Battelle are presented in Table 2.1B. When the differences in equipment and sampling procedures are considered, the agreement among the data is good.

During the program, the Battelle fume collection chamber and filter system performed well, and consistent results were obtained. In replicate sampling experiments, the fume generation rates varied by only 1 or 2 percent. The variation in calculated quantities (e.g., weight percent of electrode converted to fumes, weight of fumes per weight of deposited metal, etc.) was somewhat greater but generally less than 4 percent. These observations were confirmed by the small standard deviations associated with each group of data.

As indicated, fume samples were collected in triplicate for all of the electrodes evaluated. For any given electrode, these samples were collected one after another under the same welding conditions, and there was little variation in the individual fume sample weights. More variation would be expected if the sampling was made at random and the welding conditions were changed after each fume collection experiment.

Experimental Results

Figures and short tables are used throughout this report to summarize the data and assist in the interpretation of the results. Data on the fume generation characteristics of arc welding electrodes are presented in detailed tabular form in Appendix B. In addition to the fume generation rate for specific electrodes, these sheets contain information on (a) the conditions under which the experimental data were obtained, (b) the melting and metal deposition rates of the electrode, and (c) calculated quantities that relate the amount of fumes produced during welding to the metal deposition rate and other electrode characteristics.

Fume generation rates are expressed in "measured" and "normalized" values. The normalized value provides a means for minimizing the effects of unintentional variations in welding current on fume generation rate. Normalization was accomplished by relating the fume generation rate to the current squared. Figure 2.9 shows the relation of I^2 trend lines and data for two covered electrodes, one cored electrode, and one solid wire electrode. It is apparent that the current-squared normalization will fit a wide variety of fume generation data.

The fume generation rate is most important because it is a direct measure of the amount of the fumes produced during welding under specified conditions. This information can be used effectively by welding engineers, industrial hygienists, ventilation specialists, and others. Of equal importance is the calculated factor "weight of fumes per unit weight of deposited metal," because it can be used to estimate the amount of fumes that will be produced during production welding operations. This factor also provides a convenient means to compare welding processes on a fume-produced to weight-of-deposited-metal basis.

Table 2.1A
Welding conditions for E7018 and E70T-1 electrodes

Parameter	E7018 AWS Subcommittee on Sampling	E7018 Battelle	E70T-1 AWS Subcommittee on Sampling	E70T-1 Battelle
Current, A	140	135-145	450-470	420-450
Voltage, V	---	23-24	30	30-31
Wire feed rate, in./min (mm/s)	---	---	87.21 (206)	87.00 (205.5)
Welding speed, in./min (mm/s)	5.93 (14)	5.93 (14)	6.77 (16)	6.77 (16)
Shielding gas	---	---	CO_2	CO_2
Gas flow rate, ft^3/h (liters/min)	---	---	19-21 (40-45)	21-24 (45-50)
Electrode stickout, in. (mm)	---	---	32 (1.25)	32-35 (1.25-1.37)
Weld time, s	---	60	---	15 and 30
Electrode polarity	Positive	Postive	Positive	Positive

Table 2.1B
Fume generation characteristics of E7018 and E70T-1 electrodes
(AWS Research Committee and Battelle data)

Characteristics	E7018 AWS Subcommittee on Sampling A	B	C	E7018 Battelle	E70T-1 AWS Subcommittee on Sampling A	B	C	E70T-1 Battelle
Fume generation rate, g/min	0.50	0.52	0.43	0.51	1.15	1.05	1.27	1.26
Weight percent electrode converted to fumes	1.54	1.79	1.47	1.62	0.75	0.70	0.83	0.85
Melting rate, kg/h (lb/h)	---	---	---	---	---	9.00 (19.8)	9.17 (20.2)	8.90 (19.6)

Section IIB. Fume Generation Characteristics of Arc Welding Electrodes

Baseline Data

Research was conducted to obtain baseline data[5] on the fume generation characteristics of representative covered, flux cored, and solid electrodes used for welding carbon and low alloy steels, stainless steels, nickel and other high alloy base metals, copper alloys, and aluminum alloys. To obtain results of value to those engaged in production welding operations, welding was done at currents in the mid-to-upper portion of the range recommended by the electrode manufacturer. Some welding was done under less ideal conditions with selected electrodes to study the effect of process variables on the fume generation characteristics.

Fume Generation Rate Versus Welding Current. Experimental evidence indicates that the fume generation rate varies in accordance with a power function of current. Kobayashi, Maki, and Ohe determined the fume generation rates for a series of ilmenite, lime, and lime-titania covered electrodes with diameters ranging from 3.2 to 6.0 mm (0.128 to 0.240 inch) at appropriate current levels (Ref. 2.5). In analyzing the data, they observed that the fume generation rate (FGR) for the electrodes in each classification could be related to welding current, I, by an equation of the form

$$FGR = a\,I^b$$

where b had a value between 1.17 and 1.73. While the equations representing the fume generation behavior of the electrodes in each classification differed in accordance with the electrode covering, their similarity indicated the general dependency of fume generation rate on a power function of current.

In the present research program, this type of analysis was expanded to derive a general equation relating the fume generation rate for carbon and low alloy steel covered and flux cored electrodes to the baseline welding current. This grouping was used because there are similarities between these electrodes, particularly in the compositions of the respective coverings and fluxes. For example, Smith indicates that the high titania E6013, E7014, and E7024 covered electrodes are similar to the E70T-1 and E70T-2 flux cored electrodes in terms of the type of covering and flux, slag composition, and weld metal characteristics (Ref. 2.17).

Data on the fume generation rates of 4 mm (5/32 in.) diameter covered and 2.4 mm (3/32 in.) diameter flux cored electrodes used for welding carbon and low alloy steels are plotted in Fig. 2.2 as a function of welding current. Studies to fit the data to a power function of current showed that the fume generation rate can be expressed as a function of $I^{1.27}$ with a correlation coefficient (r^2)[6] of 0.77. The magnitude of the correlation coefficient indicates a strong dependence of fume generation rate on current. This function is in general agreement with the results obtained by Kobayashi, Moki, and Ohe (Ref. 2.5) and lends credence to the concept that approximate fume generation rates for similar covered and flux cored electrodes can be predicted by a power function of welding current.

The line shown in Fig. 2.2 represents an $I^{1.27}$ regression line. The fume generation rates associated with two E6010 electrodes were much higher than those indicated by the regression line. These high fume generation rates were a result of running tests with excessively long arc lengths. The effect of arc length and other process variables on fume generation rate is discussed later in this report.

Data obtained from several studies in the course of this program are presented in Fig. 2.3. (The figure contains an I^2 regression line for comparison.) Included are fume generation rates as a function of current for the following electrodes:

(1) E6010, E6013, and E7018 covered electrodes with diameters ranging from 2.4 mm to 4.8 mm (3/32 in. to 3/16 in.) (Tables B30 through B33)
(2) E7018 and E7024 electrodes, 6.4 mm (1/4 in.) in diameter, at high welding currents (Table B40)
(3) Stainless steel covered and flux cored electrodes with diameters of 4 mm (5/32 in.) and 2.4 (3/32 in.), respectively (Tables B7, B8, B9, and B15).

These data points fall within a fairly narrow band. It appears that the fume generation rates for such electrodes will generally lie within a narrow band approximating the regression line, if the electrodes are used in accordance with the manufacturer's recommendations.

Fume generation rates measured by Heile and Hill (Ref. 2.3) and Kobayashi, Maki, and Ohe (Ref. 2.5) were also plotted to determine the agreement between these data and the baseline data developed during this program.

5. Baseline data are those data collected when electrodes were operated at currents in the mid-to-upper part of the manufacturer's recommended operating range.

6. The correlation coefficient (r^2) is an indication of the fit between the derived curve and the experimental data. If the dependency between fume generation rate and current is strong, the correlation factor will be close to 1 or -1. If the dependency between these variables is weak, the coefficient will be close to 0.

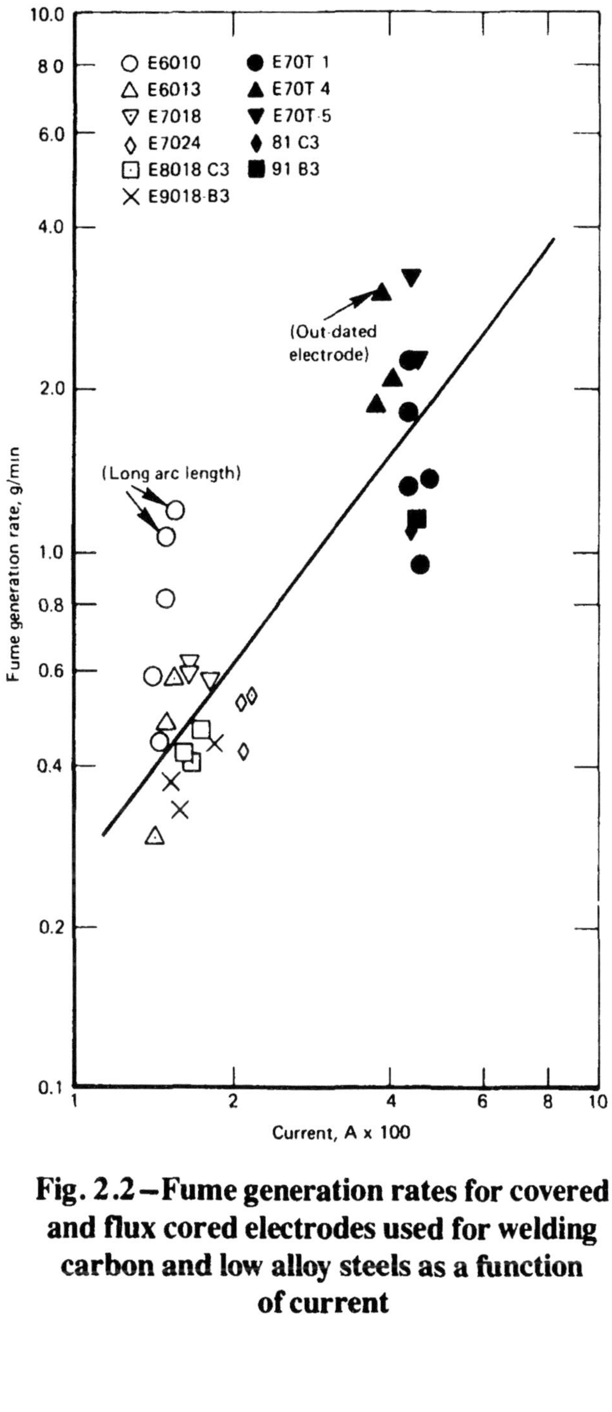

Fig. 2.2—Fume generation rates for covered and flux cored electrodes used for welding carbon and low alloy steels as a function of current

Heile and Hill evaluated the fume generation characteristics of several carbon steel covered electrodes at conventional welding currents, and the resulting data are presented in Fig. 2.4. Kobayashi, et al., investigated electrodes with ilmenite, lime, and lime-titania coverings; data associated with ilmenite coverings are included in this figure. These data also fit the I^2 regression hypothesis reasonably well. In addition, the AWS program data agree well with those developed by the other investigators.

Fume Generation Rate Relative to Electrode Type. The preceding overview of the baseline data has been primarily concerned with (a) the fume generation rates of various classifications of steel covered and flux cored electrodes and (b) an examination of the dependency of fume generation rate on welding current. Grouping of the electrodes in this manner was logical because covered and flux cored electrodes have much in common and because the covering (or flux) as well as the base metal contributes toward the formation of fumes. Data obtained from studies of solid gas metal arc electrodes were not included, because these electrodes do not incorporate a flux and the data could not be compared on an equal basis.

To examine the data obtained for all classifications of electrodes investigated during this program, the results for the following important fume generation characteristics are summarized in Fig. 2.5 and Table 2.2: (a) fume generation rate and (b) ratio of the weight of fume to the weight of deposited metal. Fume generation rate provides an immediate and direct indication of the amount of fumes to which a welder may be exposed. The ratio of the weight of fumes to the weight of deposited metal can be used to compare electrodes or welding processes on a fume exposure basis.

Under the "covered electrode" heading in Fig. 2.5, for example, the "steel" category includes electrodes that are used for welding carbon and low alloy steels: i.e., E6010, E6013, E7018, E7024, E8018 C3, and E9018 B3. The electrodes within any category can be determined from Table 2.2. Data on the fume generation characteristics of individual classifications of electrodes will be reviewed later.

A large quantity of experimental data has been compressed into Fig. 2.5 and its usefulness depends on the care with which it is interpreted. Ranges of fume generation rates and ratios of weight of fume to weight of deposited metal that may be observed when specific base metals are welded with covered, flux cored, and solid electrodes are evident in this figure. However useful this information may appear from the health and safety viewpoint, it must be used cautiously. For example, this figure indicates the fumes were produced at the highest rate when carbon and low alloy steels were welded with flux cored electrodes, and it would appear that problems associated with fumes could be minimized by using covered or solid electrodes. This may or may not be the case, since this conclusion does not consider the effects of the electrode and process variables on fume generation characteristics. When examining these data, the following should be considered:

(1) Fume generation rates are highly dependent on the welding current. Currents for flux cored arc welding were much higher than those used for welding with covered electrodes (Table 2.2). As a result, more fumes were produced during flux cored arc welding. However, since metal deposition rates were higher too, the ratio of weight of fume to weight of deposited metal was lower

for some electrodes during flux cored arc welding than it was when welding was done with certain covered electrodes. The range for this ratio even overlapped a portion of the range associated with gas metal arc welding electrodes which are recognized as low fuming types.

(2) The data in Fig. 2.5 were obtained with electrodes whose diameters were as follows: 4 mm (5/32 in.) for covered electrodes, 2.4 mm (3/32 in.) for flux cored electrodes, and 1.1 mm (0.045 in.) for most of the solid electrodes (a few of the solid electrodes had diameters of 0.8 mm [0.030 in]). Since recommended current ranges are largely based on electrode diameter, fume generation characteristics will change if diameter and current are increased or decreased. As an example, when 6.4 mm (1/4 in.) diameter electrodes were evaluated at currents near 400 A, fume generation rates were about the same as those produced by flux cored electrodes with diameters of 2.4 mm (3/32 in.) when operated at about the same current level.

(3) Other variables (arc length, electrode polarity, etc.) affect fume generation characteristics appreciably. The effects of such variables are not reflected by the data in Fig. 2.5.

Many factors must be considered when the data in Fig. 2.5 are reviewed. Electrodes are selected on the basis of economic and technical considerations (cost, weld metal properties, deposition rate, etc.), and not on the basis of their fume producing propensities. Should

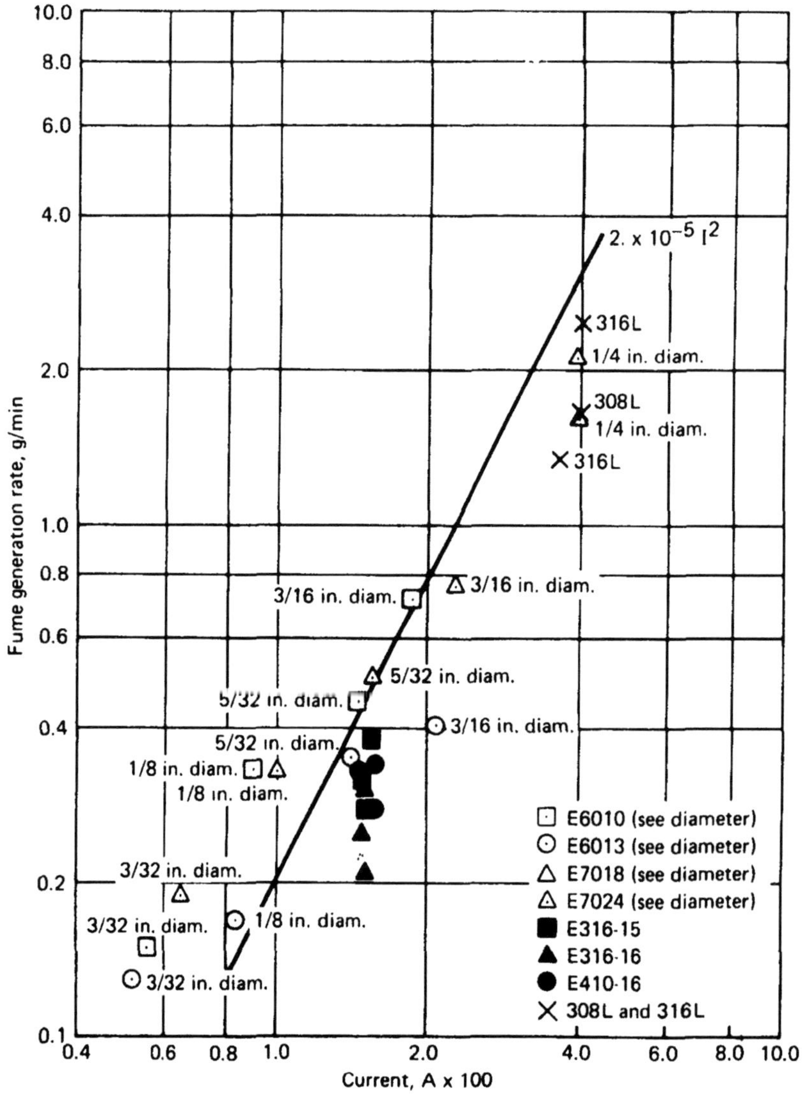

Fig. 2.3 – Fume generation rates for carbon and stainless steel covered electrodes as a function of current

the occasion arise to include fuming characteristics among the selection criteria, full use should be made of Table 2.2 (and the more detailed tables in Appendix B) to interpret the data in Fig. 2.5 properly.

Covered and Flux Cored Electrodes

In this section, data obtained at recommended operating conditions are reviewed for covered and flux cored electrodes. These electrodes are considered together because they constitute a metal-flux system, and the amount of fumes produced during welding is dependent upon the flux as well as the metallic part of the electrode.

Covered Electrodes

Fume Generation Characteristics. Data on (a) fume generation rates and (b) ratios of weight of fumes to weight of deposited metal obtained during studies of carbon steel, low alloy steel, stainless steel, nickel, and other high alloy covered electrodes are shown in bar graphs in Fig. 2.6. All of the electrodes were 4 mm (5/32 in.) in diameter and welding was done at current in the mid-to-upper part of the recommended operating range. In most instances, three electrodes made by different producers were evaluated per classification; one electrode per classification was studied in the case of the nickel and other high alloy electrodes. The data upon which this figure is based are shown in Table 2.2; detailed information on welding conditions and fume generation characteristics is contained in Tables B1 through B10.

The trends in fume generation rates and ratios of weight of fumes to weight of deposited metal shown in Fig. 2.6 are in general agreement with those obtained by other investigators (Refs. 2.3 and 2.5). The following data for electrodes with the same diameter as those examined during the AWS program were obtained by Heile and Hill:

Electrode classification	Current, A	Fume generation rate, g/min.	Weight of fume/wgt. of deposited metal, g/kg
E6010	110	0.32	19
E6010	170	0.66	24
E7018	160	0.28	8
E7018	220	0.65	18
E7024	180	0.29	8
E7024	230	0.47	8

At comparable welding currents, these data fell within or near the ranges shown in Fig. 2.6.

The data in Fig. 2.6 must be interpreted carefully because they were obtained by evaluating 4 mm (5/32 in.) diameter electrodes under conditions based on the various

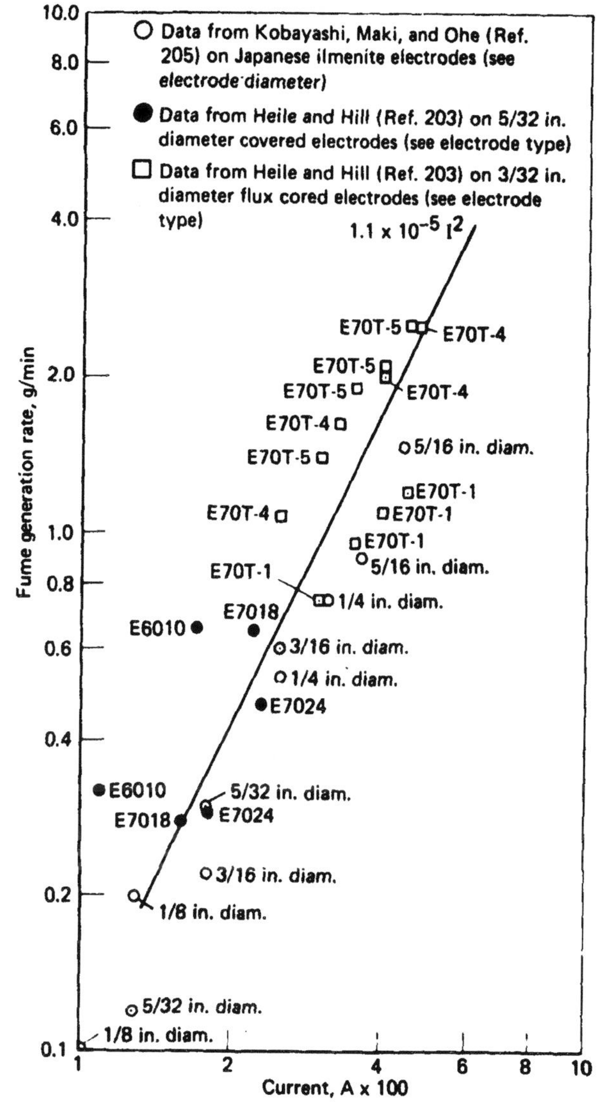

Fig. 2.4 – Fume generation rates obtained by other investigators with carbon steel covered and flux cored electrodes as a function of current

manufacturers' recommendations. Data differing from those shown in Fig. 2.6 will be obtained if the electrode diameter or welding conditions are changed.

Fume generation rates and ratios of weight of fume to weight of deposited metal were somewhat higher for E6010 electrodes than for the other covered electrodes, but there was considerable overlapping of the ranges for these quantities. Several carbon and low alloy steel electrodes had fume generation rates that fell within the lower portion of the range associated with E6010 electrodes. It is interesting to note that the fume generation

Table 2.2
Summary of baseline fume generation data for arc welding electrodes

Electrode classification	Number of electrode brands evaluated	Current range, A (nominal)	Average range: Fume generation rate, g/min	Average range: Weight of fumes/weight of deposited metal, g/kg*	Source table-App. B
		Shielded metal arc covered electrodes			
		Carbon and low alloy steel			
E6010	3	140-150	0.83 (1.20)**	35.85 (54.36)**	B1
E6013	3	145-160	0.31-0.58	14.16-25.75	B2
E7018	3	170-180	0.57-0.60	20.35-21.83	B3
E7024	3	200-230	0.43-0.55	8.92-11.11	B4
E8018 C3	3	160-175	0.43-0.47	15.92-17.82	B5
E9018 B3	3	160-180	0.36-0.46	11.19-14.94	B6
		Stainless steel and high alloy			
E316-15	3	150-155	0.28-0.38	8.02-11.08	B7
E316-16	3	145-150	0.21-0.31	6.56-11.92	B8
E410-16	3	145-160	0.28-0.34	11.75-13.97	B9
ENi-CI	1	135	0.37	12.90	B10
ENiCU-2	1	145	0.31	10.08	B10
Inconel 625	1	140-155	0.32	9.24	B10
Haynes C-276	1	130-135	0.37	14.20	B10
Haynes 25	1	135-140	0.26	8.94	B10
		Flux cored electrodes			
		Carbon and low alloy steel			
E70T-1	5	435-485	0.96-2.27	6.65-17.51	B11
E70T-4	3	370-390	1.86-2.09(2.98)**	12.76-13.83(22.70)**	B12
E70T-5	2	425-450	2.26-3.25	17.87-23.63	B13
81-C3	1	440-445	1.11	8.69	B14
91-B3	1	450	1.15	8.42	B14
		Stainless steel			
E308LT-3	1	440-445	1.64	9.11	B15
E316LT-3	2	340-405	1.34-2.48	6.97-12.32	B15
		Gas metal arc solid electrodes			
		Carbon steel			
E70S-3	3	260-290	0.41-0.46	4.97-5.68	B16
Spray w/Ar-2 O_2	3	260-290	0.41-0.46	4.97-5.68	B18
Spray w/Ar-9 CO_2	3	205-225	0.41-0.49	6.39-8.34	B16,B18
Globular w/CO_2	3	320-330	0.45-0.51	3.09-3.31	B16,B18
Short circuit w/Ar-25 CO_2	3	195-205	0.20-0.25	4.11-4.91	B16,B18
E70S-5					
Spray w/Ar-2 O_2	1	275-290	0.38	5.01	B19
Globular w/CO_2	1	325-345	0.40	2.61	B19
Short circuit w/Ar-25 CO_2	1	210-215	0.24	4.28	B19
		Stainless steel and high alloy			
ER316	1	165-175	0.04	0.58	B20
ERNiCu-7	1	250-260	0.16	2.02	B21
Inconel 625	1	190-195	0.06	0.87	B21
Haynes 25	1	200-205	0.08	1.38	B21
Haynes C-276	1	165	0.39	6.98	B21

Table 2.2 (continued)
Summary of baseline fume generation data for arc welding electrodes

Electrode classification	Number of electrode brands evaluated	Current range, A (nominal)	Average range: Fume generation rate, g/min	Average range: Weight of fumes/weight of deposited metal, g/kg**	Source table-App. B
		Aluminum			
ER4043	1	160-165	0.11-0.27	5.6-15.74	B22
ER5356	1	150-165	1.41-1.75(3.59)**	64.94-79.72(164.85)**	B23
		Copper			
ERCu	1	205-210	0.30	4.93	B24
ERCu A1-A2	1	210-215	0.47	8.12	B24

*g/kg ÷ 10 = weight of fumes expressed as a percentage of the deposit weight

**The data points in parentheses represent data obtained at non-baseline welding conditions (E6010 and ER5356) and with an E70T-4 electrode no longer in production.

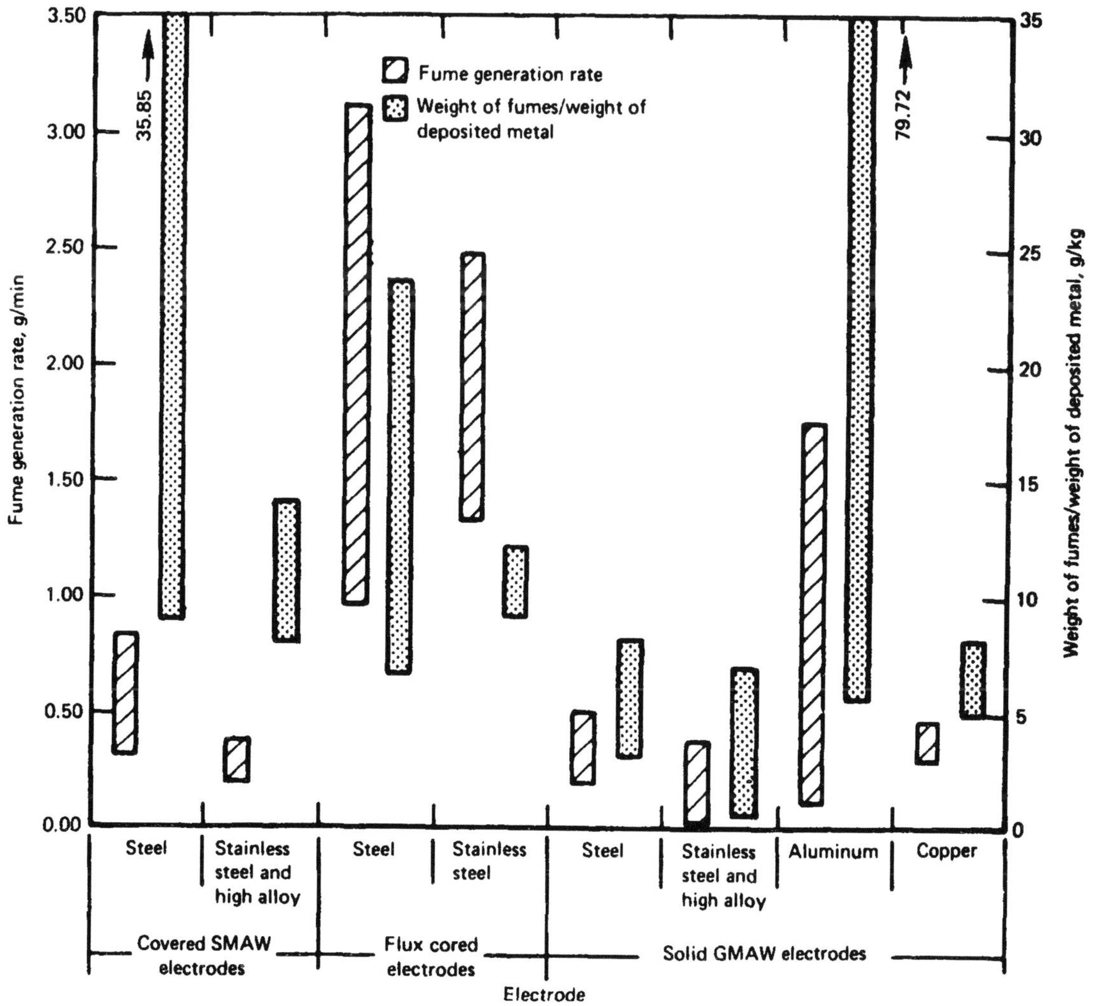

Fig. 2.5—Ranges of fume generation rates and ratios of weight of deposited metal for covered, flux cored, and solid GMAW electrodes

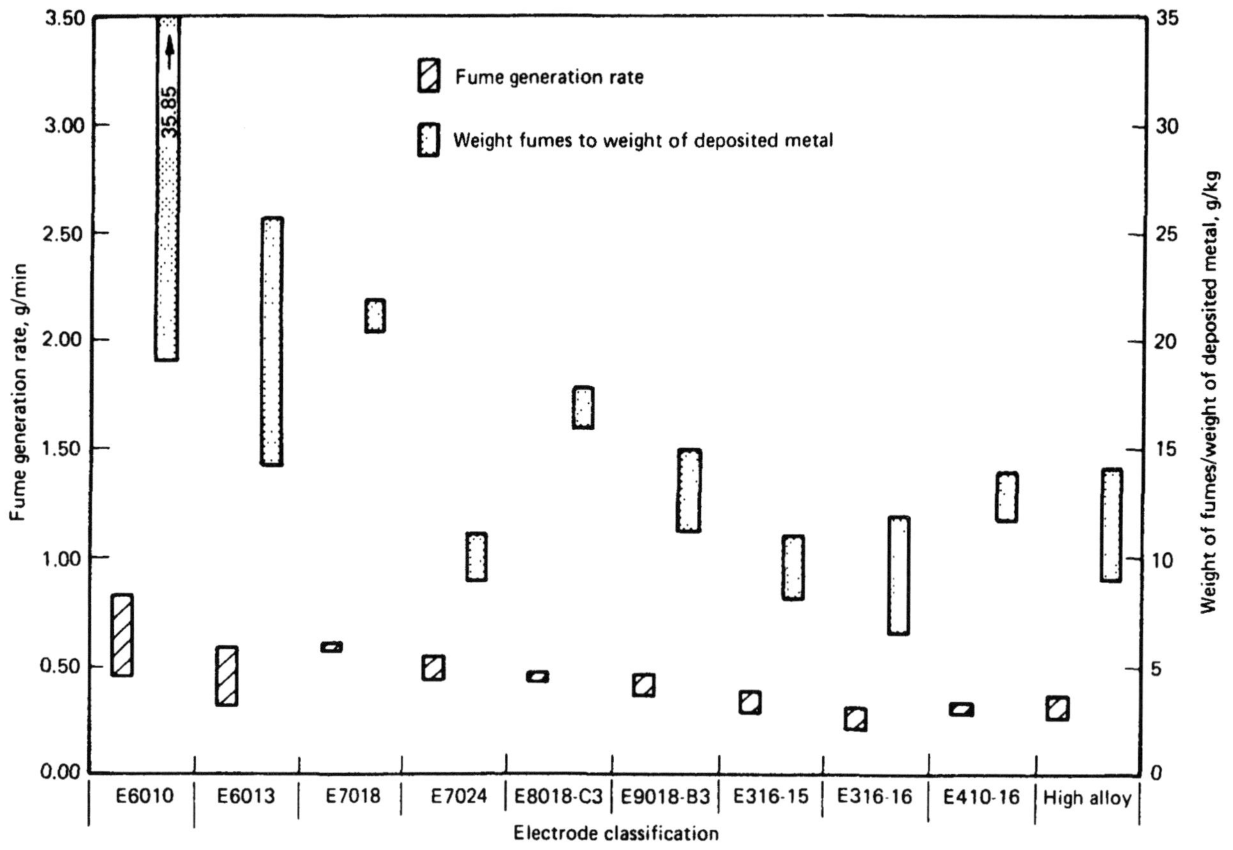

Fig. 2.6—Fume generation rates and ratios of weight of fumes to weight of deposited metal for covered electrodes

rates of the E7024 electrodes were comparable to those of other carbon and low alloy steel electrodes, even though the covering on these electrodes is much thicker than the coverings on the other electrodes. The ratios of weight of fume to weight of deposited metal for E7024 electrodes were lower than those of other electrodes. This results from the combination of a relatively low fume generation rate and the high deposition rate which results from the iron powder (up to 50 percent) in the coverings.

Fume generation rate ranges for stainless steel, nickel, and other high alloy covered electrodes were similar in magnitude; ranges for ratios of weight of fumes to weight of deposited metal generally overlapped one another.

Fume Composition. Studies to determine the complete analysis of the fumes associated with covered (and other types of) electrodes are not common; they can be quite involved and costly and may be unnecessary. Since the composition of the core wire is generally a matter of record, core wire constituents in the fumes can be readily determined by optical emission spectroscopy or atomic absorption procedures. Fume constituents attributable to the electrode coverings are more difficult to determine and a combination of analytical methods may be needed. Optical emission spectroscopy provides a means for identifying most of the constituents in the fumes, since the presence of 70 or more elements can be detected on a semiquantitative base. This method should be supplemented by (a) atomic absorption analysis for more accurately determining elements present in large amounts and (b) wet chemical analysis for determining fluoride contents. X-ray diffraction is useful if there is interest in detecting compounds in the fumes. Such detailed analyses are seldom required; industry is generally concerned only with those fume constituents that have low threshold limit values or are present in large amounts.

Since a large number of electrodes were to be evaluated during this program, the AWS Research Committee agreed that analytical efforts would be limited to the detection of the major elements in the fumes (e.g., iron, manganese, and silicon for carbon steel electrodes) and fluorides. Atomic absorption analysis was used to detect selected elements because the equipment is readily available in industry, is reasonable in cost, and produces accurate results. Wet chemistry was used to detect fluorides. Data on the core wire elements in the fumes produced by covered (and other) electrodes are presented later in this report (Table 2.21). It is noted that the analytical data do not total 100 percent because these limited studies did not include detection of fume constituents originating from the electrode covering.

Some data on expected fume constituents attributable to the coverings on the electrodes can be obtained from the literature. Compositions of the coverings of representative E6010 (high cellulose-sodium), E6013 (high titania-potassium), and E7016 (low hydrogen-potassium) electrodes have been provided by Smith and Rinehart (Ref. 2.18) and Smith (Ref. 2.19) and are shown in Table 2.3. The E6010 and E6013 electrodes are among the electrodes evaluated during this investigation. E7018 electrodes evaluated during the present program have coverings similar to that of the E7016 electrode except for the presence of iron powder in the E7018 covering. The fumes produced by the electrodes included in Table 2.3 should contain the constituents of the covering and those of the core wire, usually (but not necessarily) in the oxide form of the respective elements.

Examples of complete fume analyses of several Scandinavian electrodes, three of which appear to be low hydrogen electrodes with iron powder additions, are shown in Table 2.4.

Table 2.3
Composition of coverings on representative carbon steel electrodes (Refs. 2.18, 2.19)

Constituent	Covering composition, weight %		
	E6010	E6013	E7016
SiO_2	32.0	25.9	16.0
$TiO_2 + ZrO_2$	18.0	30.6	6.5
Al_2O_3	2.0	5.9	1.0
CaF_2	---	---	27.0
CaO	---	1.6	---
MgO	6.0	2.6	---
Na_2O	8.0	1.1	1.4
K_2O	---	6.7	1.0
CO_2	---	1.7	---
Organics	30.0	17.7	---
Fe	2.0	2.1	---
Mn	7.0	4.8	2.5
$CaCO_3$	---	---	38.0

Table 2.4
Composition of fumes produced by typical Scandinavian covered electrodes (Ref. 220)

Compound	Composition, weight %			
	Electrode 1	Electrode 2	Electrode 3	Electrode 4
SiO_2	7.0	9.5	10.0	30.5
Fe_2O_3	25.5	24.5	36.5	43.5
Al_2O_3	0.6	0.2	<0.1	<0.1
TiO_2	1.1	0.2	0.5	2.2
ZrO_2	---	0.6	---	---
MnO	4.7	7.2	8.2	9.8
ZnO	0.04	0.07	0.09	0.02
CaO	15.9	5.3	0.4	<0.1
MgO	0.1	0.1	1.0	0.1
K_2O	24.4	17.6	17.6	7.2
Na_2O	2.4	17.2	11.0	5.4
Cu	0.03	0.07	0.03	0.06
Pb	0.02	0.02	0.04	0.05
Cr	0.01	0.04	0.01	0.04
Fe	19.8	15.7	17.1	---

In the AWS studies to characterize fumes and fume particles, optical emission spectroscopy was used to determine the constituents in the fumes produced by E7024 and E410-16 covered electrodes. Data on major constituents are shown in Table 2.5.

Flux Cored Electrodes

Fume Generation Characteristics. Ranges of (a) fume generation rates and (b) ratios of weight of fumes to weight of deposited metal for representative carbon steel, low alloy steel, and stainless steel flux cored electrodes are shown in Fig. 2.7. All electrodes were 2.4 mm (3/32 in.) diameter and welding was done at currents in the mid-portion of the manufacturer's recommended operating range. Details on the welding conditions and the experimental results are contained in Tables B11 and B15. Carbon dioxide shielding was used with the E70T-1, E70T-5, 81-C3, and 91-B3 electrodes; the E70T-4 and stainless steel electrodes were self-shielding. The data upon which Fig. 2.7 is based are summarized in Table 2.2; this table should be consulted when this figure is reviewed.

(1) E70T-1 electrodes. The ranges of fume generation and ratios of weight of fumes to weight of deposited metal were broadest for this classification of electrodes. Three of these electrodes were made by the same producer; the other two were made by two different producers. The electrodes made by the same producer were designed for single- or multiple-pass welding, but the core wire and flux composition varied to achieve certain objectives: e.g., the welding of steel with different strength levels, the welding of clean or rusted steel, etc. These special-purpose electrodes defined the upper and lower limits of the ranges in Fig. 2.7. The electrode with the highest fume generation rate contained appreciable amounts of fluorides (Table 2.18); such compounds are known to enhance the production of fumes. The fluoride content of the fumes associated with the electrode having the lowest fume generation rate was very small.

(2) E70T-4 electrodes. A data point outside the indicated ranges for these electrodes is shown in Fig. 2.7. This point represents the fume generation rate for an E70T-4 electrode that was made several years ago and is no longer available on the market. Since another currently available electrode made by the same producer had a much lower fume generation rate, it appears that changes in the flux were made to reduce fume quantities.

(3) E70T-5 electrodes. The highest fume generation rates were encountered with E70T-5 electrodes. This was expected because the E70T-5 electrodes can be used with or without a shielding gas, and the basic type flux contains ingredients that produce large amounts of gas and fumes. Ratios of fume weight to weight of deposited metal were highest for these electrodes also.

Ranges for the fume generation characteristics of other flux cored electrodes overlapped one another in many instances. The experimental results are discussed below.

(1) Low alloy steel electrodes. The fume generation

Table 2.5
Optical emission spectoscopic analysis of fumes produced by two E7024 and one E410-16 covered electrodes

		Composition, weight %		
Element	(Oxide)[1]	E7024(7)[2]	E7024(8)[2]	E410-16(21)[2]
Fe	(Fe_2O_3)	20-30 (28.6-42.9)	20-30 (28.6-42.9)	10-20 (14.3-28.6)
Si	(SiO_2)	10-20 (15.7-31.4)	5-10 (7.8-15.7)	2-3 (3.1-4.6)
K	(K_2O)	8-12 (9.6-14.4)	8-12 (9.6-14.4)	10-20 (12.0-24.0)
Na	(Na_2O)	3-6 (4.0-8.1)	3-6 (4.0-8.1)	4-8 (5.4-10.8)
Mn	(MnO_2)	2-4 (3.2-6.3)	2-4 (3.2-6.3)	1-2 (1.6-3.2)
Ca	(CaO)	0.1 (0.1)	0.05 (0.1)	2-4 (2.8-5.6)
Zn	(Z_nO)	<0.1 (0.1)	<0.1 (0.1)	0.1 (0.1)
Ti	(TiO_2)	0.5 (0.8)	0.3 (0.5)	0.4 (0.7)
Al	(Al_2O_3)	0.1 (0.2)	0 05 (0.1)	0.5 (0.9)
Cr		0.01	0.01	5-10

1. Assuming that elements are completely converted to oxides.

2. The number in parentheses following the electrode designation is the code number identifying the specific electrode.

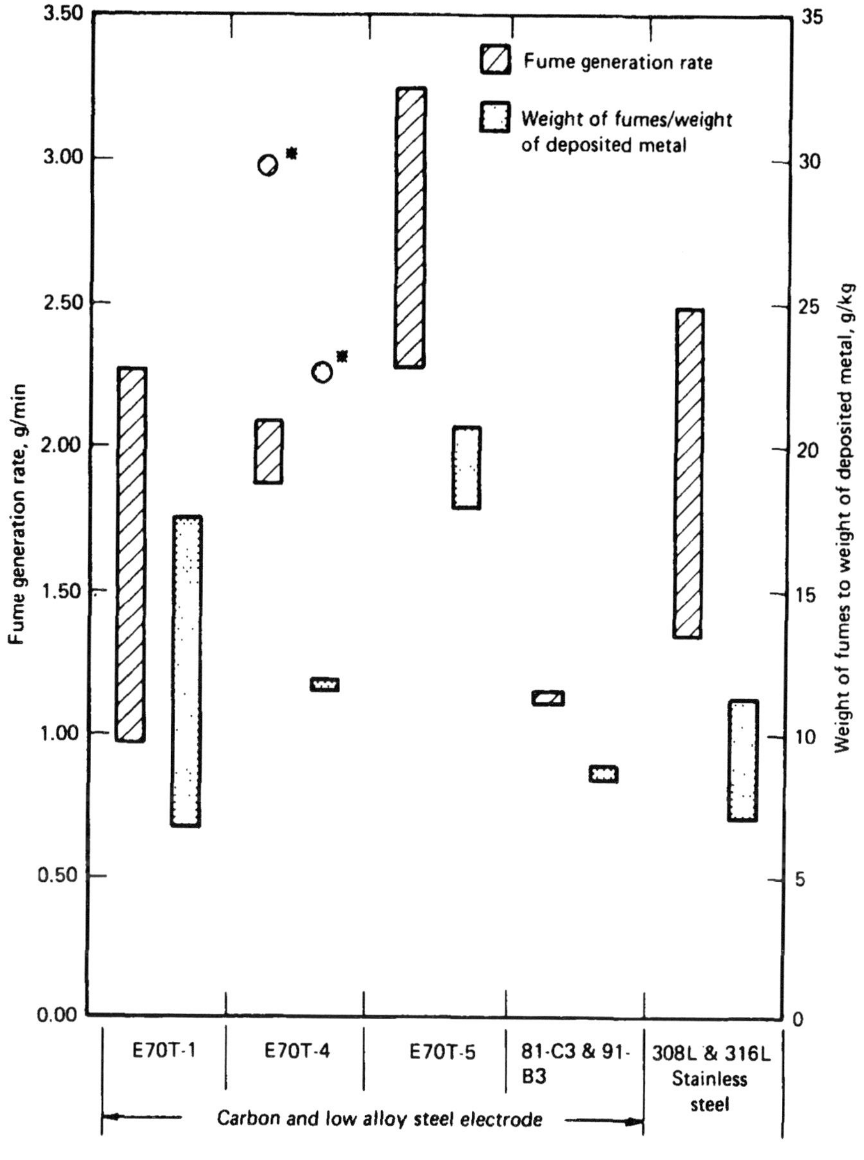

*Data from E70T-4 electrode that is no longer commercially available.

Fig. 2.7–Fume generation rates and ratios of weight of fumes to weight of deposited metal for flux cored electrodes

rates and ratios of weight of fumes to weight of deposited metal for 81-C3 and 91-B3 low alloy steel electrodes were low in comparison with those for carbon steel flux cored electrodes. These results are similar to those observed when these quantities were compared for carbon steel and low alloy steel covered electrodes.

(2) Stainless steel electrodes. The fume generation characteristics of the three self-shielded stainless steel electrodes investigated during this program varied over a wide range. This appears to be a flux-related occurrence. The E308LT-3 electrode and one of the E316LT-3 electrodes were made by the same producer, while the remaining E316LT-3 electrode was made by another producer. The fume generation rates for the electrodes made by the same producer were similar and define the low end of the range shown in Fig. 2.7; this result might be expected because the fluxes used in these electrodes were probably quite similar in composition. However, the fume generation rates and ratios of weight of fumes to weight of deposited metal varied widely for the E316LT-3 electrodes made by two producers. Since the sheath compositions are not likely to differ substantially, this variation must be associated with differences in the flux cores of the respective electrodes.

Table 2.6
Comparison of average fume generation characteristics of 2.4 mm (3/32 in.) diameter E70T-1 flux cored electrodes as a function of shielding gas

Electrode number	Shielding gas[1]	Current, A	Fume generation rate, g/min.	Weight of fumes/ weight of deposited metal, g/kg
E70T-1 (40)[2]	CO_2	475	1.35	10.40
E70T-1 (40)	Ar-25 CO_2	465	1.01	7.78
E70T-1 (42)	CO_2	440	2.27	17.51
E70T-1 (42)	Ar-25 CO_2	445	1.93	14.91

1. Ar-25 CO_2 is a convention used to designate a gas mixture of 25% CO_2, 75% Ar.
2. The number in parentheses after the AWS electrode classification is a code number identifying the specific proprietary electrode.

Table 2.7
Typical flux compositions of the three carbon dioxide shielded flux cored electrode types, percent (Ref. 2.17)

	Composition, weight percent		
	Type 1	Type 2	Type 3
Compound or element	Titania type (non-basic) flux	Lime-titania type (basic or neutral) flux	Lime type (basic) flux
SiO_2	21.0	17.8	7.5
Al_2O_2	2.1	4.3	0.5
TiO_2	40.5	9.8	- - -
ZrO_2	- - -	6.2	- - -
CaO	0.7	9.7	3.2
Na_2O	1.6	1.9	- - -
K_2O	1.4	1.5	0.5
CO_2 (as carbonate)	0.5	- - -	2.5
C	0.6	0.3	1.1
Fe	20.1	24.7	55.0
Mn	15.8	13.0	7.2
CaF_2	- - -	18.0	20.5
AWS classification	E70T-1 or E70T-2	E70T-1	E70T-1 or E70T-5

As noted previously, the E70T-1 and E70T-5 electrodes were used with CO_2 shielding. To investigate the premise that fume generation rates can be affected by the type of shielding gas used during welding, the characteristics of two E70T-1 flux cored electrodes were determined with Ar-25 CO_2 shielding. Other than shielding gas, the welding conditions were essentially the same as the baseline conditions. The resulting data are summarized in Table 2.6; additional details are contained in Appendix B (Table B25).

The fume generation rates for each electrode were reduced between 15 and 25 percent when Ar-25 CO_2 shielding was used instead of CO_2. This is because the argon-based shielding gas has a lower oxidation potential than CO_2, and oxidation processes contributing to the formation of fumes at and near the tip of the electrode were decreased as a consequence. The metal deposition rates for these electrodes were unaffected by the type of gas shielding used. Thus, the ratios of weight of fumes to weight of deposited metal were also smaller when Ar-25 CO_2 was used for shielding.

Fume Composition. As in the case of covered electrodes, the composition of fumes produced by flux cored electrodes is determined by composition of the electrode sheath, flux core, and, to a minor extent, by the base metal. Since the composition of the sheath for electrodes within a classification is unlikely to vary much from producer to producer, variations in fume compositions for such electrodes are caused primarily by differences in the flux core ingredients.

To provide an insight into the constituents that may be detected in the fumes associated with welding operations, data on the composition of flux cores and slags associated with representative flux cored electrodes are shown in Tables 2.7 and 2.8 (Ref. 2.17). The compositions of the fluxes used in three CO_2 shielded flux cored electrodes are shown in Table 2.7. Electrodes incorporating these fluxes can be included in one or two AWS classifications; classification criteria are discussed in the article from which these data were taken (Ref. 2.17). The type of flux also determines the classification of self-shielded flux cored electrodes as indicated in Table 2.8.

Table 2.8
Typical flux compositions of the four types of self-shielded flux cored electrodes, percent (Ref. 2.17)

	Composition, weight percent			
	Type 1	Type 2	Type 3	Type 4
Compound or element	Fluorspar-aluminum flux	Fluorspar-titania flux	Fluorspar-lime-titania flux	Fluorspar-lime flux
SiO_2	0.5	3.6	4.2	6.9
Al	15.4	1.9	1.4	---
Al_2O_3	---	---	---	0.6
TiO_2	---	20.6	14.7	1.2
CaO	---	---	4.0	3.2
MgO	12.6	4.5	2.2	---
K_2O	0.4	0.6	---	---
Na_2O	0.2	0.1	---	0.6
C	1.2	0.6	0.6	0.3
CO_2 (as carbonate)	0.4	0.6	2.1	1.3
Fe	4.0	50.0	50.5	58.0
Mn	3.0	4.5	2.0	7.9
Ni	---	---	2.4	---
CaF_2	63.5	22.0	15.3	22.0
AWS classification	E70T-4 E60T-7 E60T-8	E70T-3	E70T-6	E70T-5

Most of the flux constituents shown in these tables will be present as elements and compounds in the fumes produced by electrodes incorporating them. They will not be present in the same proportions as in the flux because some constituents are transferred by the arc to the slag and, in some cases, to the weld metal. Elemental compositions of the fumes produced by typical E70T-1, E70T-4, and E70T-5 electrodes are shown in Table 2.9 along with calculated oxide contents (assuming that the elements are converted to oxides (Ref. 2.3). These electrodes were classified as follows: E70T-1, rutile-base; E70T-4, fluorspar-base; and E70T-5, silica-base.

Fume compositions determined by optical emission spectroscopy are shown in Table 2.10 for the E70T-1, E70T-4, and E70T-5 electrodes investigated during this program. These data were acquired during studies to characterize fumes and fume particles. From these data, it appears that the E70T-4 electrodes evaluated by Battelle and by Heile and Hill had fluxes based on similar ingredients (e.g., aluminum and fluorspar). Fluxes for the respective E70T-1 and E70T-5 electrodes differed considerably.

Solid Electrodes

The fume generation characteristics of solid electrodes and rods used for gas metal arc welding and gas tungsten arc welding are reviewed in this section. In comparison with covered and flux cored electrodes, these are low fuming electrodes and rods whose fume generation tendencies are a direct function of the amount of electrode or rod consumed during welding. Other factors which influence the rate at which fumes are produced by the solid electrodes include the type of shielding gas, the metal transfer characteristics of the arc, and the welding conditions.

Gas Metal Arc Electrodes

Fume Generation Characteristics. Ranges for fume generation rates and ratios of weight of fume to weight of deposited metal associated with carbon steel, stainless steel, high alloy, copper, and aluminum gas metal arc welding electrodes investigated during this program are shown in Fig. 2.8. The data upon which these ranges are based are summarized in Table 2.2; detailed information on the fume generation characteristics of individual electrodes are contained in Tables B16 to B24 along with the welding conditions.

(1) Steel Electrodes. The results obtained with carbon steel electrodes are reviewed separately from those obtained with other types of solid electrodes. Caution is required in interpreting the data in Fig. 2.8 because of the dependency of the fume generation characteristics on welding current and other process variables. The fume generation characteristics of three 1.1 mm (0.045 in.) E70S-3 electrodes, each made by a different manufacturer, were determined at appropriate current levels in the following transfer modes: spray transfer with Ar-2 O_2 or Ar-9 CO_2 shielding; globular transfer with CO_2 shielding; and short circuiting transfer with Ar-25 CO_2 shielding. Although the electrodes were made by different producers, there was little variation in composition. The thickness of the copper coating on the electrode surface varied from zero to several microinches. A single E70S-5 electrode was also evaluated in various transfer modes: spray transfer with Ar-2 O_2 shielding; globular transfer with CO_2 shielding; and short circuiting transfer with Ar-25 CO_2 shielding.

Fume generation rates for the three E70S-3 electrodes were similar in magnitude during spray transfer welding with Ar-2 O_2 or AR-9 CO_2 shielding, and during globular transfer welding with CO_2 shielding. Under these conditions, the ratios of weight of fume to weight of deposited metal varied somewhat because of differ-

Table 2.9
Composition of fumes from flux cored arc welding (Ref. 2.3)

Element	(Oxide)	Composition, weight percent					
		E70T-1		E70T-4		E70T-5	
Fe	(Fe_2O_3)	28	(40.0)	13	(18.6)	23	(32.9)
Mn	(MnO_2)	5	(7.9)	2	(3.2)	4	(6.3)
Si	(SiO_2)	25	(39.3)	4	(6.3)	20	(31.4)
Ca	(CaO)	---		13	(18.2)	8	(11.2)
Ti	(TiO_2)	2	(3.3)	---		---	
Mg	(MgO)	3	(5.0)	9	(14.9)	---	
Al	(Al_2O_3)	---		9	(17.0)	---	
F		7		13		7	

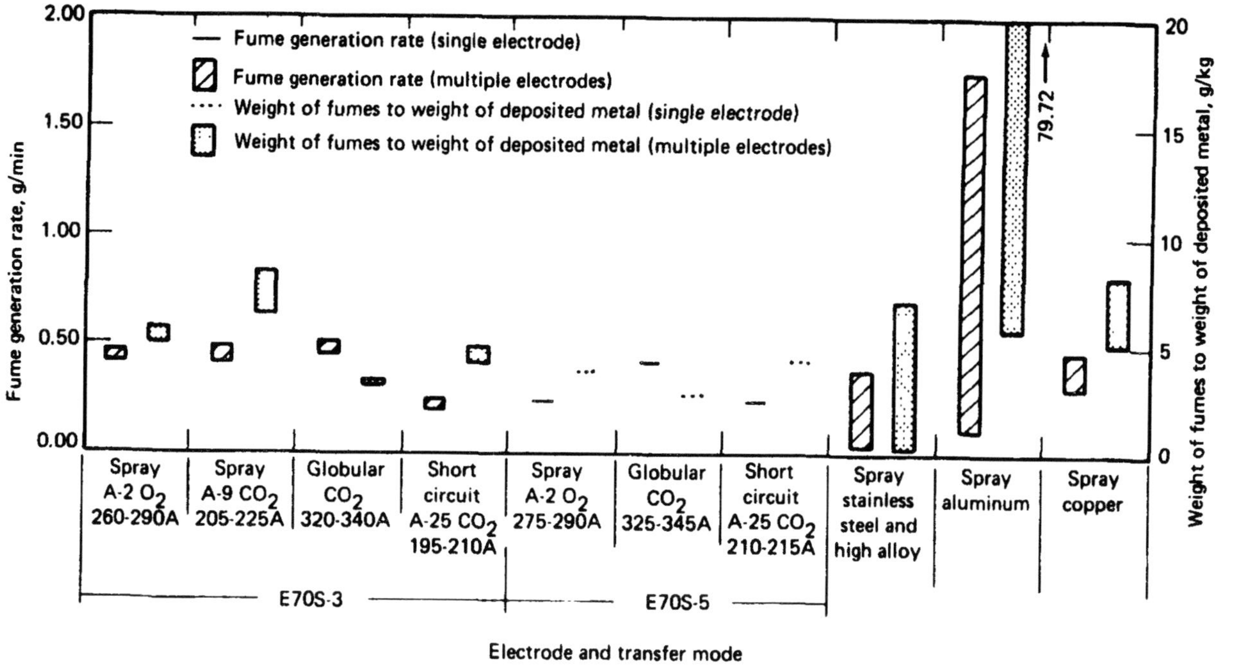

Fig. 2.8–Fume generation rates and ratios of weight of fumes to weight of deposition metal for gas metal arc electrodes

Table 2.10
Optical emission spectroscopic analysis of fumes produced by E70T-1, E70T-4, and E70T-5 flux cored electrodes

		Composition, weight %					
Element	(Oxide)[1]	E70T-1 (42)[2]		E70T-4 (49)[2]		E70T-5 (50)[2]	
Fe	(Fe_2O_3)	30-40	(42.9-57.2)	15-25	(21.4-35.8)	30-40	(42.9-57.2)
Si	(SiO_2)	2-3	(3.1-4.6)	0.1	(0.2)	2-3	(3.1-4.6)
K	(K_2O)	1.0	(1.2)	2.0	(2.4)	4-6	(4.8-7.2)
Na	(Na_2O)	4-6	(5.4-8.1)	0.1	(0.1)	1.0	(1.4)
Mn	(MnO_2)	4-6	(6.3-9.5)	2-3	(3.2-4.7)	4-6	(6.3-9.5)
Ca	(CaO)	0.1	(0.1)	15-25	(21-35)	8-12	(11.2-16.8)
Zn	(ZnO)	0.1	(0.1)	0.1	(0.1)	0.1	(0.1)
Ti	(TiO_2)	0.5	(0.8)	0.01	(0.02)	0.2	(0.3)
Al	(Al_2O_3)	0.4	(0.8)	7-10	(13.2-18.9)	1-2	(1.9-3.8)
Mg	(MgO)	0.02	(0.03)	7-10	(11.6-16.6)	0.2	(0.3)

1. Assuming that elements are completely converted to oxides.
2. The number in parentheses following the electrode designation is the code number identifying the specific electrode.

ences in the metal deposition rates for the respective electrodes. Under short circuiting conditions with Ar-25 CO_2 shielding, fume generation rates were lower than they were with other metal transfer modes and other shielding gases.

The effects of type of shielding gas and mode of metal transfer on fume generation characteristics are not clearly evident from the data in Fig. 2.8 or the tabular data mainly because of the differences in the currents used for welding. If the effects of current are taken into account by normalizing[7] the data to a selected value of current (multiplying the normalized fume generation rate for each electrode by current squared), the effects of shielding gas and metal transfer mode are more readily observed. To facilitate interpreting the data, the average fume generation rates were normalized to a current of 250 A. These data together with the measured rates are shown in Table 2.11.

As indicated in Table 2.11, fumes were produced at the highest rate during spray transfer welding with Ar-9 CO_2 shielding, followed approximately in descending order by spray transfer welding with AR-2 O_2, short circuiting transfer welding with Ar-25 CO_2, and globular transfer welding with CO_2. The effects of shielding gas on fume generation rates can be best observed by comparing the rates at which fumes were produced during spray transfer welding with Ar-2 O_2 or Ar-9 CO_2 shielding. When Ar-9 CO_2 was used, the fume generation was much higher than when Ar-2 O_2 shielding was used. This reflects the higher oxidation potential of the Ar-9 CO_2 shielding gas, and the fume generation rate presumably would increase further if the content of CO_2 in the argon-based shielding gas was increased.

The E70S-5 electrode is similar to the E70S-3 electrode except that it contains aluminum as well as manganese and silicon as a deoxidizer. If the same welding conditions are used, the fume generation characteristics of the E70S-3 and E70S-5 electrodes should be similar, as indeed they were (see Tables B16 through B19).

(2) Other Solid Electrodes. Representative stainless steel, high alloy, copper, and aluminum solid electrodes, 1.1 mm (0.045 in.) in diameter,[8] were also investigated. In each instance, a single electrode per classification was evaluated and the welding current was adjusted to produce acceptable spray transfer conditions.

(3) High Alloy Electrodes. Five high alloy electrodes (one stainless steel electrode, three nickel base electrodes, and one cobalt-base electrode) were investigated during this program, and all but one had fume generation rates and ratios of weight of fumes to weight of deposited metal well below those of carbon steel electrodes. This trend was also observed with covered electrodes of the same types. The exception, a nickel-base electrode, had fume generation rates approaching

7. In general, fume generation rates for electrodes studied in this and other investigations varied in accordance with current to a power that ranged from greater than one to less than three (current equals I^x where $1 < x < 3$). With current squared accepted as a reasonable power function compromise, a normalized fume generation rate that includes the effect of current on the rate at which fumes are produced can be established. Then, if an electrode is used at one current, it is possible to use the normalized rate (fume generation rate per unit of current squared) and predict the approximate fume generation rate when the electrode is used at a higher or lower rate (provided other welding variables are controlled). Normalized fume generation rates are shown in the tabular data in Appendix B along with measured rates.

8. Because of availability, electrodes with diameters other than 0.045 in. (1.1 mm) were used in a few instances; such deviations are noted in the tabular data in Appendix B.

Table 2.11
Measured and normalized (Ref. 2.1) fume generation rates for E70S-3 electrodes

Metal transfer mode	Shielding gas	Fume generation rate g/min					
		E70S-3(54)[2]		E70S-3(57)[2]		E70S-3(58)[2]	
		Measured	Normalized	Measured	Normalized	Measured	Normalized
Spray	AR-2 O_2	0.41	0.35	0.46	0.36	0.45	0.35
Spray	AR-9 CO_2	0.41	0.58	0.44	0.62	0.49	0.61
Globular	CO_2	0.51	0.29	0.46	0.28	0.45	0.27
Short circuit	Ar-25 CO_2	0.20	0.33	0.25	0.38	0.24	0.38

1. Fume generation rates normalized to a current of 250 A.

2. The number in parentheses following the electrode designation is the code number identifying the specific electrode.

those of the carbon steel electrodes. It was a small diameter electrode, 0.09 mm (0.035 in.), and the current density was higher than that used with the other electrodes; as a result, arc temperatures were probably higher and more fumes were produced.

(4) Aluminum Electrodes. The aluminum electrodes presented mixed results. The ER4043 electrode produced fumes at low rates and the ratio of weight of fumes to weight of deposited metal was low also. The ER5356 electrode produced fumes at much higher rates than the ER4043 electrode. High fume generation rates were not unexpected, because the ER5356 electrode contains magnesium, a metal that oxidizes easily and has a high vapor pressure. In contrast, the ER4043 electrode contains silicon, an element with a much lower vapor pressure. Using a contact tube-to-work distance of 12.5 mm (0.50 in.), the fume generation rate for the ER5356 electrode was similar to that obtained by Heile and Hill when differences in welding current are taken into account (Ref. 2.3). Much higher fume generation rates were produced at longer contact tube-to-work distances, and there was evidence of lack of shielding of the arc and pool of weld metal. These results point to the need for careful welding torch setup control and good shielding to minimize fumes as well as to insure the production of high quality welded joints.

(5) Copper Electrodes. Fume generation rates and ratios of weight of fumes to weight of deposited metal for an ERCu and an ERCuAl-A2 electrode were comparable in magnitude to those associated with carbon steel electrodes.

Fume Compositions of Gas Metal Arc Electrodes. In contrast to covered and flux cored electrodes, the composition of the fumes produced by gas metal arc welding electrodes is controlled by the composition of the electrode and any coating that might be on the electrode surface. This can be illustrated by considering the composition of the fumes produced by the three E70S-3 electrodes investigated during this study:

	Fume composition, weight percent						
Electrode no.	Fe	(Fe_2O_3)	Mn	(MnO_2)	Si	(SiO_2)	Cu
E70S-3 (54)	63.7	(91.1)	5.3	(8.4)	0.05	(0.1)	0.11
E70S-3 (57)	65.7	(93.9)	3.8	(6.0)	0.79	(1.2)	0.60
E70S-3 (58	62.5	(89.4)	8.5	(13.4)	0.53	(0.8)	1.00

Each of the fume samples contained iron, manganese, silicon, and copper (or their oxides). The amount of copper in the fumes produced by the electrode that was not coated with copper, E70S-3 (54), was higher than expected. Subsequent analyses of short lengths of each electrode from which the copper coating was stripped showed that they all contained a small (0.02 to 0.03) weight percent of copper as a residual element in the electrodes themselves. Apparently, the residual copper in the E70S-3 (54) electrode plus perhaps some copper picked up from contact tube and wire drive was the source of copper in the fumes from this electrode.

Because of its low threshold limit value, there is more concern about the presence of copper fume in the welding environment than about iron oxide fume. Table 2.12 combines data on the baseline fume generation rates of three E70S-3 electrodes with the compositional data on

Table 2.12
Results of studies on copper in the fumes produced by E70S-3 electrodes used for gas metal arc welding

	Electrode		
Characteristic	E70S-3(54)*	E70S-3(57)	E70S-3(58)
Copper coating on electrode			
Thickness, μ in.	0	18	24
weight percent	0	0.19	0.24
Copper content of fumes, weight %	0.11	0.60	1.00
Iron (oxide) content of fumes, weight %	63.7 (91.1)	65.7 (93.9)	62.5 (89.4)
Total fume generation rate, g/min.	0.41	0.46	0.45
Copper fume generation rate, g/min.	0.0004	0.0028	0.0045
Iron (oxide) fume generation rate, g/min.	0.26 (0.37)	0.30 (0.43)	0.28 (0.40)

*The number in parentheses following the electrode designation is a code number identifying the specific electrode.

the fumes. The ACGIH time-weighted average threshold limit value for iron oxide fumes (5 mg/m³) is 25 times as large as that for copper fumes (0.2 mg/m³). However, the results of the experimental studies indicate that the iron oxide fume generation rates for the E70S-3 (54), E70S-3 (57), and E70S-3 (58) electrodes were 925, 154, and 89 times as great, respectively, as the copper fume generation rates. Thus, when welding with these electrodes, the allowable limit for iron oxide fumes would be exceeded long before reaching the limit for copper fumes.

Compositions of the fumes produced by the gas metal arc electrodes used for welding other base metals (stainless steels, nickel base alloys, etc.) are determined by the composition of the respective electrodes.

Gas Tungsten Arc Welding

Fume Generation Characteristics. Two filler rods used for gas tungsten arc welding were also evaluated: a 3.2 mm (1/8 in.) diameter E308L stainless steel wire and a 2.4 mm (3/32 in.) diameter 5356 aluminum alloy wire. The welding conditions are indicated below:

308L Stainless Steel Filler Metal. Welding was done with a torch equipped with a 2.4 mm (3/32 in.) diameter 2 percent thoriated tungsten electrode (AWS EWTh-2). The welding current was 200-215 A (dc), the electrode polarity was negative, and argon was used for shielding.

5356 Aluminum Filler Metal. Welding was done with a torch equipped with a 2.4 mm (3/32 in.) diameter tungsten electrode (AWS EWP). The welding current was 250 A (ac) and argon was used for shielding.

Under these conditions, the fume generation rates were 0.0025 and 0.0065 g/min for the stainless steel and aluminum filler wires, respectively. Such rates are a small fraction of the rates associated with other processes investigated during this program.

Fume Compositions. No analyses were made.

Section IIC. Effects of Process Variables on Fume Generation Rates

While baseline data on the fume generation characteristics of covered, flux cored, and solid electrodes provide a wealth of information that is useful to those concerned with the health and safety of welding personnel, they are by no means the entire story. The term "baseline" itself implies that these data were obtained when the electrodes were used in general accordance with recommended conditions, and questions concerning the effects of the process variables on fume generation characteristics naturally arise. In reviewing the baseline data, reference to the effect of current on these characteristics has been frequently made because it appears to be the dominant variable..Other welding variables also have a bearing on the rate at which fumes are produced during welding. Some of the other variables affecting fume production are:

(1) Voltage
(2) Electrode polarity
(3) Shielding gas (flux cored and solid electrodes)
(4) Contact tube-to-work distance (flux cored and solid electrodes)
(5) Metal transfer mode
(6) Electrode characteristics (manufacturer, diameter, core wire composition, covering or flux composition, etc.).

All of these variables in addition to current and their effects on fume generation characteristics will be discussed in this section. It should be noted that the effects of many of these variables are interrelated.

Current Effects

Current is acknowledged to have a major effect on the rate at which fumes are produced during welding. In the present investigation, current effects were studied with representative covered, flux cored, and solid electrodes used for welding carbon steels. Trends similar to those observed should be obtained with other classifications of electrodes.

Covered Electrodes. Fume generation rate as a function of current is shown in Fig. 2.9 for 4 mm (5/32 in.) E6010 and E7018 electrodes; the data forming the basis for this figure are contained in Tables B26 and B27. In each instance, these rates varied nonlinearly with current. Studies to fit the data to a power function indicated that the fume generation rate (FGR) relation to current was different for each type of electrode. For example, the regression equations for two of the electrodes were

$$FGR_{E6010} = 0.000011\, I^{2.24} \text{ with } r^2 = 0.93$$
$$FGR_{E7018} = 0.00017\, I^{1.54} \text{ with } r^2 = 0.88$$

The regression lines in the figure are I^2 lines included for comparison purposes.

The results obtained are similar to those obtained by other investigators. Kobayashi, Maki, and Ohe noted that fume generation rates for ilmenite, lime, and lime-titania covered electrodes varied with current to powers of 1.17 to 1.74 (Ref. 2.5). Agreement with their results was good considering differences between domestic and foreign electrodes and differences in fume sampling methods. In work done by Heile and Hill with E6010 and E7018 electrodes, similar trends in fume generation rates were observed (Ref. 2.3).

Flux Cored Electrodes. The effects of welding current on fume generation rates for a 2.31 mm (3/32 in.) diameter E70T-1 electrode using CO_2 shielding are also shown in Fig. 2.9; details on the welding conditions and the results of this investigation are provided in Table B28. Above 450 A, the fume generation rate increased almost linearly with current; below 450 A, the behavior of the fume generation rate with current was nonlinear. In particular, there was a well-defined and reproducible minimum at low current levels that may have been caused by

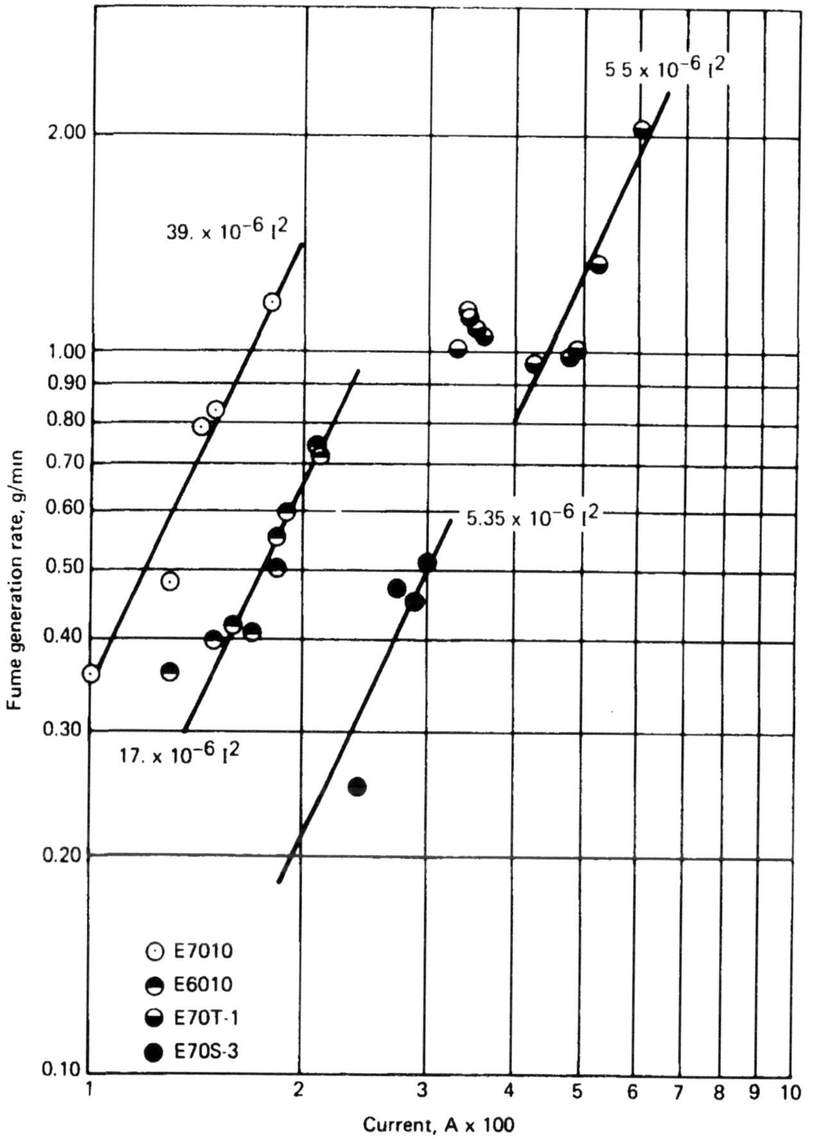

Fig. 2.9–Fume generation rates for selected covered, flux cored, and solid GMAW electrodes as a function of current

changes in arc length, changes in the sizes of globules being transferred across the arc, short-circuiting, etc. Attempts to fit a power function to the data obtained over the entire current range were unsuccessful because of the minimum in the curve at low currents. A reasonable fit with the I^2 line was obtained for data obtained at currents above 425 A. Thus, at currents between 425 and 600 A, there was a strong dependency of fume generation rate on current; at currents below 425 A, the fume generation rate was dependent upon variables other than current. It is emphasized that the indicated relationship between fume generation rate and current is valid only for the data obtained during this study.

A video camera was used to study the arc at currents between 350 and 500 A. At currents around 500 A, the arc was steady and appeared to have a length of about 3.2 mm (1/8 in.). At currents in the 350 A range, the arc was turbulent and its length varied rapidly; at times, the arc was buried. Apparently, the minimum in the curve and the increase in fume generation rate at lower currents were associated with this turbulence and the changes in type of metal transfer.

Solid GMAW Electrodes. The effect of current on the fume generation rates for a 1.1 mm (0.045 in.) diameter E70S-3 electrode is shown in Fig. 2.9. The shielding gas was Ar-2 O_2, and welding was done in the spray transfer range to avoid introducing variations in data associated with the change from globular to spray transfer. The welding conditions and experimental data are contained in Table C29.

As indicated in Fig. 2.9, the fume generation rate increased gradually with current over the range investigated. Lower fume generation rates in this range were observed by Heile and Hill (Ref. 2.3). Such differences could be associated with a lower voltage (shorter arc length), shorter contact tube-to-work distance, better shielding, etc. Their current range extended downward to 100 A and a minimum in the fume generation rate was observed where the transfer mode changed from spray to globular.

Covered Electrode Welding at High Currents. Studies were undertaken to determine if the fume generation characteristics of covered electrodes were similar in magnitude to those obtained with flux cored electrodes when welding was done at comparable current levels. Two 6.4 mm (1/4 in.) diameter electrodes were included in this investigation: a low hydrogen, iron powder electrode (E7018) with a recommended operating range of 300 to 350 A, and a high titania, iron powder electrode (E7024) with a recommended operating range of 385 to 400 A. The resulting data are contained in Table B40.

Table 2.13 was prepared to facilitate comparison of the characteristics of these electrodes with those flux cored electrodes. The fume generation rates for the E7018 and E7024 electrodes at high welding currents were similar in magnitude to those of the flux cored electrodes. Only one electrode, E70T-5(50), had a substantially higher fume generation rate than either of the covered electrodes. The ratio of weight of fumes to weight of deposited metal was higher for the E7018 electrode than for any of the flux cored electrodes. This reflects the relatively low metal deposition rate of the E7018 electrode in comparison to those of the flux cored electrodes. For an electrode with a much higher deposition rate, E7024, this ratio was similar to those of several flux cored electrodes. The results indicate that the fume generation rates for covered and flux cored electrodes are similar when the electrodes are used at similar current levels.

Current Density Effects

Since current density also has a bearing on the rate at which fumes are produced, this variable was studied to obtain a better understanding of its effects when welding was done with electrodes having different diameters.

Variable Current Density. The fume generation characteristics of three electrodes with diameters of 3.2 mm (1/8 in.), 4 mm (5/32 in.), and 4.8 mm (3/16 in.) were investigated; all electrodes were made by the same producer. Welding with all electrodes was done at a nominal current of 150 A; both direct current (electrode positive) and direct current (electrode negative) were used. Under these conditions, the approximate current densities were as follows: 12 200 A/in.^{2} (1890 A/cm^2) for the 1/8 in.

Table 2.13
Comparison of fume generation characteristics of covered and flux cored electrodes obtained at comparable current levels

Electrode (no.)	Current, A	Fume generation rate: Measured, g/min	Fume generation rate: Normalized, µg/min/I^2	Weight of fumes/ weight of deposited metal, g/kg*
		Covered electrodes		
E7018 (72)	390-400	1.60	10.3	27.91
E7024 (73)	385-395	2.13	14.0	19.28
		Flux cored electrodes		
E70T-1 (40)	470-485	1.36	5.8	10.40
E70T-4 (41)	455-465	0.96	4.5	6.65
E70T-4 (42)	435-445	2.27	11.5	17.51
E70T-1 (46)	435-445	1.33	6.9	9.97
E70T-1 (47)	435	1.83	9.7	14.17
E70T-4 (56)	370-390	1.86	12.5	12.76
E70T-4 (49)	400-410	2.09	12.9	13.83
E70T-5 (44)	425-445	2.26	11.8	17.87
E70T-5 (50)	440-450	3.25	16.6	23.63

*g/kg ÷ 10 = weight of fumes expressed as a percentage of deposit weight

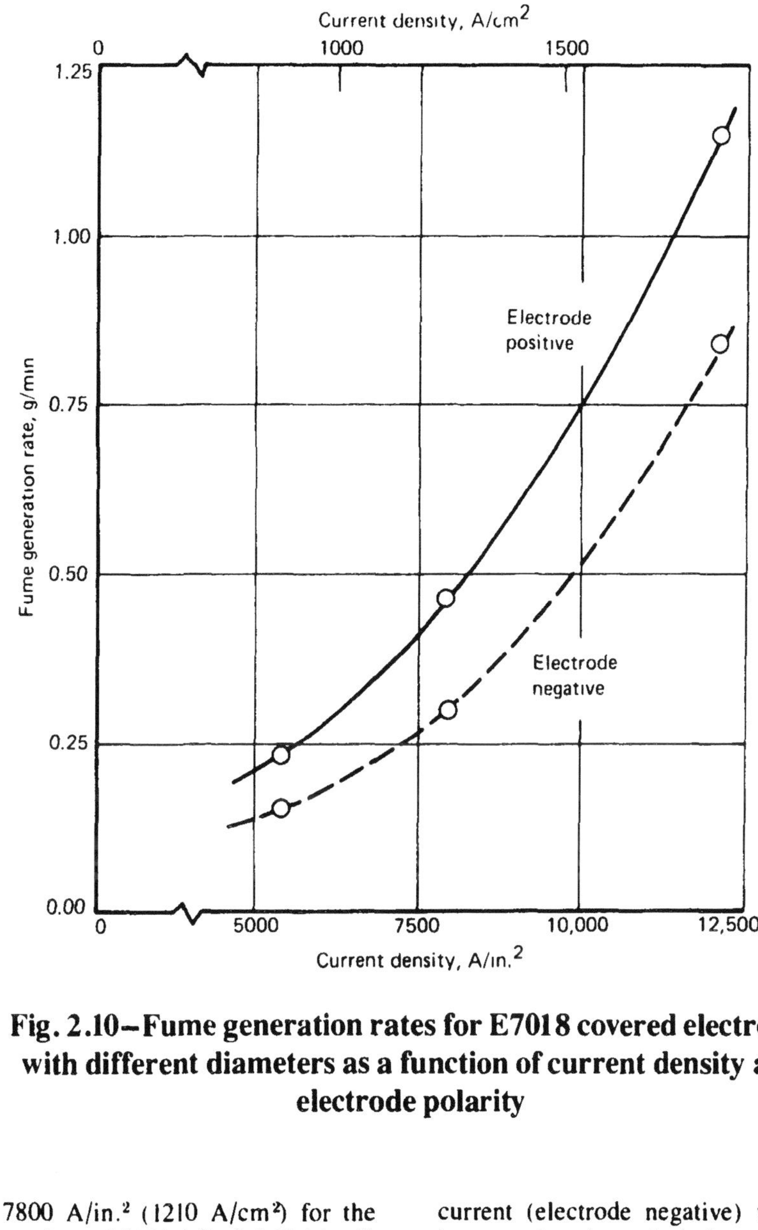

Fig. 2.10–Fume generation rates for E7018 covered electrodes with different diameters as a function of current density and electrode polarity

diameter electrode; 7800 A/in.^{2} (1210 A/cm^2) for the 5/32 in. diameter electrode; and 5400 A/in.^{2} (840 A/cm^2) for the 3/16 in. diameter electrode. Note that there are non-ideal operating conditions, since the current density for the 1/8 in. diameter electrode is high and the current density for the 3/16 in. diameter electrode is low. However, the conditions used were useful in studying current density effects. The results of this investigation are shown in Fig. 2.10; experimental details appear in Tables B30 and B31.

In Fig. 2.10, the fume generation rate increased with current density in what appears to be a parabolic relationship. The current density was highest in the 1/8 in. diameter electrode. As a result, the arc temperature was higher than that for larger electrodes and a resultant higher fume generation rate was observed. When direct current (electrode positive) was used, the fume generation rate was higher than that observed when direct current (electrode negative) was used. This occurred because the electrode is heated to higher temperatures when the anode spot is on the electrode.

Constant Current Density. The fume generation characteristics of four E7018 electrodes with diameters ranging from 2.4 mm (3/32 in.) to 4.8 mm (3/16 in.) were studied to determine the effects of constant current density. The welding currents were selected to produce current densities of approximately 8000 to 8500 amperes per square inch. The resulting data are shown in Fig. 2.11 and 2.12 and in Table B32. With these welding conditions, the fume generation rate varied approximately with the square of electrode diameter (Fig. 2.11). The electrode melting rate was also approximately proportional to the square of the diameter (Table B32). The tabular data indicate that the amount of fumes produced during welding was largely dependent upon the amount of electrode consumed. Figure 2.12 shows the dependence

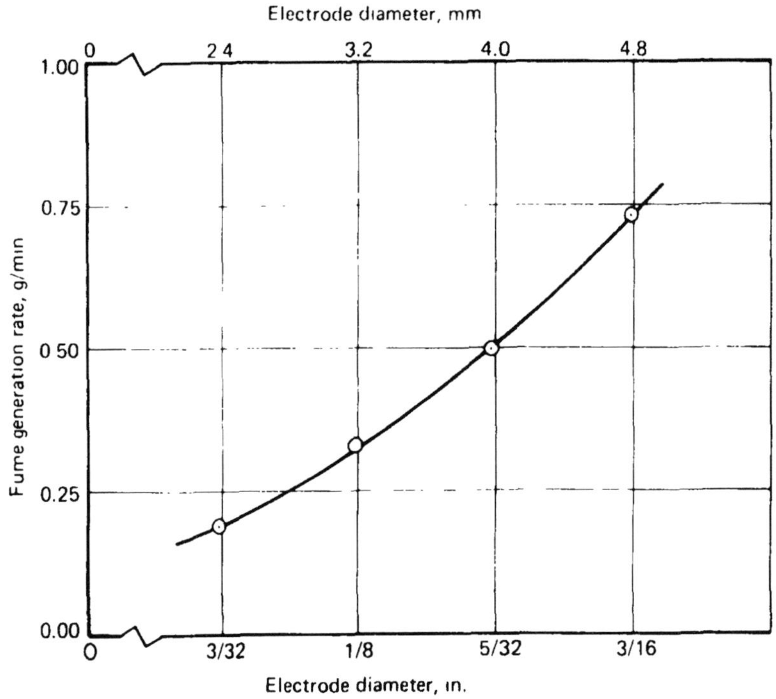

Fig. 2.11–Fume generation rates for E7018 covered electrodes with different diameters at constant current densities

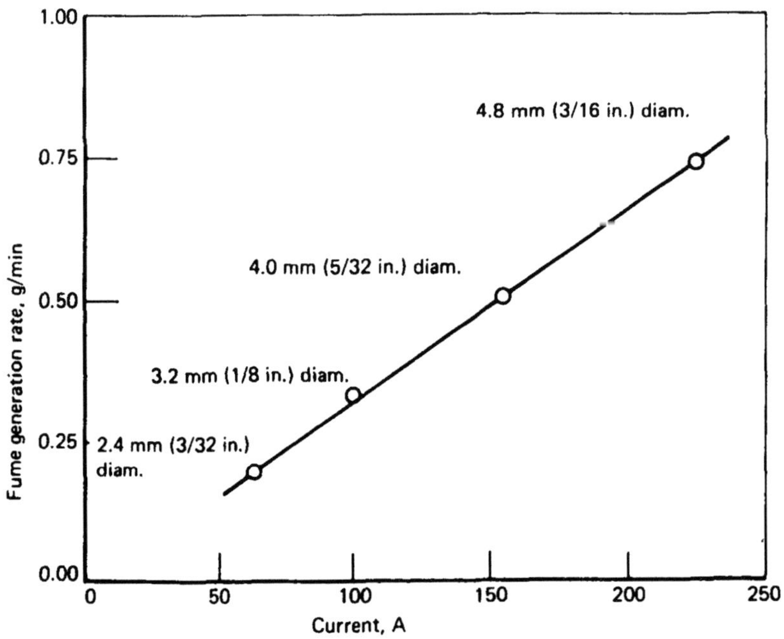

Fig. 2.12–Fume generation rates for E7018 covered electrodes with different diameters as a function of current

of fume generation rate on welding current. With approximately equal current densities in each electrode, the variation of fume generation rate with current was almost linear. Four E6010 and four E6013 electrodes were evaluated in a similar manner and similar results were obtained; these data are shown in Tables B33 and B34 respectively.

Arc Voltage—Arc Length Effects

Arc voltage and arc length are closely related welding parameters: for any given arc length, there is a corresponding value of voltage whose magnitude is determined by the characteristics of the electrode (resistivity, melting rate, etc.) and those of the welding process and power supply. During the investigation, the fume generation characteristics of selected electrodes as a function of voltage (measured across the electrodes) and arc length were studied. Because of the interrelationship between these variables, their effects on the rate at which fumes were produced were similar.

Previously discussed studies to evaluate the fume generation characteristics of a flux cored electrode, E70T-1 (41), were extended to include the effects of voltage. To investigate the dependency of fume generation rate on voltage, welding was done at high and low voltages near the low, middle, and high portions of the current range used for this electrode (330 to 600 amperes). For any given current, the "high" voltage was slightly below that at which the arc became unstable and difficult to control; the "low" voltage was just higher than that at which stubbing of the electrode occurred. The results obtained at the middle and high portions of the current range are shown in Fig. 2.13; tabular data are provided in Table B28. The results showed an increase in voltage was accompanied by an increase in the fume generation rate. Other investigators have reported similar results. In this instance, the effect of voltage was most pronounced at higher current levels.

The effects of arc voltage (or arc length) on the fume generation characteristics of an E6010 covered electrode were also examined. Welding was done at a nominal current of 150 amperes over a voltage range that extended from about 20 to 40 volts. Data on fume generation rates for this electrode are also plotted in Fig. 2.13; complete data on the results of this study are shown in Table B35. Although fumes were produced at different rates by the E6010 and E70T-1 electrodes, the trend toward increasing quantities of fumes at higher voltages was common to both.

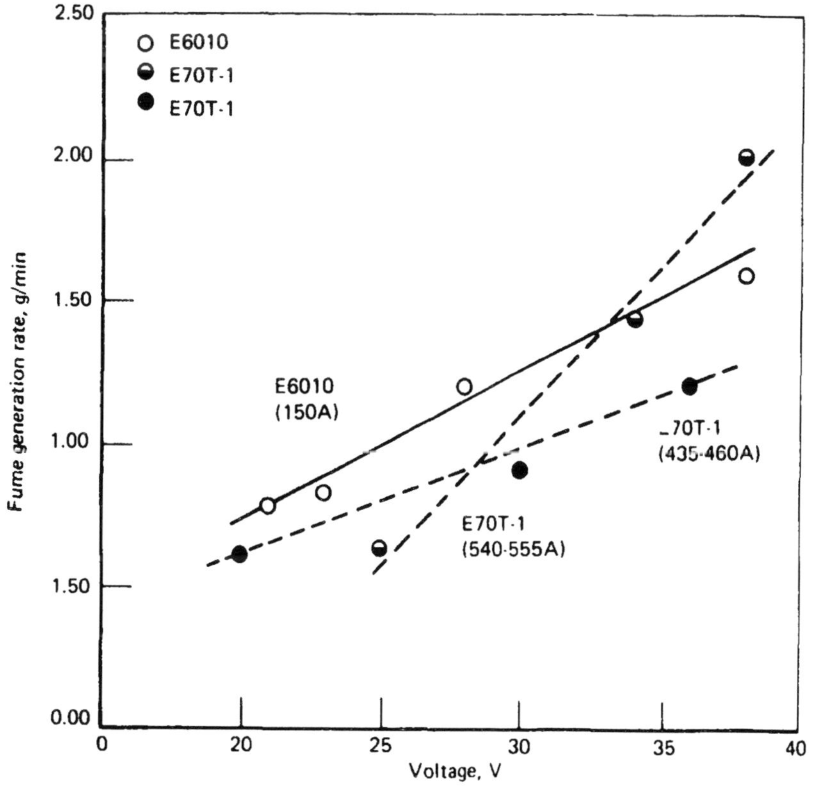

Fig. 2.13—Fume generation rates for an E6010 covered electrode and an E70T-1 flux cored electrode as a function of voltage at selected current levels

Arc length effects were studied with 4 mm (5/32 in.) diameter E6010, E6013, and E7018 electrodes. A video camera connected to a television unit and a video tape recorder were used to display and record arc length; current and voltage were monitored with a strip chart recorder.

Fume samples were obtained when welding was done with each electrode at selected arc lengths. To obtain the desired data, a welding arc was established and the power unit was adjusted to provide current at about 150 amperes. The arc was photographed with the video camera and the result was concurrently displayed on the television screen and stored on video tape for future use. Then, as the welder monitored the screen to maintain as constant an arc length as possible, a bead-on-plate weld was deposited and the fume sample was collected. The welding time was about 30 seconds. This process was repeated for each arc length.

The resulting data were analyzed to determine the effects of arc length on fume generation rates. To accomplish this objective, the data stored on the video tape were displayed on the television screen and the average arc length over the welding interval was measured from the plate surface to the electrode tip. It should be emphasized that this was a subjective measurement, because the arc length varied during welding. Since current varied also, the fume generation rates were normalized at a current of 150 amperes to provide a common basis for comparison. The results of this investigation are summarized in Fig. 2.14; supporting data are provided in Table B36.

Fume generation rates increased with increasing arc length for all of the electrodes included in this investigation. The slope of the lines relating these variables indicated that arc length had slightly more effect on the rates at which fumes were produced by the E6010 and E6013 electrodes than on the rate at which fumes were produced by the E7018 electrode. The results of this investigation (i.e., the increase in fume generation rate with increasing arc length) are in general agreement with those observed by Kobayashi, Maki, and Ohe (Ref. 2.5).

In a related experiment, the welding current was

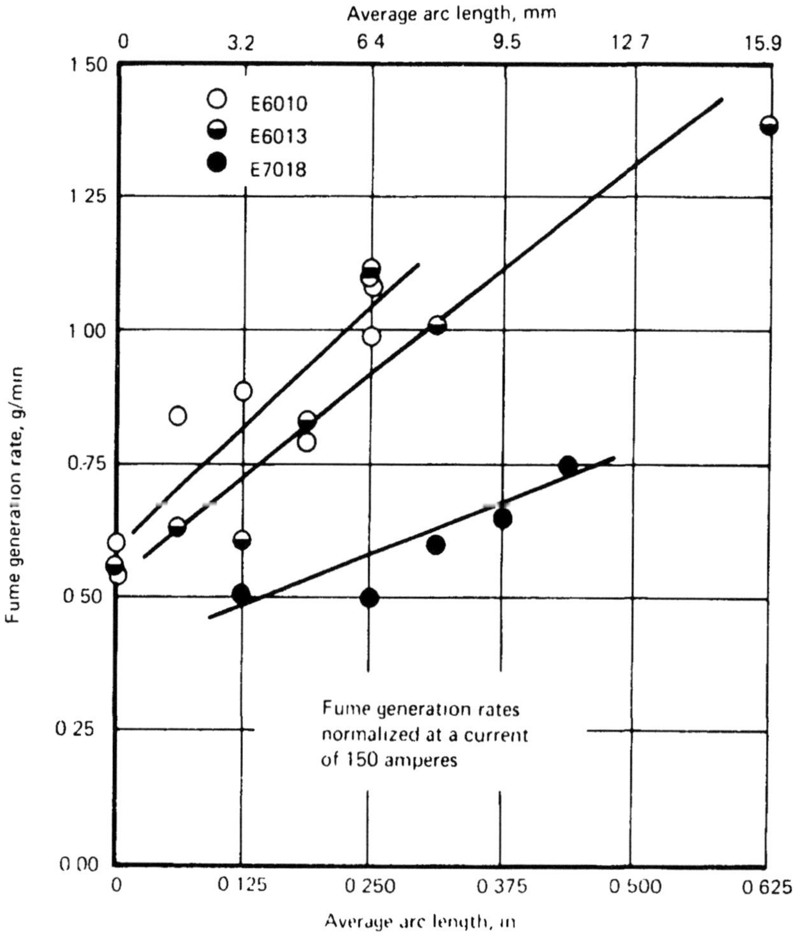

Fig. 2.14 – Fume generation rates for selected covered electrodes as a function of arc length

initially established at 150 amperes for an arc length that would normally be used during welding. The current was not adjusted thereafter, but was allowed to vary with different arc lengths. For example, when the arc length was increased, voltage increased and current decreased. Fume samples were collected at selected arc lengths for E6010, E6013, and E7018 electrodes during welding intervals of about 30 seconds. Arc length data were recorded on video tape and analyzed later in the manner described previously. All fume generation rates were normalized to a common current of 150 amperes for comparison purposes. The results were similar to those discussed previously; that is, fume generation rates increased with increasing arc length (see Table B36).

Contact-Tube-To-Work Distance Effects

The distance between the end of the contact tube and the surface of the base plate is a variable that affects the melting and metal deposition rates in flux cored arc welding and gas metal arc welding. Since fume generation rates are closely associated with melting rates, the effects of contact tube-to-work distance were included in this investigation.

The fume generation characteristics of an E70T-1 flux cored electrode as a function of contact tube-to-work distance were determined. Welding was done at two current levels, nominally 450 and 520 amperes, and the contact tube-to-work distance was varied from 19 mm (0.75 in.) to 38 mm (1.50 in.). As the contact tube-to-work distance was varied, the wire feed rate was adjusted to maintain the selected current level. The power supply voltage remained constant during these studies, even though it would be normally increased or decreased with changes in contact tube-to-work distance so that arc length could be controlled.

The following observations are based on an examination of the data presented in Tables B42 and B43 and in Fig. 2.15:

(1) At each current level, there was a gradual increase in metal deposition rate with increasing contact tube-to-work distance (Tables B42 and B43). When the contact tube-to-work distance increases, the wire feed speed must be increased to maintain a constant current level. The increased melting rate is accompanied by a higher deposition rate.

(2) For each current level, the fume generation rate remained essentially constant for each contact tube-to-work distance.

(3) The ratio of weight of fumes to weight of deposited metal decreased gradually with increasing contact tube-to-work distance.

(4) The effect of current on fume generation rate is evident in Fig. 2.15. At 450 amperes, the fume generation rate was about 1 g/min; at 520 amperes, the rate was about 1.3 g/min.

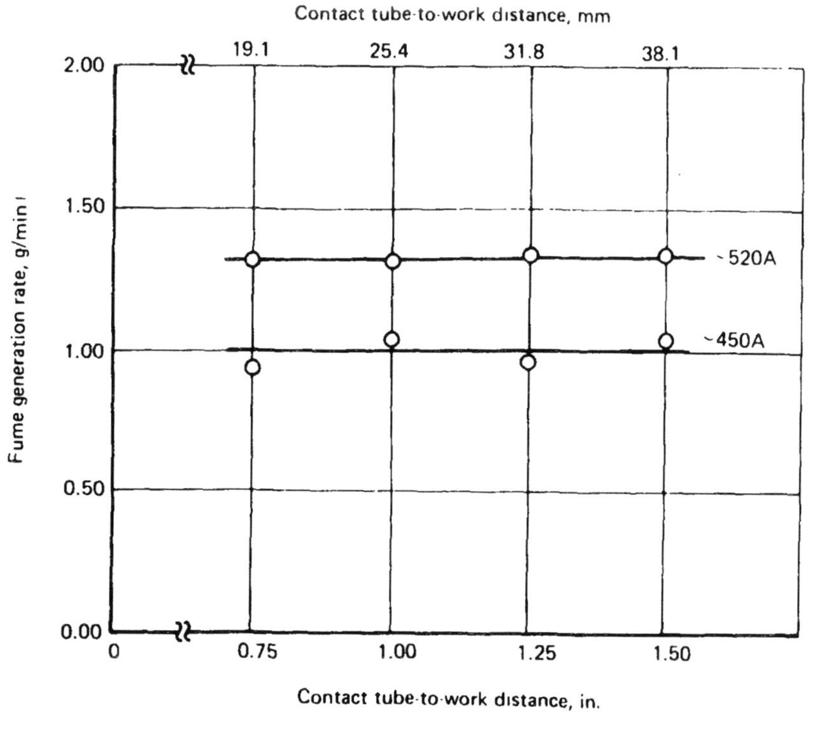

Fig. 2.15—Fume generation rate of a 2.4 mm (3/32 in.) E70T-1 flux cored electrode as a function of contact tube-to-work distance at selected constant current levels

Shielding Gas Effects

The shielding gas is a variable associated with the flux cored and gas metal arc welding processes, and it affects the rate at which fumes are produced during welding and the composition of the fumes. The type of shielding gas also affects the kind of gases to be found in the welding area.

The effects of shielding gases on fume generation rates have already been discussed in the review of the baseline data for flux cored and gas metal arc welding electrodes. These rates were highest when CO_2 or gas mixtures containing CO_2 were used to shield the welding arc.

Since the oxidation potential of the shielding gas affects the rate at which specific elements oxidize as they are vaporized from the tip of the electrode, studies were conducted to determine the effects of shielding gas type on the composition of the fumes produced by E70S-3 gas metal arc electrodes used for welding carbon steels. Each of the three E70S-3 electrodes included in this investigation was evaluated using welding conditions to produce the following transfer modes: (a) spray transfer with Ar-2 O_2 shielding, (b) spray transfer with Ar-9 CO_2 shielding, and (c) globular transfer with CO_2 shielding. The fume samples were collected on cellulose membrane filters and analyzed by atomic absorption procedures. The resulting data plus the calculated weight percentages of each oxide are shown in Table 2.14 and the apparent trends are discussed below:

(1) Iron (iron oxide) contents were highest when welding was done in the spray transfer mode with either Ar-2 O_2 or Ar-9 CO_2 shielding and lowest when welding was done in the globular transfer mode with CO_2 shielding These results are in agreement with those of Heile and Hill (Ref. 2.3).

(2) Manganese (manganese oxide) contents were highest when welding was done in either the spray transfer mode with Ar-9 CO_2 shielding or in the globular transfer mode with CO_2 shielding; this element (or elemental oxide) content was lowest when welding was done in the spray transfer mode with Ar-2 O_2 shielding. Here, the oxidation potential of the shielding gas had a slight effect on the contents of manganese (manganese oxide) in the fumes.

(3) Silicon (silicon dioxide) contents were highest when welding was done in the globular transfer mode with an oxidizing gas, CO_2. These results are also in agreement with those of Heile and Hill (Ref. 2.3). There was no apparent explanation for the low silicon content in the fumes produced by the E70S-3(54) and E70S-3(57) electrodes.

(4) Copper contents in the fumes appeared to be less

Table 2.14
Effects of shielding gas on fume composition

Electrode[1]		Shielding gas	Fume sample weight, g	Composition, weight %						
				Fe	(Fe_2O_3)	Mn	(MnO_2)	Si	(SiO_2)	Cu
E70S-3	(54)	Ar-2 O_2[2]	0.23	61.7	(88.2)	4.6	(7.3)	1.0	(1.6)	0.09
E70S-3	(54)	Ar-9 CO_2	0.55	62.5	(89.4)	6.1	(9.6)	0.5	(0.8)	0.09
E70S-3	(54)	CO_2	0.42	56.3	(80.5)	6.3	(10.0)	2.3	(3.6)	0.14
E70S-3	(57)	Ar-2 O_2[2]	0.32	62.7	(89.7)	4.4	(7.0)	1.1	(1.7)	0.85
E70S-3	(57)	Ar-9 CO_2	0.44	62.2	(88.9)	6.5	(10.3)	0.4	(0.6)	0.70
E70S-3	(57)	CO_2	0.33	55.4	(79.2)	6.8	(10.7)	1.5	(2.4)	1.20
E70S-3	(58)	Ar-2 O_2[2]	0.41	62.1	(88.8)	5.6	(8.8)	1.1	(1.7)	1.29
E70S-3	(58)	Ar-9 CO_2	0.54	62.0	(88.7)	4.6	(7.3)	1.5	(2.4)	0.99
E70S-3	(58)	CO_2	0.40	52.5	(75.1)	5.5	(8.7)	2.5	(3.9)	1.00

1. Welding Conditions:

For spray transfer welding with A-2 O_2 shielding
Electrode No. 54: 35 V; 260-280 A
Electrode No. 57: 35 V; 270-280 A
Electrode No. 58: 34.5 V; 290-300 A

For globular transfer welding with CO_2 shielding
Electrode No. 54: 36 V; 330-340 A
Electrode No. 57: 35 V; 320 A
Electrode No. 58: 32 V; 330 A

For spray transfer welding with A-9 CO_2 shielding
Electrode No. 54: 34.5 V; 210-220 A
Electrode No. 57: 35 V; 205-215 A
Electrode No. 58: 35 V; 215-225 A

2. Average of three analyses.

a function of the shielding gas than were the other core wire elements and their oxides.

Effects of Iron Powder Additions

Investigation was made into the effects of iron powder additions on the fume generation characteristics of the two groups of mild steel electrodes listed below:

(1) High Titania Electrodes. This group included electrodes with high titania coverings and iron powder additions ranging from none (E6013) to moderate (E7014) to high (E7024).

(2) Low Hydrogen Electrodes. This group included electrodes with low hydrogen coverings and iron powder additions ranging from none (E7016) to moderate (E7018) to high (E7028).

These studies were conducted with 4 mm (5/32 in.) diameter electrodes, all of which were made by one producer. Welding was done at currents near the high end of the operating range. The results of this investigation are summarized in Table 2.15; detailed data are contained in Tables B37 and B38.

High Titania Electrodes. At the measured welding current, the fume generation rates for the E6013, E7014, and E7024 electrodes were very similar in magnitude (0.68 to 0.72 g/min) and appeared to be unaffected by the presence or absence of iron powder additions to the electrode coverings. This is not the case if the fume generation rates for these electrodes are compared at equivalent welding currents. When these rates were normalized to a current of 175 A, for example, the fume generation rates varied inversely with the amount of iron powder in the electrode covering.

The ratio of the weight of fumes to weight of deposited metal also varied inversely with the amount of iron powder in the covering when these electrodes were used at recommended current levels. Under these conditions, the E7024 electrode had the lowest ratio, because the fume generation rate for this electrode was low and the metal deposition rate was high.

Low Hydrogen Electrodes. At recommended current levels, the fume generation rates for the E7016 (no iron powder addition) and E7018 (moderate iron powder addition) electrodes were almost identical; the rate for the E7028 electrode (high iron powder addition) was twice as large. When the normalized rates for these electrodes were compared, they were very similar in magnitude (0.50 to 0.58 g/min). The rates at which fumes were produced appeared to be controlled by ingredients other than iron powder in the electrode covering. Calcium fluoride, a compound common to these low hydrogen electrodes, probably exerted a controlling influence on the rate at which fumes were produced.

Ratios of weight of fumes to weight of deposited metal for these electrodes were similar in magnitude.

Other Studies

Electrode Angle Effects. The effect of electrode angle on the rate at which fumes are produced by covered electrodes was studied with 4 mm (5/32 in.) diameter E6010, E6013, and E7018 electrodes. Each electrode was evaluated at angles of 30, 60, and 90 degrees to the workpiece; a tubular guide for the electrode was used to establish and maintain the desired electrode angle. Bead-on-plate welds were deposited with each electrode at a nominal current of 150 A. The results are shown in Fig. 2.16; supporting data are presented in Table B39.

For all electrodes examined, the data indicated a gradual decrease in the fume generation rate as the angle between the electrode and workpiece increased toward 90 degrees. This result agrees with the findings of other investigators (Refs. 2.5 and 2.18). However, care should be observed in interpreting the data presented strictly in terms of an electrode angle effect, because a lengthening of the arc probably occurred as the electrode angle decreased. As indicated in the section on arc length, fume generation rates are affected substantially by changes in this variable.

Two sets of operating conditions that produced arc lengths differing by about 1.6 mm (1/16 in.) were used with the E6010 electrode; data obtained under those conditions are also included in Fig. 2.16. The curves representing fume generation rates at various angles were separated by an amount largely attributable to the effect of

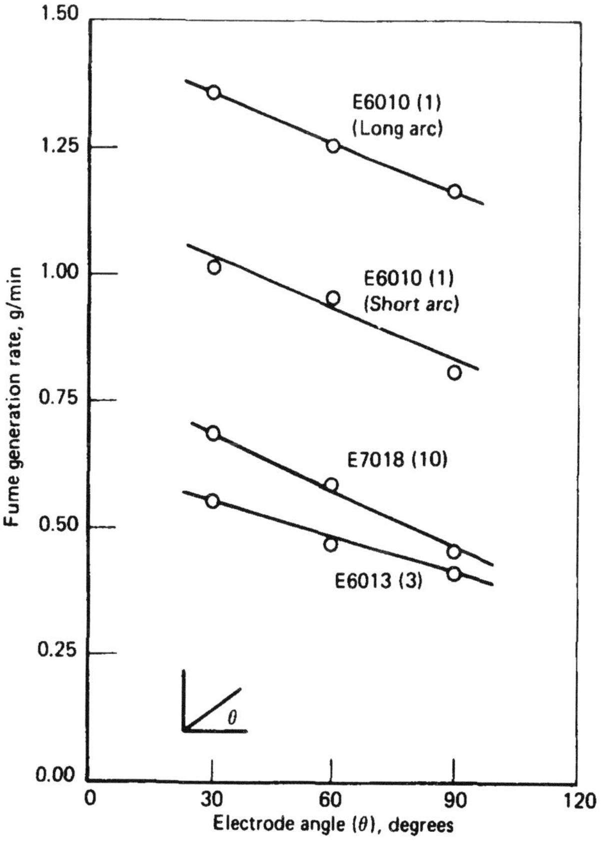

Fig. 2.16–Effect of electrode angle on the fume generation rates of covered electrodes

Table 2.15
Summary of data on effects of iron powder on fume generation characteristics of high titania and low hydrogen covered electrodes

Electrode	Covering	Current, A (nominal)	Fume generation rate: Measured, g/min	Fume generation rate: Normalized[1], g/min	Weight of fume/wgt. of deposited metal, g/kg[2]
E6013	High titania	165	0.73	0.81	29.86
E7014	High titania	175	0.72	0.75	23.72
E7024	High titania	210	0.68	0.47	12.13
E7016	Low hydrogen	170	0.50	0.51	20.84
E7018	Low hydrogen	175	0.51	0.50	18.39
E7024	Low hydrogen	230	1.03	0.58	21.27

1. Fume generation rates normalized to 175 A.
2. g/kg ÷ 10 = weight of fume expressed as a percentage of deposit weight.

arc length differences; however, in both instances, the rates decreased with increasing electrode angle by about the same amount.

Welding with covered electrodes is done mostly at large (>60 degree) angles between the electrode and workpiece, so problems with fumes are minimized. It is important to be aware that increased amounts of fumes can be expected when welding is done under confined conditions where the electrode angle might be restricted to small values.

Base Plate Effects. Fume generation characteristics for the covered, flux cored, and solid electrodes investigated during this program were determined with base plates whose composition was similar to that of the core wire or sheath. That is, E6010 and E70T-1 electrodes were evaluated with carbon steel plate, E410-15 electrodes with 400-series stainless steel plate, and so on. Since the base plate has relatively little effect on the production of fumes, the need to match the base plate to the electrode was questioned, particularly with respect to the evaluation of stainless steel electrodes. To resolve this question, fume samples for gravimetric and chemical analyses were collected during the deposition of stainless steel bead-on-plate welds on carbon steel and stainless steel plates. Welding was done with a 4 mm (5/32 in.) diameter E316-16 covered electrode at baseline conditions. The experimental results are discussed below with the aid of Tables 2.16 and 2.17.

(1) As indicated in Table 2.16, fume generation rates and ratios of weight of fumes to weight of deposited metal were unaffected by the type of base metal used during welding. Thus, either carbon steel or stainless steel could be used with equal facility to evaluate stainless steel electrodes in terms of their fume generation rates. Caution in extrapolating these results to other electrodes is advised.

(2) The composition of the fumes did reflect the composition of the base metal upon which the welds were deposited as well as that of the electrode. When the E316-16 electrode was evaluated on stainless steel, the fumes contained higher contents of manganese, nickel, and chromium than when this electrode was used on carbon steel (Table 2.17). The effects of the electrode itself on fume composition can be estimated by examining the composition of the fumes produced when welds were deposited on carbon steel. The difference between the contents of manganese, nickel, and chromium in this sample and in the sample collected when welding was done on stainless steel represents the effects of the stainless steel base plate on fume composition. It was concluded that electrode and base plate should be matched if fume compositions are to be determined.

Diameter Effects with Flux Cored Electrodes. The fume generation characteristics of a 1.6 mm (1/16 in.) diameter E70T-1 electrode were compared to those of a 2.4 mm (3/32 in.) diameter electrode made by the same manufacturer. Welding was done at current levels appropriate for the respective electrodes and CO_2 was used for shielding (Table B41). Data on the "as measured" and "normalized" fume generation rates are presented below:

Electrode	Electrode diam., in. (mm)	Welding current, A	Fume generation rate, g/min.: Measured	Fume generation rate, g/min.: Normalized*
E70T-1 (40)	3/32 (2.4)	480	1.36	1.36
E70T-1 (45)	1/16 (1.6)	330	1.01	2.14

*Fume generation rates normalized to a current of 480 A.

The measured fume generation rate for the small diameter electrode was less than that associated with the large diameter electrode. However, when the effect of current on this characteristic was taken into account by normalizing the fume generation rates to a common current (480 A

Table 2.16
Effect of base plate on average fume generation characteristics of 4 mm (5/32 in.) diameter E316-16 covered electrode

Characteristic	Base plate	
	Carbon steel	Stainless steel
Fume generation rate, g/min.	0.24	0.24
Weight of fume/weight of deposited metal, g/kg*	9.96	9.87

*g/kg ÷ 10 = %

Table 2.17
Composition of carbon steel and stainless steel base plate and composition of fume samples

Type	Composition, weight %				
	Fe	Mn	Si	Cr	Ni
		Base plate			
Carbon steel	Bal.	0.68	<0.01	---	---
Stainless steel (Type 304)	Bal.	1.70	0.41	19.43	8.31
		Fumes			
Electrode/base plate					
E316-16/carbon steel	7.85	7.53	<0.01	4.85	1.03
E316-16/stainless steel	8.16	7.80	<0.01	5.60	1.21

in this case), more fumes were produced by the small diameter electrode than by the large diameter one. Fume generation rates are dependent upon the rate at which the electrode is consumed. If two electrodes with different diameters are used at the same current level, the melting rate for the smaller of the two electrodes will be greater in the small diameter electrode. The effect of increased I^2R heating in the smaller electrode probably also contributed to increased melting rate. The effect of I^2R heating on deposition rates was studied by Wilson, Claussen, and Jackson (Ref. 2.22).

Section IID. Analytical Studies

Various analytical methods were used to determine the presence of selected constituents in the fumes produced during arc welding operations. Some of the characteristics of these methods are discussed as follows:

(1) Optical emission spectroscopy yields a wealth of analytical data because 70 or more elements can be detected simultaneously and their content can be estimated on a semiquantitative basis. It is best suited for accurately determining the presence of elements that are present in small amounts. This method is most valuable when the elements in question are present in concentrations of 1 percent or less. If elements are present in large amounts, concentrations are often expressed in percentage ranges, and other techniques (e.g., atomic absorption) must be used to obtain greater accuracy. In the case of welding fumes, optical emission spectroscopy can provide an overview of most of the elements present in the fume sample, regardless of their origin; i.e., from the core wire or sheath, electrode covering or flux, or base plate. This method was used during this program in studies to characterize the nature of welding fumes, and the results have been presented in a previous section of this report. It was not used more extensively because capital equipment costs for optical emission spectroscopy are high, and such equipment is not as readily available in industry as that used with other analytical methods.

(2) Atomic absorption analysis is also capable of determining the presence of about 70 elements, but the concentration of each element must be detected individually. Analytical costs per individual element are not high, but they can be appreciable if the sample contains

several elements (e.g., the fumes produced by stainless steel or nickel-base electrodes). This method is better suited than optical emission spectroscopy for the detection of elements that are present in large amounts, and the detection accuracy is excellent. Equipment for atomic absorption analysis is reasonable in cost and readily available in industry. Optical emission spectroscopy combined with atomic absorption analyses of elements present in large concentrations provide reliable data on fume sample compositions.

(3) Classical chemical analysis procedures have been developed to detect the presence of most elements and compounds. Insofar as welding fumes are concerned, these methods are most useful in determining constituents whose presence cannot be detected by optical emission spectroscopy or atomic absorption analysis (e.g., fluorides, chlorides, gases, etc.).

Analysis of Fume

In this program, atomic absorption analysis was selected as the primary analytical technique for detecting elemental fume constituents because equipment for conducting such analyses is commonly available and relatively simple to use. However, by agreement with the AWS Research Committee, this method was used only to detect the major constituents that originated in the core wire or electrode sheath (i.e., iron, manganese, and silicon in the case of fumes produced by low carbon steel covered or flux cored electrodes) because of (a) the large number of samples to be analyzed and (b) the need to detect constituents on an element-by-element basis. As a result, analytical data on the composition of fumes associated with covered or flux cored electrodes do not total 100 percent because they do not include all of the contents of constituents originating in the electrode covering or flux (e.g., sodium, calcium, titanium, etc.). In some instances, these constituents represent a major fraction of the constituents in the fumes. In the case of fumes produced by solid gas metal arc electrodes, the analytical data more nearly approach 100 percent as a total, assuming that most elements are present in the oxide form.

To supplement these data, fluoride contents in the fumes produced by many of the low hydrogen electrodes were determined by wet chemical analysis. Different analytical techniques are needed to determine total fluorides and water-soluble fluorides. Total fluoride contents in the fumes produced by most of the low hydrogen covered electrodes investigated during this program were obtained. Water-soluble fluorides (but not total fluorides) were determined for several other electrodes. In the course of these studies, both total and water-soluble fluorides were determined in samples of the fumes produced by representative E7018 covered and E70T-1 flux cored electrodes provided by the AWS Subcommittee on Sampling. The results are shown below.

	Fluoride content of fumes, wgt. %		
Electrode	Total	Water-soluble	Water-insoluble*
E7018	22.1	5.9	16.2
E70T-1	0.23	0.16	0.07

*Obtained by subtraction.

The data indicate that a large fraction of the fluorides in the fumes produced by the E7018 electrode was insoluble in water; this result may be attributable to the presence of calcium fluoride in the fumes. This compound, a common and major ingredient in low hydrogen electrode coverings and fluxes, is soluble in water to a limited degree only. The fluoride content of the fumes associated with the E70T-1 flux cored electrode was negligible from the hygienic viewpoint; the fumes generated by the E7018 electrode contained appreciable fluorides.

Concentrations of selected elements in the fumes produced by representative covered, flux cored, and solid electrodes investigated during this program were determined by atomic absorption analysis of fume samples collected on cellulose membrane filters. The analytical results are shown in Table 2.18 along with supplementary data on fluorides in the fumes produced by some of these electrodes. With few exceptions (e.g., carbon steel electrodes used for gas metal arc welding), the data represent the results obtained from the analysis of one sample per electrode.

The following comments are based on a review of the data contained in Table 2.18:

(1) When converted to their common oxide form, the constituents of fumes produced by solid gas metal arc welding electrodes generally totalled in the 90 percent range. Lower totals were observed for constituents associated with the fumes generated by ER5356, ERNiCu-7, Inconel 625, and ERCu electrodes.

(2) The constituents (in oxide form) of fumes produced by covered and flux cored electrodes did not approach 100 percent as a total because concentrations of elements associated with the electrode covering or flux were not determined by analysis and, as a result, could not be included in such compilations. Fume constituent totals were highest for electrodes with relatively thin coverings (E6010 and E6013) and lowest for electrodes with thick coverings (E7018 and E7024). This is an indirect indication of the contribution of the electrode covering to the overall fume generation rate.

(3) The fluoride content of the fumes produced by the E8018 C3, E9018 B3, and E316-15 covered electrodes was high in comparison to that observed with other electrodes. This may be characteristic of these electrodes since they were all produced by the same manufacturer.

(4) The fluoride content (water-soluble) of the fumes associated with four of the E70T-1 electrodes included in this investigation was small (0.06 to 0.31 percent). Much higher fluoride contents (6.33 and 8.68 percent) were observed in the fume produced by two

Table 2.18
Concentrations of selected constituents in the fumes produced by arc welding electrodes (Analyses by atomic absorption technique)

Electrode	Code	Fume sample weight, g	Concentration, weight percent									Other
			Fe	Mn	Si	Ni	Cu	Cr	Mo	Al	Mg	F	
Covered electrodes													
E6010 (IP)	1	0.49	47.5	3.0	5.7	---	---	---	---	---	---	---	---
E6010 (IP)	2	0.35	49.6	3.4	6.5	---	---	---	---	---	---	---	---
E6010 (IP)	11	0.56	41.6	3.4	3.7	---	---	---	---	---	---	---	---
E6013	4	0.16	61.4	5.1	12.2	---	---	---	---	---	---	---	---
E6013	6	0.08	45.0	5.5	15.0	---	---	---	---	---	---	---	---
E6013	3	0.27	62.2	4 1	11.2	---	---	---	---	---	---	---	---
E7018	5	0.26	26.1	4.5	<0.2	---	---	---	---	---	---	13.1[1]	---
E7018	12	0.30	28.7	4.1	<0.2	---	---	---	---	---	---	14.0[1]	---
E7018	10	0.26	24.2	3.6	<0.2	---	---	---	---	---	---	15.0[1]	---
E7024	7	0.30	30.2	5 3	18.3	---	---	---	---	---	---	---	---
E7024	8	0.31	29.5	5.6	19.7	---	---	---	---	---	---	---	---
E7024	9	0.31	33.3	7.8	13.2	---	---	---	---	---	---	---	---
E6013[3]	34	0.25	41.4	---	---	---	---	---	---	---	---	---	---
E7014[3]	30	0.23	39.0	---	---	---	---	---	---	---	---	---	---
E7024[3]	35	0.27	27.8	---	---	---	---	---	---	---	---	---	---
E7016[3]	29	0.21	32.7	---	---	---	---	---	---	---	---	15.2[1]	---
E7018[3]	36	0.22	27.8	---	---	---	---	---	---	---	---	12.7[1]	---
E7028[3]	32	0.42	38.1	---	---	---	---	[illegible]	---	---	---	11.6[1]	---
E8018 C3	14	0.12	45.2	7.2	---	0.3	---	0.1	<0.1	---	---	35.8[1]	---
E9018 B3	17	0.16	21.9	5.9	---	0.1	---	1.6	<0.1	---	---	28.1[1]	---
E316-15	22	0.15	8.4	7.7	---	1.1	---	5.8	<0.1	---	---	---	---
E316-16	20	0.17	10.0	8.8	---	1.5	---	6.5	<0.1	---	---	17.2[1]	---
E410-16	21	0.11	33.1	5.2	---	<0.1	---	---	---	---	---	---	---
ENi-CI	28	0.35	2.5	0.3	---	6.9	<0.1	---	---	---	---	10.0[1]	---
EniCu-2	33	0.31	0.1	2.1	---	4.2	6.2	---	---	---	---	---	---
Inconel 625[7]	62	0.25	0.6	---	---	4.6	0.7	5.9	2.1	---	---	---	---
Haynes C-276[8]	69	0.35	0.3	0.3	---	1.1	---	2.5	0.6	1.0	1.4	5.9[1]	0.1Co-0.1W-2.4Ca-4.9Ti
Haynes 25[8]	70	0.25	---	4.6	---	1.8	---	6.9	---	1.1	0.1	7.7[1]	7.7Co-1.3W-2.4Ca-5.3Ti
Flux cored electrodes													
E70T-1	40	0.53	34.4	7.1	3.7	---	---	---	---	---	---	0.06[2]	---
E70T-1	41	0.60	25.2	13.5	7.5	---	---	---	---	---	---	0.06[2]	---
E70T-1	42	0.91	33.6	9.7	2.2	---	---	---	---	---	---	8.68[2]	---
E70T-1	45	0.51	46.1	12.4	5.2	---	---	---	---	---	---	0.17[2]	---
E70T-1	46	0.63	41.2	6.2	5.4	---	---	---	---	---	---	0.31[2]	---
E70T-1	47	0.89	37.9	6.4	1.0	---	---	---	---	---	---	6.33[2]	---

Table 2.18 (cont.)
Concentrations of selected constituents in the fumes produced by arc welding electrodes
(Analyses by atomic absorption technique)

Electrode	Code	Fume sample weight, g	Concentration, weight percent: Fe	Mn	Si	Ni	Cu	Cr	Mo	Al	Mg	F	Other
Flux cored electrodes (cont.)													
E70T-4	48	1.45	11.5	3.3	<0.05	<0.01	---	---	---	---	---	2.73[2]	---
E70T-4	49	0.98	16.2	4.6	<0.05	<0.01	---	---	---	---	---	0.84[2]	---
E70T-4	56	0.95	15.1	1.0	<0.05	<0.01	---	---	---	---	---	1.84[2]	---
E70T-5	44	0.90	26.7	11.3	<0.05	---	---	---	---	---	---	2.63[2]	---
E70T-5	50	0.61	29.2	10.9	0.09	---	---	---	---	---	---	4.80[2]	---
316L	55	0.65	12.4	7.3	0.05	1.06	---	12.5	0.34	---	---	11.5 [2]	---
Gas metal arc electrodes													
E70S-3[4]	54	---	63.7	5.3	0.05	---	0.11	---	---	---	---	---	---
E70S-3[4]	54	0.54	61.0	3.6	1.1	---	0.07	---	---	---	---	---	---
E70S-3[4]	54	0.55	60.4	5.0	1.9	---	0.09	---	---	---	---	---	---
E70S-3[5]	54	0.55	62.5	6.1	0.5	---	0.09	---	---	---	---	---	---
E70S-3[6]	54	0.42	56.3	6.3	2.3	---	0.14	---	---	---	---	---	---
E70S-3[4]	57	0.23	65.7	3.8	0.8	---	0.60	---	---	---	---	---	---
E70S-3[4]	57	0.29	61.2	3.5	0.6	---	1.20	---	---	---	---	---	---
E70S-3[4]	57	0.44	61.1	5.8	1.8	---	0.74	---	---	---	---	---	---
E70S-3[5]	57	0.44	62.2	6.5	0.4	---	0.70	---	---	---	---	---	---
E70S-3[6]	57	0.33	52.2	6.8	1.5	---	1.00	---	---	---	---	---	---
E70S-3[4]	58	0.23	62.5	8.5	0.5	---	1.00	---	---	---	---	---	---
E70S-3[4]	58	0.49	60.1	3.4	1.0	---	1.80	---	---	---	---	---	---
E70S-3[4]	58	0.50	63.6	4.8	1.7	---	1.08	---	---	---	---	---	---
E70S-3[5]	58	0.54	62.0	4.6	1.5	---	0.99	---	---	---	---	---	---
E70S-3[6]	58	0.40	55.4	5.5	2.5	---	1.20	---	---	---	---	---	---
E70S-5	59	0.11	61.7	5.8	0.93	---	1.75	---	---	---	---	---	---
ER5356	60	1.06	---	---	---	---	---	---	---	38.0	3.8	---	---
ER4043	61	0.10	---	---	1.7	---	---	---	---	46.2	---	---	---
ERNiCu-7	64	0.13	5.0	1.1	0.65	22.1	44.4	<0.01	---	---	---	---	---
Inconel 625[7]	53	0.07	1.9	---	---	27.2	0.69	15.4	2.1	---	---	---	---
Haynes C-276[8]	68	0.25	3.6	1.0	---	32.5	---	8.2	16.2	0.4	---	---	3.2W
Haynes 25[8]	71	0.07	---	15.4	---	7.1	---	14.9	---	0.3	---	---	1.0W-33.5 Co
ERCuAl-A2	75	0.38	---	---	---	---	70.5	---	---	3.4	---	---	---
ERCu	76	0.31	---	---	---	---	66.0	---	---	---	---	---	---

Notes.

1. Total fluorides
2. Water-soluble fluorides
3. These electrodes were used to study the effect of iron powder additions on fume generation rates. Thus, only iron was determined in the analysis.
4. Shielding gas: A-2 O_2
5. Shielding gas: A-9 CO_2
6. Shielding gas: CO_2
7. TM, International Nickel Company
8. TM, Stellite Division, Cabot Corporation

E70T-1 electrodes, each of which was made by a different producer. As discussed by Smith (Ref. 2.17), the fluxes of some E70T-1 electrodes contain appreciable amounts of calcium fluoride to form a basic type slag.

Fume Characterization

A study was undertaken to characterize the fumes associated with various welding operations in terms of composition, the presence or absence of crystalline phases, and particle morphology. The fumes produced by electrodes used for shielded metal arc welding (E7024 and E410-16) and flux cored arc welding (E70T-1, E70T-4, and E70T-5) were analyzed for elemental composition by optical emission spectroscopy (OES) and for crystalline phases by x-ray diffraction (XRD). Scanning electron microscopy (SEM) was used to study particle morphology. Semiquantitative results of the OES analysis of the fume samples are shown in Table 2.19.

Optical emission spectroscopy is capable of detecting the presence of 70 or more elements on a semiquantitative basis. Thus, totals of data for the individual fume samples listed in Table 2.19 may approach 100 percent, if the elemental fume constituents are converted to their common oxide forms. In some instances, elemental concentrations are expressed in percentage ranges, if the element is present in large amounts. Another analytical method (atomic absorption, wet chemistry, etc.) must be used to achieve more accuracy. Only one crystalline phase, iron oxide (Fe_3O_4, or magnetite), was detected by XRD in the fumes produced by the covered electrodes; manganese was probably present in this phase also since it can replace iron in the crystal lattice. Crystalline phases containing silicon were not detected. However, since silicon was detected in the fumes produced by the E7024 and E410-16 electrodes, it was probably present as a glassy SiO_2 phase along with potassium and sodium. The x-ray diffraction pattern from the fumes produced by the E410-16 electrode had an extra diffraction line whose presence was attributed to a chromium compound.

In the fumes produced by the flux cored electrodes,

Table 2.19
Elemental composition of fumes produced by various electrodes[1] as determined by optical emission spectroscopy

Element	Composition, weight percent					
	E7024 (7)	E7024 (8)	E410-16 (21)	E70T-1 (42)	E70T-4 (49)	E70T-5 (50)
Fe	20.0-30.0	20.0-30.0	10.0-20.0	30.0-40.0	15.0-25.0	30.0-40.0
Si	10.0-20.0	5.0-10.0	2.0-3.0	2.0-3.0	0.1	2.0-3.0
K	8.0-12.0	8.0-12.0	10.0-20.0	1.0	2.0	4.0-6.0
Na	3.0- 6.0	3.0-6.0	4.0-8.0	4.0-6.0	<0.1	1.0
Mn	2.0-4.0	2.0-4.0	1.0-2.0	4.0-6.0	2.0-3.0	4.0-6.0
Ca	0.1	0.5	2.0-4.0	0.1	15.0-25.0	8.0-12.0
Zn	≤ 0.1	<0.1	0.1	<0.1	<0.1	<0.1
Ti	0.5	0.3	0.4	0.5	<0.01	0.2
Pb	0.2	0.2	0.01	<0.01	0.2	0.02
Al	0.1	0.05	0.5	0.4	7.0-10.0	1.0-2.0
Sn	0.1	0.1	0.2	0.01	<0.01	<0.01
Cr	0.01	0.01	5.0-10.0	0.01	<0.01	<0.01
B	0.01	0.5	0.2	<0.01	<0.01	<0.01
Mg	0.01	0.03	0.3	0.02	7.0-10.0	0.2
Mo	0.005	0.005	0 05	<0.01	<0.01	<0.01
V	0.01	<0.01	0.03	<0 01	<0.01	0.02
Cu	0.05	0.1	0.1	0.1	0.01	0.02
Ni	0.005	0.01	0.05	0.01	<0.01	<0.01
Co	---	<0 01	0 01	<0 01	<0 01	<0.01
Ba	---	<0 01	0.03	0.01	0 05	0.03
Zr	---	---	---	0.1	<0.01	0.03
Sr	---	---	---	<0.01	<0.01	<0.01

1. The number in parentheses following the electrode designation is the code number identifying the specific electrode.

the major crystalline phase detected was Fe_3O_4; manganese was probably present in this phase as well. In addition, a crystalline calcium fluoride phase (CaF_2, or fluorite) was detected in the fumes associated with the E70T-4 and E70T-5 electrodes. Although magnesium was present as a major constituent in the fumes produced by the E70T-4 electrode, crystalline phases containing magnesium were not detected.

The initial SEM examination of samples of the fumes produced by E7024 and E410-16 covered electrodes revealed the presence of rounded particles 4 to 40 μm in diameter; structures that were obviously crystalline were not evident in the SEM micrographs (Figs. 2.17a and 2.17b). Very fine fume particles were not visible in the micrographs, probably because of the manner in which the samples were prepared for examination. To overcome this difficiency, silver membrane filters capable of capturing and retaining particles less than one micrometer in diameter were used to obtain additional samples of E7024 fumes for study by SEM. SEM micrographs of some very fine fume particles are shown in Fig. 2.18a and 2.18b. Semiquantitative analyses of individual fume particles in an area containing extremely fine particles are presented in Table 2.20. Morphologically, silicon was not present in the form of crystalline silica.

In SEM analysis, the relative intensity of the x-ray energy from individual elements is detected in counts/second, and the results are displayed on a cathode ray tube screen. This analytical method is semiquantitative unless standards are available and the effects of absorption and intensity are taken into account. In the SEM analyses, the contents of individual elements totalled 100 percent because the relative intensities were normalized.

SEM studies of the fumes associated with flux cored electrodes were also conducted. Many relatively small, irregularly-shaped particles were observed, but crystalline structures were not detected.

Gas Detection

Studies to detect the presence (or absence) of selected gases (carbon monoxide and nitrogen dioxide) produced during arc welding operations were conducted using gas detector tubes containing ready-to-use stable chemicals which provide a colorimetric indication when specific gases are present. In practice, a measured volume of air is drawn through the tube and the resulting strain length is measured and related to the gas concentration by means of a calibration chart.

Two gases, carbon monoxide and nitrogen dioxide, were of specific interest because they are frequently present during arc welding operations. While ozone was also of interest, its detection with gas detector tubes is subject to interference by other airborne contaminants, and attempts to detect ozone in this manner were not made. Ozone is best detected by instruments equipped to monitor the chemiluminescent reaction between ethylene and ozone. The AWS publication, "The Welding Environment," lists several references concerned with ozone in the vicinity of arc welding processes (Refs. 2.10 and 2.12).

For this work, the appropriate gas detector tube was mounted in the apex of the fume collection chamber. Welding was done with representative covered, flux cored, and solid gas metal arc electrodes at conditions similar to those used in obtaining baseline fume generation data for these electrodes. During the welding and sampling interval, the gases were permitted to rise in the chamber by convection and escape through the open filter holder.

The following comments are based on a study of the data in Table 2.21:

(1) Unless ventilation is provided, carbon monoxide (CO) may present a problem during welding with processes in which the arc is shielded by CO_2 or gas mixtures containing CO_2; e.g., flux cored arc welding and gas metal arc welding. Carbon monoxide in inconsequential amounts may also be present during shielded metal arc welding.

(2) Nitrogen dioxide presents a potential problem based on published ACGIH threshold limit values when welding is done with processes in which the arc is shielded by argon or gas mixtures containing large amounts of argon.

The trends observed in this study were in agreement with the results obtained by other investigators. The

Table 2.20
Composition of fine fume particles from an E7024 electrode by scanning electron microprobe analysis

Particle	Composition, percent of counts						
	Fe	Mn	Si	Ti	Al	K	Ca
A	25	7	19	35	---	8	6
B	7	5	13	47	3	16	9
C	98	---	2	---	---	---	---
Scan area	34	8	39	---	---	20	---

Fig. 2.17A—Scanning electron micrograph of fume from E7024 (8) electrode (magnification x 500)

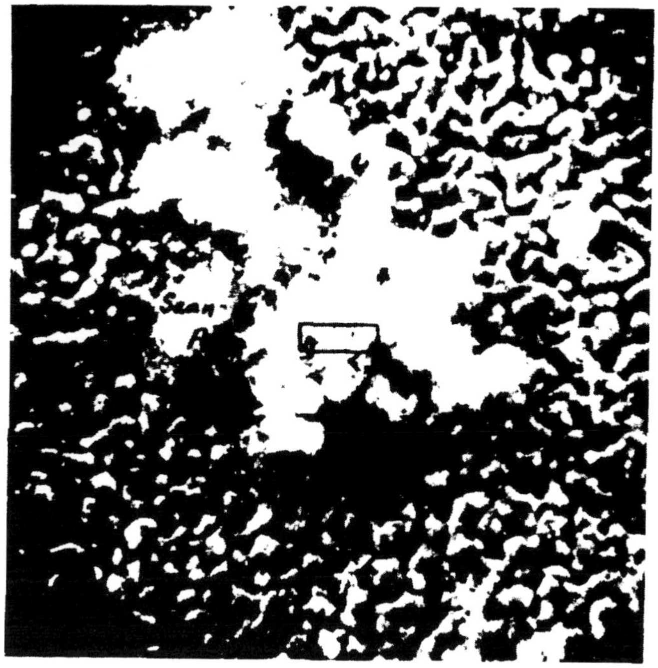

Fig. 2.18A—Fine fume from E7024 Electrode
(Background is Flotronic silver membrane)

Fig. 2.17B—Scanning electron micrograph of fume from E410-16 (21) electrode (magnification x 500)

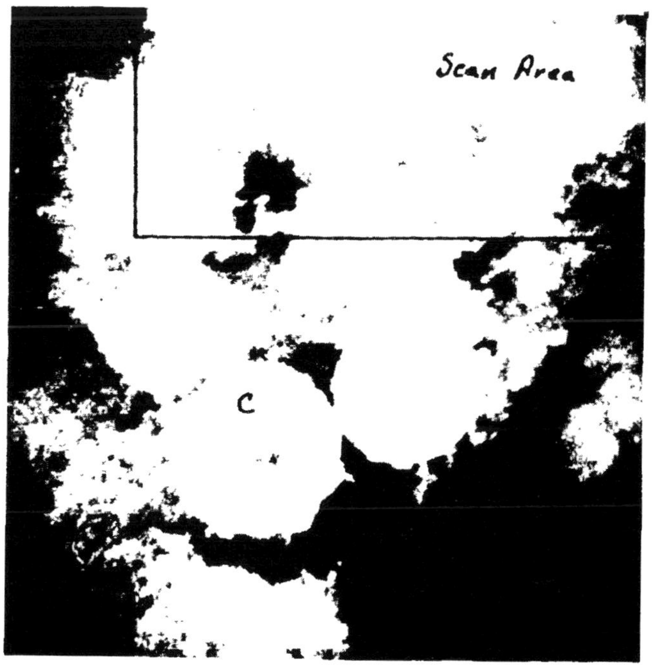

Fig. 2.18B—Fine fume from E7024 electrode

Table 2.21
Analyses of gases associated with arc welding

Electrode (code)[2]		Gases (ppm)[1] CO	NO_2	Supplementary shielding
E6010	(11)	5	---	None
E6010	(1)	5	---	None
E6013	(3)	5	---	None
E6013	(4)	3	---	None
E7018	(10)	7.5	---	None
E7018	(12)	7	---	None
E7024	(7)	3	---	None
E7024	(9)	10	---	None
E8018 C3	(13)	10	---	None
E8018 C3	(15)	5	---	None
E9018 B3	(18)	5	---	None
E9018 B3	(16)	5	---	None
E316-15	(27)	8	0.25	None
E316-16	(27)	5	0.25	None
ENiCu 2	(33)	3	0.25	None
E70T-1	(40)	100	0.50	CO_2
E70T-1	(40)	50	0.75	Ar-25 CO_2
E70T-1	(41)	80	0.50	CO_2
E70T-1	(41)	50	0.75	Ar-25 CO_2
E70T-4	(50)	---	0.50	None
E70T-5	(44)	50	0.50	CO_2
E70S-3	(58)	80	0.75	Ar-9 CO_2
E70S-3	(58)	20	1.50	Ar-2 O_2
E70S-3	(58)	150	0.25	CO_2
E5356	(60)	---	1.00	Ar
Inconel 625		---	7.00	Ar

1. A dash (- - -) signifies that CO or NO_2 was not present in detectable amounts.
2. The number in parentheses following the electrode designation is the code number identifying the specific electrode.

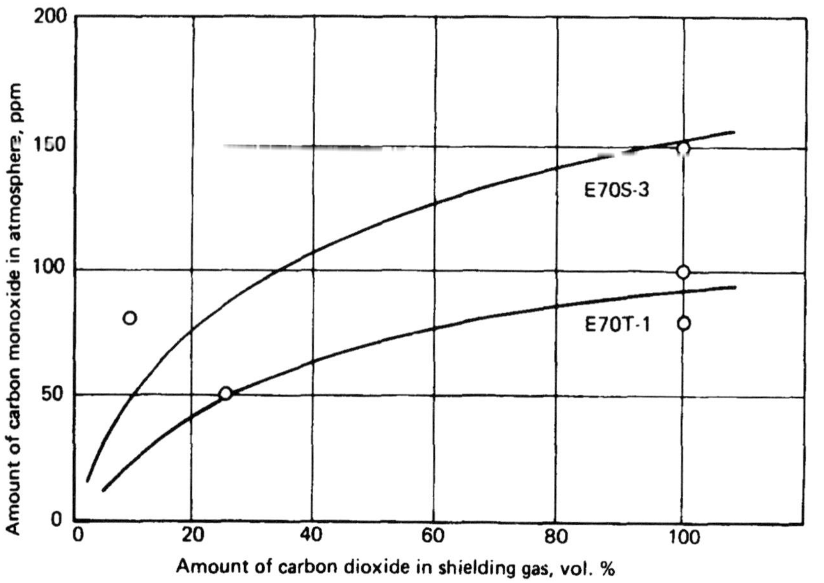

Fig. 2.19—Content of carbon monoxide in the atmosphere as a function of the amount of carbon dioxide in the shielding gas during gas metal arc and flux cored arc welding

concentrations of carbon monoxide and nitrogen dioxide in the vicinity of production welding operations may differ substantially from those obtained in the fume collection chamber because of the dilution of these gases by natural or artificial ventilation.

In gas shielded welding operations, problems with carbon monoxide can be minimized by the use of argon or argon-based shielding gases (Ar-2 O_2, Ar-9 CO_2, etc.).

The data in Table 2.18 can be used to establish a relationship between the amount of carbon dioxide in the shielding gas and the amount of carbon monoxide in the welding atmosphere (Fig. 2.19). As indicated, carbon monoxide concentrations rose gradually as the amount of carbon dioxide in the shielding gas increased from 0 to 100 percent. Additional data points would be needed to define this relationship with more accuracy.

Section IIE. Effect of Humidity on Fume Sampling Results

Early during Task 1 on arc welding fumes and gases of the American Welding Society research program "Improving the Welding Environment," the effect of humidity changes in the sampling area on the consistency of fume sampling results was examined. These studies were done at the request of the AWS Subcommittee on Sampling. The objective was to verify the procedures established by the working document "AWS Standard Method for Shop Sampling the Welding Environment" (July 30, 1975). As outlined, the study involved the collection of fume samples under low, medium, and high humidity conditions. The welding electrodes used were:

(1) E70S-3, 0.045 in. (1.1 mm) diameter (low fuming type)
(2) E70T-4, 0.120 in. (3 mm) diameter (high fuming type)

Welding was done at currents near the high end of the operating range (i.e., at about three-fourths of maximum recommended current for each electrode). For each humidity condition, replicate fume samples (3) were collected using a helmet sampler and a lapel sampler. The results of the effect of humidity on fume sampling results are reviewed hereinafter.

Sampling Area

Fume sampling was done in an environmentally controlled laboratory meeting Federal Clean Room Specifications. Since there was a laminar flow of air through this laboratory, an area near the air inlet was curtained off to produce a space for locating the welding equipment so that welding was done under still air conditions. Smoke tests were conducted to verify the existence of such conditions. During welding, the fumes rose vertically toward the ceiling; however, they billowed and drifted randomly about the welder's helmet as welding progressed.

The controls for this area could be adjusted to produce relative humidities ranging from less than 20 to more than 95 percent. Samples were collected under the following easily obtainable conditions of relative humidity:

(1) Low (23 to 25 percent R.H.)
(2) Medium (51 to 53 percent R.H.)
(3) High (>95 percent R.H.)

Experimental Procedures

Welding. Conditions for welding with E70S-3 and E70T-4 electrodes were established in accordance with recommendations from the electrode manufacturers as follows:

	E70S-3	E70T-4
Current, A	320-330	550
Voltage, V	31-32	33
Wire feed rate, in./min (mm/s)	400-425 (169-180)	230 (97)
Shielding gas	Ar-2 O_2	—
Shielding gas flow rate, ft^3/h (liters/m)	40 (18.88)	
Contact tube-to-work distance, in. (mm)	3/4 (19)	2-3/4 (70)
Travel speed, in./min (mm/s)	~20 (8.5)	~20 (8.5)
Polarity	Electrode positive	Electrode positive

Sampling. Sampling was conducted in the manner detailed in the previously referenced working document. Sampling was done at a location 13-1/2 in. (0.34 m) above and 12 in. (0.30 m) away from the arc. Some variations in this position occurred as the welder shifted his position to make adjustments and ease discomfort during periods of continuous welding. These movements affected the specific location of the helmet and lapel filter holders with respect to the fume stream; however, the orientation of one holder with respect to the other remained constant.

The filters were weighed on an analytical balance capable of weighing to 0.001 mg. This balance was located in a temperature and humidity controlled area in a building adjacent to that in which sampling was done. The filters were kept in a desiccator in this area until they were required. The manner in which the filters were handled is outlined below.

(1) The filters were weighed, placed between watch glass covers, and put in a desiccator.

(2) The filters were transported to the sampling area in the desiccator, and only enough filters for the required tests were removed from the weighing area. The filters were removed from the watch glass covers with smooth-faced tweezers and placed in filter holders.

(3) After the fume samples were collected, the filters were removed from the holders and placed between the watch glass covers again and returned to the desiccator.

(4) The filters were returned to the weighing area,

desiccated overnight, and then weighed to determine the quantity of fume collected.[9]

Experimental Results

Results obtained during the sampling of fumes associated with E70T-4 and E70S-3 electrodes are shown in Tables 2.22 and 2.23, respectively. Calculated fume concentrations for individual sampling experiments are shown in these tables. Also included are averages of the results of tests performed with the same electrode, filter holder location, and humidity level. These averages are neither threshold limit values nor time-weighted averages.

The following comments are based on a review of the experimental data:

(1) For a given filter location and humidity level, the weight of fumes collected during welding with the E70T-4 electrode was generally a direct function of the sampling time. The variations that occurred can be attributed to the random manner in which the fumes billowed and drifted and to movements of the welder during welding. The dependence of fume weight on sampling time was less evident when fumes produced by the E70S-3 electrodes were collected.

(2) The quantity of fumes produced by each electrode must also be considered in addition to the behavior of the fume stream and possible movements of the welder during welding. Fume quantities associated with the E70S-3 electrode were very small and the fumes tended to drift mainly upward as they were produced. In contrast, large quantities of fumes were produced by the E70T-4 electrode. These fumes drifted outward and upward as they were produced. Thus, the filters were more likely to be exposed to a consistent quantity of fumes when the E70T-4 electrode was used for welding than when the E70S-3 electrode was used.

(3) It was expected that concentrations of fumes at the welder's lapel would be consistently larger than those behind the welder's helmet, but this was not always the case, particularly when the E70S-3 electrode was used. This is explained by considering the respective locations of the helmet and lapel filters and the fume characteristics of the E70S-3 and E70T-4 electrodes. Because of the orientation of the welder with respect to the arc, the lapel filter was located slightly lower and further from the arc than the helmet filter. As a result of their specific location with respect to the fume stream, the helmet and lapel filters were exposed to different amounts of fumes. In previous work at Battelle, large variations in fume concentrations were observed for quite small variations in the location of the filter.

(4) The humidity level in the welding area did not significantly affect the results when fumes produced by the E70S-3 electrode were sampled. There was an appreciable increase in fume concentration with increasing humidity level for fumes produced by the E70T-4. The implications of this observation are discussed below.

Significant implications may be applied to evidence that the humidity level in the welding area had an apparent effect on fume concentrations associated with the E70T-4 electrode but not on those associated with the E70S-3 electrode. Among them are the following:

(1) Concentrations of fumes produced by E70T-4 electrodes (and possibly covered electrodes and other flux cored electrodes) may vary in accordance with the ambient humidity.

(2) Gravimetric analyses of fumes produced by such electrodes may not produce valid results unless humidity effects are taken into account.

Fume sample weight would be increased if atmospheric moisture is (a) absorbed (or adsorbed) by the filter, or the fume sample, or both; (b) chemically combined with constituents in the fumes; or (c) both absorbed and chemically combined. If increased fume sample weight was caused by the absorption of moisture, a valid weight could be obtained by desiccating or vacuum drying the sample to remove absorbed moisture. These procedures would be ineffective if moisture was chemically combined with one or more fume constituents. The data tend to favor absorption as shown by these results:

(1) Five filters used to collect fumes produced by the E70S-3 electrode under high humidity conditions (>95 percent R.H.) were weighed immediately after use and again after overnight desiccating. A weight loss averaging 14 μg was observed between weighing periods. Presumably, this weight loss represented moisture absorbed by the filter and fume sample during the sampling interval and subsequently removed by the desiccant. In this instance, absorption appeared to be a reasonable cause for moisture pickup.

(2) Two filters used to collect fumes produced by the E70T-4 electrode under the same high humidity conditions were handled in the same manner, and a weight gain of 7 to 8 μg was observed. This implies that additional moisture was absorbed by the fume sample during desiccation; e.g., some constituent in the fume sample was a better desiccant than the one being used in the desiccator (anhydrous calcium sulfate). The results do not shed any light on the possible presence of water that is chemically combined with a fume constituent. In this case, however, chemical combination of water with one or more flux constituents appears possible.

To resolve this problem, samples of the fumes produced by E70T-4 electrodes were collected under high and medium humidity conditions. These samples were collected directly in the fume stream at a constant distance above the arc. The fume samples were weighed immediately after collection and after various conditioning treatments (desiccating, vacuum drying, etc.) to detect the presence (or absence) of absorbed moisture. The samples were also analyzed using a thermal gravi-

9. Several filters used to sample fumes under high humidity conditions (>95 percent R.H.) were weighed immediately after sampling and before desiccating. These filters were weighed again after overnight desiccating to determine if moisture was picked up during sampling.

Table 2.22
Fume sampling results with E70T-4 electrode

Filter location	Relative humidity, %	Fume wgt., mg	Sampling time, min.	Concentration,[1] mg/m^3	Average concentration, mg/m^3
	23-25	0.624	30	11.1	
Helmet	23-25	0.292	10	15.7	13.6
	23-25	0.157	6	14.0	
	23-25	1.621	30	22.9	
Lapel	23-25	0.959	10	40.8	34.8
	23-25	0.577	6	40.9	
	51-53	1.555	30	27.9	
Helmet[2]	51-53	0.438	6.75	34.9	31.4
	51-53	1.398	24	31.3	
	51-53	0.307	5.5	30.0	
Lapel[2]	51-53	0.578	10	31.1	30.2
	51-53	0.682	12	30.5	
	95	1.072	12	48.0	
Helmet	95	0.628	6	68.3	51.9
	95	0.881	12	39.5	
	95	1.755	12	62.2	
Lapel	95	0.762	6	54.0	57.1
	95	1.553	12	55.1	

1. Concentrations are not time-weighted averages.
2. The Bendix pump was used to collect all samples under medium humidity conditions. For low and high humidity conditions, the Bendix pump was used to collect the helmet samples and the MSA pump was used to collect the lapel samples.

metric procedure capable of detecting the presence of chemically combined water.

Thermal Gravimetric Studies. Thermal gravimetric analysis (TGA) is a technique that is well-suited to detecting the temperatures at which absorbed and combined moisture is released by various materials. The equipment incorporates a sensitive balance that can detect weight changes in the microgram range when a sample is heated at a constant rate.

The samples for thermal gravimetric analysis were collected with Gelman DM Metricel filters, 37 mm in diameter with a pore size of 0.8 micron. Positioning the filter in the fume stream about 20 in. (0.51 m) above the arc permitted collection of a large fume sample within a reasonable welding time. Sampling was done at a rate of 2.5 to 3.0 liters/minute. Samples of E70T-4 fumes were collected when the humidity in the welding area while sampling was either greater than 95 percent or about 40 percent.

Thermal gravimetric analyses of the fume samples were conducted using a heating rate of 10°C/min. The results of individual analyses are discussed below:

(1) Initial attempts to analyze the entire sample (filter plus fume sample) were not successful because decomposition of the filter masked the weight loss from the fume sample.

(2) Analyses of the fume samples (obtained by scraping the fume deposit from the filter) produced positive results (Fig. 2.20). The sample collected at greater than 95 percent relative humidity had several well-defined temperature ranges over which significant weight losses were observed. The sample collected at 40 percent humidity also lost weight but at a more gradual rate. Both samples experienced about a 0.8 percent weight loss at 100°C due to absorbed (or adsorbed) moisture on the fume particles. The sample collected under high humidity conditions lost weight sporadically as the temperature increased. The sample collected at 40 percent lost weight uniformly in these regions. Presumably, these weight loss patterns are associated with the volatilization of moisture combined with different compounds in the fume sample.

Of particular note was the fact that there was little difference in the magnitude of the total weight loss over the temperature range examined, regardless of the humidity at which the samples were collected; 3.3 percent

Table 2.23
Fume sampling results with E70S-3 electrode

Filter location	Relative humidity, %	Fume wgt., mg	Sampling time, min.	Concentration,[1] mg/m^3	Average concentration, mg/m^3
	23-25	0.531	26.5	10.8	
Helmet	23-25	0.249	30	4.5	5.7
	23-25	0.084	26	1.7	
	23-25	0.437	26.5	7.0	
Lapel	23-25	0.230	30	3.3	3.9
	23-25	0.087	26	1.4	
	51-53	0.219	30	3.9	
Helmet	51-53	0.163	27.5	3.2	3.7
	51-53	0.118	16	3.9	
	51-53	0.254	30	3.6	
Lapel	51-53	0.176	27.5	2.7	3.2
	51-53	Note 2.	---	---	
	95	0.116	15	4.2	
Helmet	95	0.058	15	2.1	3.2
	95	Note 3.	---	---	
	95	0.364	15	10.3	
Lapel	95	0.158	15	4.5	6.8
	95	0.216	16.5	5.6	

1. Not time-weighted averages.
2. No sample reported, filter holder leaked.
3. No sample reported, moisture condensed in flowmeter tube.

for the sample collected at 95 percent humidity as opposed to 2.9 percent for the one collected at 40 percent humidity. If a fume sample collected at 20 to 25 percent relative humidity were analyzed by this technique, it is expected that the weight loss curve would resemble that obtained with the sample collected at 40 percent humidity, but the total weight loss would be less.

Discussion of Results. The results of this investigation indicate that the humidity in the welding area does not affect the weight of the fume samples; however, the amount of moisture picked up by absorption (or adsorption) and by combination with one or more constituents of the fumes is small (about 3 percent). Errors of this magnitude should have little bearing on the validity of gravimetric fume sampling results.

While the weight loss data obtained by thermal gravimetric analyses indicated that moisture was picked up by the fume sample, the amount of moisture detected was not large enough to explain the observed increase in the weight of E70T-4 fume samples as the relative humidity in the welding area increased. The thermal gravimetric data appeared to be valid, because comparable weight losses were observed independently by a member of the AWS Subcommittee on Sampling when fume samples were heated to 1200° F (649° C).

Other factors that may contribute to the observed increase of E70T-4 fume weights with increasing relative humidity are the effects of humidity on the (a) fume generation rate, (b) fume composition, (c) collection efficiency of the filter, and (d) the electrostatic attraction between particles. These factors were not examined during the course of the research program.

Summary of Part II

Principal Fumes and Gases Present

From the literature (Ref. 2.20), ranges for the major constituents in the fumes produced by four typical Scandinavian covered electrodes are as listed below. Three of these electrodes appear to have been low hydrogen electrodes with iron powder additions (individual results in Table 2.4).

Particle	Range, wgt.%
Fe_2O_3	25 to 44
SiO_2	7 to 30
K_2O	7 to 24
F^-	0 to 20
Na_2O	2 to 17
CaO	0 to 16
MnO	5 to 10
TiO_2	0.2 to 2
MgO	0.1 to 1
Al_2O_3	0 to 0.6
ZrO_2	0 to 0.6
ZnO	0.02 to 0.09
Cu	0.03 to 0.07
Pb	0.02 to 0.05
Cr	0.01 to 0.04

In the present study the major constituents in the fumes produced by one E410-16 and two E7024 covered electrodes were determined by optical emission spectroscopy to be as follows (complete data in Table 2.5). The tabulation assumes that the elements detected were completely converted to oxides.

		Range, wgt.%	
Element	Oxide	E7024	E410
Fe	Fe_2O_3	28 to 43	14 to 29
Si	SiO_2	8 to 31	3 to 5
K	K_2O	10 to 14	12 to 24
Na	Na_2O	4 to 8	5 to 11
Mn	MnO_2	3 to 6	2 to 3
Ca	CaO	0.1	3 to 6
Zn	ZnO	0.1	0.1
Ti	TiO_2	0.5 to 0.8	0.7
Al	Al_2O_3	0.1 to 0.2	0.9
Cr	—	0.01	5 to 10
	(Total oxides)	(89 to 103)	(47 to 90)

From the literature (Ref. 2.3), the elemental compositions of fumes produced by typical E70T-1, E70T-4, and E70T-5 electrodes, assuming an oxide form of these fumes, have been reported as follows:

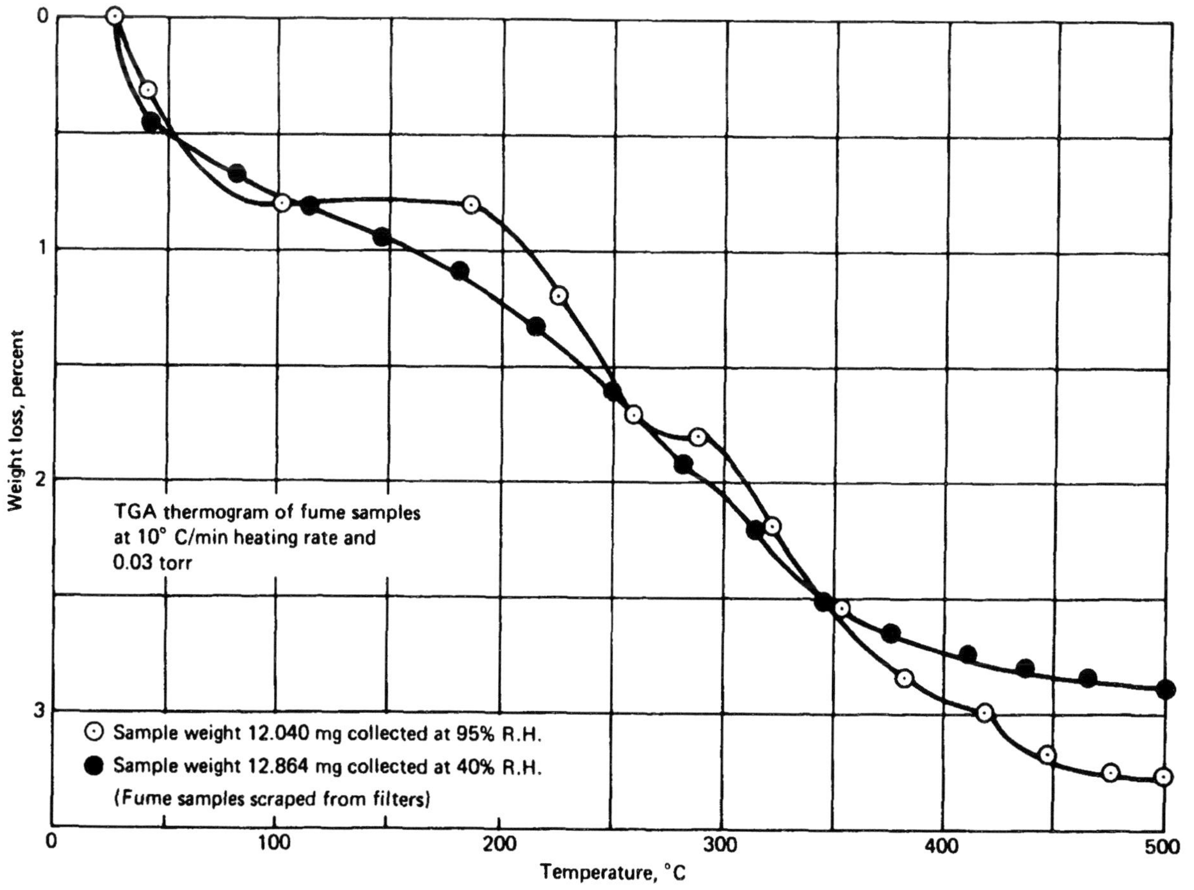

Fig. 2.20–Thermal gravimetric analysis of fume samples from E70T-4 steel electrodes

Element	Oxide	Composition, oxide weight %		
		E70T-1	E70T-4	E70T-5
		Rutile-base	Fluorspar-base	Silicon-base
Fe	Fe_2O_3	40	19	33
Mn	MnO_2	8	3	6
Si	SiO_2	39	6	31
Ca	CaO	—	18	11
Ti	TiO_2	3	—	—
Mg	MgO	5	15	—
Al	Al_2O_3	—	17	—
F	F^-	7	13	7
	Totals	102	91	87

In this study, the fumes from three flux covered electrodes were identified by optical emission spectroscopy to have the compositions that are listed below. The E70T-1 and E70T-5 electrodes studied in this program appear to have differed considerably (at least in silica content) from those reported above from the literature, while the E70T-4 electrode seems to have had an alumina-and-fluorspar flux similar to that reported above (see also Table 2.10).

Element	Oxide	Range, oxide weight %		
		E70T-1	E70T-4	E70T-5
Fe	Fe_2O_3	43 to 57	21 to 36	43 to 57
Mn	MnO_2	6 to 10	3 to 5	6 to 10
Si	SiO_2	3 to 5	0.2	3 to 5
Ca	CaO	0.1	21 to 35	11 to 17
Ti	TiO_2	0.8	0.2	0.3
Mg	MgO	0.03	12 to 17	0.3
Al	Al_2O_3	0.8	13 to 19	2 to 4
Na	Na_2O	5 to 8	0.1	1
K	K_2O	1	2	5 to 7
Zn	ZnO	0.1	0.1	0.1
	Total	58 to 80	72 to 114	71 to 101

Figures and short tables are used throughout this report to summarize the data and assist in the interpretation of the results. Data on the fume generation characteristics of specific arc welding electrodes are presented in detailed tabular form in Appendix B. These tables contain information on (a) the conditions under which the experimental data were obtained, (b) the melting and metal deposition rates of the electrode, and (c) calculated quantities that relate the amount of fumes produced during welding to the metal deposition rate and other electrode characteristics.

Fume generation rates have been expressed in "measured" and "normalized" values. The normalized value minimizes the effects of unintentional variations in welding current on fume generation rate. Normalization was accomplished by relating the fume generation rate to the current squared. The current squared normalization will fit a wide variety of fume generation data.

The fume generation rate can be used effectively by welding engineers, industrial hygienists, ventilation specialists, and others. The calculated factor "weight of fumes per unit weight of deposited metal" approximates the amount of fumes produced during production welding operations. The factor compares welding processes on the basis of fumes produced per weight of deposited metal.

Covered Electrodes

Fume Generation Characteristics. Generation rates and fume ratios of carbon steel, low alloy steel, stainless steel, nickel, and other high alloy covered electrodes are shown in bar graphs in Fig. 2.6.

The data in Fig. 2.6 must be interpreted carefully. They were obtained by evaluating 4 mm (5/32 in.) diameter electrodes from various producers under conditions recommended by the manufacturers. Data differing from those shown will be obtained if the electrode diameter or welding conditions are changed.

Fume generation rates and ratios of weight of fume to weight of deposited metal were somewhat higher for E6010 electrodes than for the other covered electrodes. The E7024 electrode rates were comparable to those of other carbon and low alloy steel electrodes. The ratios of fume to weight of deposited metal for E7024 electrodes were lower than those of other electrodes. This results from the combination of a relatively low fume generation rate and the high deposition rate contributed by the iron powder (up to 50 percent) in the coverings.

Fume generation rate ranges for stainless steel, nickel, and other high alloy covered electrodes were similar in magnitude; ranges for ratios of weight of fumes to weight of deposited metal generally overlapped one another.

Fume Composition. Data for the major elements in the fumes (e.g. iron, manganese, and silicon for carbon steel electrodes) and fluorides produced by covered (and other) electrodes are presented in Table 2.18. The analytical data do not total 100 percent, principally because the limited studies did not include detection of fume constituents originating from the electrode covering.

Major constituents in the fumes produced by E7024 and E410-16 covered electrodes are shown in Table 2.5.

Flux-Cored Electrodes

Fume Generation Characteristics. Ranges of (a) fume generation rates and (b) ratios of weight of fumes to weight of deposited metal for representative carbon steel, low alloy steel, and stainless steel flux cored electrodes are shown in Fig. 2.7. The data upon which Fig. 2.7 is based are summarized in Table 2.2.

E70T-1 Electrodes. The ranges of fume generation and ratios of weight of fumes to weight of deposited metal were broadest for this classification of electrodes. The electrode with the highest fume generation rate contained appreciable amounts of fluorides (Table 2.18); such compounds are known to enhance the production of fumes.

E70T-4 Electrodes. High fume generation rates were experienced with the self-shielded E70T-4 electrodes. Profuse fume generation is shown for an E70T-4 electrode made several years ago that is no longer available on the market. Since another currently available electrode made by the same producer had a much lower fume generation rate, it appears that changes in the electrode flux have been made to reduce fume quantities.

E70T-5 Electrodes. The highest fume generation rates were encountered with E70T-5 electrodes; the E70T-5 electrodes can be used with or without a shielding gas, and the basic type flux contains ingredients that produce large amounts of gas and fumes. Ratios of fume weight to weight of deposited metal were highest for these electrodes also.

Low Alloy Steel Electrodes. The fumes generated by 81-C3 and 91-B3 low alloy steel electrodes were low compared to those produced by carbon steel flux cored electrodes. A similar comparison holds for carbon steel and low alloy steel covered electrodes.

Stainless Steel Electrodes. The fumes generated by three self-shielded stainless steel electrodes varied widely; this characteristic appears to be flux-related.

Effect of Shielding Gas. Fume generation rates can be affected by the type of shielding gas used during welding. The data summarized in Table 2.24 show a reduction of 15 to 25 percent in fume generation rates when Ar-25 CO_2 shielding is used instead of CO_2. The argon-based shielding gas has a lower oxidation potential than CO_2, and oxidation processes contributing to the formation of fumes at and near the tip of the electrode are consequently decreased. The metal deposition rates for these electrodes are unaffected by the type of gas shielding used. Thus, the ratio of weight of fumes to weight of deposited metal are also smaller when Ar-25 CO_2 is used for shielding.

Fume Composition. The composition of fumes produced by flux cored electrodes is determined by the compositions of the electrode sheath, flux core, and (to a minor extent) base metal. Since the composition of the sheath varies but little from producer to producer, variations in fume compositions for such electrodes are mostly caused by differences in the flux core ingredients.

The constituents that may be detected in flux cores and slags associated with representative flux cored electrodes are shown in Tables 2.7 and 2.8. Most of these flux constituents will be present as elements and compounds in the fumes produced by electrodes incorporating them. The proportions will differ because some constituents are transferred by the arc to the slag or to the weld metal. Elemental compositions of the fumes and postulated oxides produced by typical E70T-1, E70T-4, and E70T-5 electrodes are shown in Table 2.9.

Fume compositions determined by optical emission spectroscopy are shown in Table 2.10.

Solid Electrodes

Fume Generation Characteristics. The fume generation characteristics of solid electrodes and rods used for gas metal arc welding and gas tungsten arc welding are a direct function of the amount of electrode consumed during welding, modified by the type of shielding gas, the metal transfer characteristics of the arc, and the welding conditions.

Gas Metal Arc Electrodes. Fume generation characteristics for carbon steel, stainless steel, high alloy, copper, and aluminum gas metal arc welding electrodes are shown in Fig. 2.8.

Steel Electrodes. The fume generation characteristics of steel electrodes made by three different manufacturers are shown for spray transfer with Ar-2 O_2 or Ar-9 CO_2 shielding, globular transfer with CO_2 shielding, and short circuiting transfer with Ar-25 CO_2 shielding. There was little variation in composition. A single E70S-5 electrode was also evaluated in the same transfer modes.

Other Solid Electrodes. Representative stainless steel, high alloy, copper, and aluminum solid electrodes were investigated under spray transfer conditions. The data in Fig. 2.8 show strong dependency on welding current and other process variables.

High Alloy Electrodes. Four high alloy electrodes (one stainless steel electrode, two nickel-base electrodes, and one cobalt-base electrode) had fume generation rates and ratios of weight of fumes to weight of deposited metal well below those of carbon steel electrodes. A third nickel-base electrode had a fume generation rate approaching those of the carbon steel electrodes. It was a small diameter electrode; therefore, the current density was high and the higher arc temperature produced more fumes.

Aluminum Electrodes. The aluminum electrodes varied in performance. The ER4043 electrode produced fumes at low rates with a low ratio of weight of fumes to weight of deposited metal. The ER5356 electrode produced fumes at much higher rates. Both electrodes had much higher fume generation rates at longer contact tube-to-work distances with evidence of lack of shielding of the arc and weld pool. Good shielding and careful control of the welding torch setup are needed to minimize fumes, resulting in the production of high quality welded joints as an added benefit.

Copper Electrodes. Fume generation rates and ratios of weight of fumes to weight of deposited metal for an ERCu and an ERCuA1-A2 electrode were comparable in magnitude to those associated with carbon steel electrodes.

Fume Compositions. The composition of the fumes produced by solid gas metal arc welding electrodes is controlled by the composition of the electrode and any coating that might be on the electrode surface. Each of the fume samples contained iron, manganese, silicon, and copper (or their oxides).

Because of its low threshold limit value, there is more concern about the presence of copper fume in the welding environment than about iron oxide fume. Table 2.12 combines data on the baseline fume generation rates of three E70S-3 electrodes with the compositional data on the fumes. The ACGIH time-weighted average threshold limit value for iron oxide fumes (5 mg/m^3) is 25 times as large as that for copper fumes (0.2 mg/m^3). However, the results of the experimental studies indicate that the iron oxide fume generation rates for the E70S-3 (54), E70S-3 (57), and E70S-3 (58) electrodes were 925, 154, and 89 times as great, respectively, as the copper fume generation rates. Thus, when welding with these electrodes, the allowable limit for iron oxide fumes would be exceeded long before reaching the limit for copper fumes.

Compositions of the fumes produced by the gas metal arc electrodes used for welding other base metals (stainless steels, nickel-base alloys, etc.) are determined by the composition of the respective electrodes.

Fume Generation Characteristics of Gas Tungsten Arc Welding Rods

The fume generation rates for an E308L stainless steel filler rod and a 5356 aluminum filler rod were a small fraction of the rates associated with other processes investigated during this program.

Process Variables

Some of the variables affecting fume production are:

(1)) Welding process (summarized above)

(2) Current

(3) Voltage

(4) Electrode polarity

(5) Shielding gas (flux cored and solid electrodes)

(6) Contact tube-to-work distance (flux cored and solid electrodes)

(7) Metal transfer mode

(8) Electrode characteristics (manufacturer, diameter, core wire composition, covering or flux composition, etc.)

Current Effects. Fume generation rate, as shown in Fig. 2.9 (based on Tables B26 and B27), varies nonlinearly with current. Studies to fit the data to a power function indicated that the fume generation rate (FGR) relation to current was different for each type of electrode. The simplest regression equation seems to be

$$FGR = aI^2$$

where a is a constant and I is the current.

Current Density Effects. Current density also has a bearing on the rate at which fumes are produced.

Variable Current Density. The fume generation characteristics of three E6013 electrodes, all made by the same producer, are shown in Fig. 2.10.

The fume generation rate increased with current density in what appears to be a parabolic relation. With direct current, the fume generation rate is higher with electrode positive than that observed with electrode negative because the electrode is heated to higher temperatures when the anode spot is on the electrode.

Constant Current Density. The fume generation characteristics of four E7018 electrodes are shown in Figs. 2.11 and 2.12. The fume generation rate (Fig. 2.10) and the electrode melting rate (Table B32) were approximately proportional to the square of the diameter. The amount of fumes produced during welding was largely dependent upon the amount of electrode consumed. Fig. 2.12 shows the dependence of fume generation rate on welding current. With approximately equal current densities in each electrode, the variation of fume generation rate with current was almost linear.

Arc Voltage—Arc Length Effects. Arc voltage and arc length are closely related welding variables. The fume generation characteristics of selected electrodes increased with increasing arc length for all of the electrodes included in this investigation.

Contact Tube-To-Work Distance Effects. The distance between the end of the contact tube and the surface of the base plate affects the melting and metal deposition rates in flux cored arc welding and gas metal arc welding. The fume generation characteristics of an E70T-1 flux cored electrode were determined as a function of contact tube-to-work distance. Two current levels were studied with the following results:

(1) At each of the current levels there was a gradual increase in metal deposition rate with increasing contact tube-to-work distance (Tables B42 and B43). When the contact tube-to-work distance increases, the wire feed rate must be increased to maintain a constant current level. The increased melting rate is accompanied by a higher deposition rate.

(2) For each current level, the fume generation rate remained essentially constant for each contact tube-to-work distance.

(3) The ratio of weight of fumes to weight of deposited metal decreased gradually with increasing contact tube-to-work distance.

(4) The effect of current on fume generation rate is evident in Fig. 2.15. At 450 A, the fume generation rate was about 1 g/min.; at 520 A, the rate was 1.3 g/min.

Shielding Gas Effects. The shielding gas in flux cored and gas metal arc welding affects (a) the rate at which fumes are produced during welding, and (b) the composition of the fumes. Copper contents in the fumes appeared to be less a function of the shielding gas than were the other core wire elements and their oxides.

Metal Transfer Mode. The effects of shielding gas type on the composition of the fumes produced by E70S-3 gas metal arc electrodes used for welding carbon steels were evaluated using welding conditions to produce the following transfer modes: (a) spray transfer with Ar-2 O_2 shielding, (b) spray transfer with Ar-9 CO_2 shielding, and (c) globular transfer with CO_2 shielding. The fume samples were collected on cellulose membrane filters and analyzed by atomic absorption procedures. The resulting data plus the calculated weight percentages of each oxide are shown in Table 2.14. The apparent trends were these:

(1) Iron (iron oxide) contents were highest in the spray transfer mode with either Ar-2 O_2 or Ar-9 CO_2

shielding and lowest in the globular transfer mode with CO_2 shielding. These results are in agreement with those of Heile and Hill (Ref. 2.3).

(2) Manganese (manganese oxide) contents were highest in either the spray transfer mode with Ar-9 CO_2 shielding or in the globular transfer mode with CO_2 shielding; the Mn (or MnO_2) content was lowest in the spray transfer mode with Ar-2 O_2 shielding. Here, the oxidation potential of the shielding gas had a slight effect on the contents of manganese (manganese oxide) in the fumes.

(3) Silicon (silicon dioxide) contents were highest in the globular transfer mode with an oxidizing gas, CO_2. These results are also in agreement with those of Heile and Hill (Ref. 2.3). There is no apparent explanation for the low silicon contents in the fumes produced by electrodes 54 and 57.

Effects of Iron Powder Additions. The fume generation characteristics of two groups of mild steel iron powder electrodes are summarized in Table 2.15.

High-Titania Electrodes. At the measured welding current, the fume generation rates for E6013, E7014, and E7024 electrodes are similar in magnitude (0.68 to 0.72 g/min.) and appear unaffected by the presence or absence of iron powder additions to the electrode coverings. Rates normalized to a current of 175 A varied inversely with the amount of iron powder in the electrode covering.

The ratio of the weight of fumes to weight of deposited metal also varied inversely with the amount of iron powder in the covering when these electrodes were used at recommended current levels. The E7024 electrode had the lowest ratio because its fume generation rate was low and its metal deposition rate was high.

Low Hydrogen Electrodes. At recommended current levels, the fume generation rates for E7016 (no iron powder addition) and E7018 (moderate iron powder addition) electrodes were almost identical; the rate for an E7028 electrode (high iron powder addition) was twice as large. The normalized rates for these electrodes were similar in magnitude (0.50 to 0.58 g/min.). The rates at which fumes were produced appeared to be controlled by ingredients other than iron powder in the electrode covering. Calcium fluoride, a compound common to these low hydrogen electrodes, probably exerted a controlling influence on the rate at which fumes were produced.

Ratios of weight of fumes to weight of deposited metal for these electrodes were similar.

Other Studies

Electrode Angle Effects. The effect of electrode angle on the rate at which fumes are produced by covered electrodes was studied at angles of 30, 60, and 90 degrees to the workpiece; a tubular guide was used to establish and maintain the desired electrode angle. Bead-on-plate welds were deposited with each electrode at a nominal current of 150 A. The results are shown in Fig. 2.16.

The fume generation rate gradually decreased as the angle between the electrode and workpiece increased toward 90 degrees. Care should be observed in interpreting these data strictly in terms of an electrode angle effect because a lengthening of the arc probably occurred as the electrode angle decreased. Fume generation rates are affected substantially by changes in arc length.

Base Plate Effects. Fume generation characteristics for covered, flux cored, and solid electrodes were determined with a base plate having composition similar to that of the core wire or sheath. These studies included fume samples for gravimetric and chemical analyses collected from stainless steel bead-on-plate welds deposited on carbon and stainless steel plate. The experimental results are summarized in Tables 2.16 and 2.17 and as follows:

(1) Fume Generation Characteristics. Fume Generation rates and ratio of weight of fumes to weight of deposited metal were unaffected by the type of base metal used during welding.

(2) Fume Composition. The composition of the fumes did reflect the composition of the base metal upon which the welds were deposited as well as that of the electrode. On stainless steel, the fumes from an E316-16 electrode contained higher contents of manganese, nickel, and chromium than they did when that electrode was used on carbon steel (Table 2.17). It is concluded that electrode and base plate should be matched if fume compositions are to be determined.

Diameter Effects with Flux Cored Electrodes. The measured fume generation rate for a small diameter electrode is less than that for a large diameter electrode. However, if the fume generation rates are normalized to a common current, the data indicate that more fumes will be produced by the small diameter electrode than by the large diameter electrode at the same current level. Fume generation rates are dependent upon the rate at which the electrode is consumed. At the same current level, the melting rate for the smaller of the two electrodes will be greater than that for the larger because of the greater current density in the small diameter electrode and the effect of increased I^2R heating in the smaller electrode.

Analytical Studies

Optical emission spectroscopy yields a wealth of analytical data because 70 or more elements can be detected simultaneously and their content can be estimated on a semiquantitative basis. This method provides an overview of most of the elements present in the fume sample, regardless of their origin; i.e., from the core wire (sheath), electrode covering (flux), or flux plate.

Atomic absorption analysis is also capable of determining the presence of about 70 elements, but each element must be detected individually. This method is better-suited for the detection of elements present in large amounts and the detection accuracy is excellent. This method, combined with optical emission spectroscopy, provides reliable data on fume sample compositions.

Classical chemical analysis procedures are most useful

in determining constituents whose presence cannot be detected by optical emission spectroscopy or atomic absorption analysis (e.g., fluorides, chlorides, gases, etc.).

Analysis of Fume. In this study, AA analysis was used to detect only the major constituents that originated in the core wire or electrode sheath (i.e., iron, manganese, and silicon in the case of fumes produced by low carbon steel covered or flux cored electrodes). As a result, analytical data on fumes associated with covered or flux cored electrodes do not total 100 percent, because they do not include certain constituents originating in the electrode covering or flux (e.g., sodium, calcium, titanium, etc.).

To supplement these data, fluoride contents in the fumes produced by many of the low hydrogen electrodes were determined by wet chemical analysis.

Fume Characterization. The fumes produced by electrodes used for shielded metal arc welding (E7024 and E410-16) and flux cored arc welding (E70T-1, E70T-4, and E70T-5) were analyzed for elemental composition by optical emission spectroscopy (OES) and for crystalline phases by x-ray diffraction (XRD). Scanning electron microscopy (SEM) was used to study particle morphology. Semiquantitative results of the OES analysis of the fume samples are shown in Table 2.19.

Only one crystalline phase, iron oxide (Fe_3O_4, magnetite), was detected by XRD in the fumes produced by the covered electrodes; manganese was probably present in this phase also since it can replace iron in the crystal lattice. Crystalline phases containing silicon were not detected; however, since silicon was detected in the fumes produced by the E7024 and E410-16 electrodes, it was probably present as a glassy SiO_2 phase along with potassium and sodium. The x-ray diffraction pattern from the fumes produced by the E410-16 electrode had an extra diffraction line whose presence was attributed to a chromium compound. The study did not determine the valence of chromium in chromium compounds detected in fume samples.

In the fumes produced by the flux cored electrodes, the major crystalline phase detected was Fe_3O_4; manganese was probably present in this phase also. In addition, a crystalline calcium fluoride phase (CaF_2, fluorite) was detected in the fumes associated with E70T-4 and E70T-5 electrodes. Although magnesium was present as a major constituent in the fumes produced by the E70T-4 electrode, crystalline phases containing magnesium were not detected. Crystalline phases containing silicon were not detected.

Gas Detection. Studies to detect the presence (or absence) of selected gases (carbon monoxide and nitrogen dioxide) produced during arc welding operations were conducted using gas detector tubes.

Two gases, carbon monoxide and nitrogen dioxide, were of specific interest because they are frequently present during arc welding operations. Literature references (Refs. 2.22 and 2.23) also report ozone in the vicinity of arc welding processes.

Unless ventilation is provided, carbon monoxide (CO) may present a problem during welding with processes in which the arc is shielded by CO_2 or gas mixtures containing CO_2; e.g., flux cored arc welding and gas metal arc welding. Carbon monoxide in inconsequential amounts may also be present during shielded metal arc welding. In gas shielded welding operations, problems with carbon monoxide can be minimized by the use of argon or argon-based shielding gases (Ar-2 O_2, Ar-9 CO_2, etc.).

The data in Table 2.18 can be used to establish a relationship between the amount of carbon dioxide in the shielding gas and the amount of carbon monoxide in the welding atmosphere (Fig. 2.19). Carbon monoxide concentrations rose gradually as the amount of carbon dioxide in the shielding gas increased from 0 to 100 percent. Additional data would be needed to define this relationship more accurately.

Nitrogen dioxide is potentially a problem, based on published ACGIH threshold limit values, when welding is done with the arc shielded by argon or gas mixtures containing large amounts of argon (probably by interaction between excited argon atoms and nitrogen molecules that penetrated the gaseous shield). This study did not find nitrogen dioxide in the unshielded arc.

Effect of Humidity or Fume Sampling. The studies of fume sampling collected under low, medium, and high humidity conditions indicate that the humidity in the welding area does affect the weight of the fume samples; however, the amount of moisture picked up by absorption (or adsorption) and by combination with one or more constituents of the fumes is small (about three percent). Fume weights collected from E70T-4 welds increased with increasing relative humidity more than their measured moisture contents would explain. It is felt that other factors not studied in this program may contribute to this anomaly.

Part III. Brazing Fumes and Gases

Introduction

As part of the American Welding Society research program "Improving the Welding Environment," Battelle-Columbus examined the fumes generated during brazing. The broad objective of this task was to show experimentally the relationship of the fumes given off to the brazing filler metal used.

Early in the program a literature search was conducted to learn the extent and usefulness of the bibliographic data available. The major conclusions drawn were:

(1) The cadmium oxide fume hazard has received by for the most attention, but even this takes the form of "case histories" without significant attention to the basic causes of the problem. Only a few investigators have chosen to examine the effects of process and procedure variables on the magnitude of the hazard.

(2) The possibility of harmful fluoride concentration in the brazing environment is covered briefly in the literature. As in the case of cadmium, there is a dearth of useful information that would permit adequate definition of the hazard under varying conditions of use.

(3) Almost no information exists that can be used to define the brazing environment and the safe practices to be followed.

Based on these conclusions and the experiences of members of the AWS Research Committee, an experimental research program on brazing fumes was agreed upon. The experimental approach was developed without benefit or knowledge of any published prior experience by other investigators. Its magnitude and scope were restrained by the availability of funds assignable for brazing fume studies.

Scope

The research plan was designed to minimize the variables caused by changes in flame shape, character, and position that occur during torch brazing. Thus, procedures similar to those used during welding fume studies were suggested. Total fume samples collected in the AWS-Battelle welding fume collection chamber would be used to find the influence of brazing variables on the fume generation rates and quantities. Also the fume sample would be analyzed by atomic absorption methods to determine how brazing variables affect fume compositions. Unfortunately, it increasingly became evident as the program progressed that the procedures chosen did not provide the precision required for accurate measurement of filler metal fumes.

The filler metals chosen were: Ag-15Cu-16Zn-24Cd (BAg-1) and Ag-22Cu-20Zn-7Cd-1Sn. These filler metals have similar melting characteristics and are used extensively in industry for torch brazing.

Limitations

The investigation was limited to the BAg-class filler-metals and was directed toward determining the fume generation characteristics of two filler metals having widely different cadmium contents.

As the data presented in this report show, the quantity of fume developed during brazing with filler metals containing cadmium and zinc is quite small. They also show that when a flux is used, it is the major contributor to the fume produced. In addition, the results clearly indicate the need for the prior development and validation of precision procedures for use during any experimental study of the fumes produced by brazing filler metals and fluxes.

Program

The experimental program covered the following tests, conducted in random order:

(1) BAg-1 filler metal and Type 3A flux, heated to 1200° F (649° C), to 1250° F (677° C) and to 1350° F (732° C); five tests at each temperature.

(2) BAg-1 filler metal under CO_2 shielding, heated to 1200° F, to 1350° F, and to 1500° F (816° C); five tests at each temperature plus three extra tests at 1500° F.

(3) Low cadmium filler metal and Type 3A flux, heated to 1250° F, to 1350° F, and to 1500° F; four tests at each temperature plus two extra tests at 1500° F.

(4) Low cadmium filler metal under CO_2 protection, heated to 1250° F, to 1350° F, and to 1500° F; four tests at each temperature plus one extra at 1250° F (of questionable accuracy and therefore discarded).

Equipment, Materials, Procedure

Crucibles (boats)

All brazing filler metal melting was carried out in shallow pure alumina boats having these nominal internal dimensions: 1.75 x 1.13 x 0.38 in. (44 x 29 x 9.7 mm, approximately). The internal dimensions varied because the wall thickness was not uniform from boat to boat.

The possible surface area variable was neglected and not recorded during the study because not enough boats were available to permit sorting for uniformity.

During analysis of the results, the surface area variation was found to range between −4 and +12 percent from the nominal area; thus it was considered a significant possible source of error in the data collected.

Filters

Initial experiments were run using two types of filters: glass fiber absolute filters (MSA, Cat. No. 75428) when weight determinations were desired and cellulose membrane filters (Millipore Type RA, 1.2μm pore size) for samples to be analyzed.

It was found, however, that the absolute filters gave excessive variation in results. A satisfactory explanation of these variations was not found; therefore, all fume data used for study of the results were taken from samples collected on membrane filters.

The active filter area was 33 square in. (21 290 mm^2) in all tests. No prefilter was used. Each filter was dried for 1 hour at 220°F (105°C) immediately prior to weighing and use.

Melting Furnace and Fume Collection Chamber

A resistance furnace capable of heating the brazing filler metal to melting temperature in less than 10 minutes and to 1500° F (816° C) in less than 15 minutes was built for the program. The melting furnace was enclosed in the AWS-Battelle fume collection chamber when in use. It is shown in position in Fig. 3.1. Closeup views of the furnace prepared for melting under a flux and under a CO_2 gas blanket are shown in Figs. 3.2 and 3.3 respectively. The CO_2 gas outlet was a flare fitting with an opening measuring 0.050 x 1.813 in. (1.27 x 46 mm).

A standard temperature control system utilizing a chromel-alumel thermocouple was used. For tests under CO_2, the gas flow was regulated through a flow meter at 8 ft^3 per hour (3.8 liters per minute, approximately).

Filler Metals

The composition of the filler metals chosen for study was given above.

The standard BAg-1 filler metal was furnished in 1.50 x 1.25 x 0.38 in. (approx. 38 x 32 x 9.7 mm) blocks. These were cut to 1.5 x 1.00 x 0.38 in. (approx. 38 x 25 x 9.7 mm) in order to fit the crucible (boat). The excess pieces were used in check tests near the end of the study. The weight of the cut-down blocks varied between 84.7 and 82.6 grams. Melts made with the excess pieces varied between 74.7 to 76.6 grams. Atomic absorption analyses of these blocks made at Battelle gave these results 25.4 percent Cd and 16.5 percent Zn. The liquidus 1145° F (618° C) was taken as that given in the literature.

The low cadmium filler metal was furnished as 1.50 x 1.00 x 0.13-in. (approx. 38 x 25 x 3.2 mm) blocks. Three blocks were used for each melt, the weight of which varied between 85.0 and 85.8 grams. Battelle atomic absorption analysis of these blocks gave these results: 7.02 percent Cd, 19.6 percent Zn. The liquidus of this filler metal was determined at Battelle and found to be 1210° F (654° C).

Other Materials

The flux used was commercial AWS Type 3A and was weighed to give 3 to 4 grams for each melt. Standard welding grade CO_2 gas having a −50° F (−46° C) specified dew point was used. The actual dew point was not checked.

Test Procedure

The standard procedure used for brazing filler metal melting and fume collection was as follows:

(1) Weigh ingot

(2) Add 3 to 4 g flux, or start CO_2 gas flow, as required

(3) Place in cool furnace in fume collection chamber

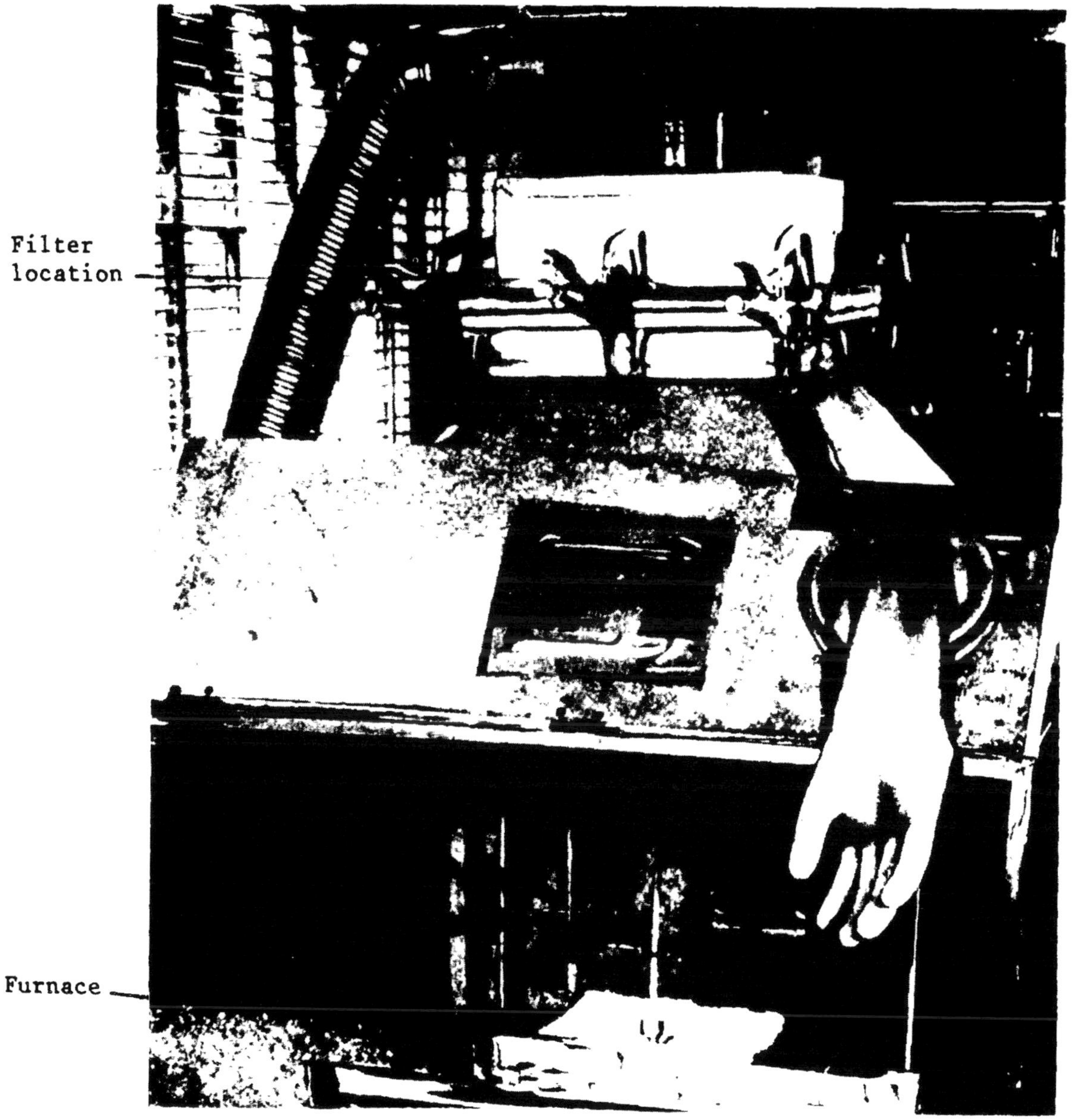

Fig. 3.1—Brazing filler metal melting furnace in AWS-Battelle welding fume collection chamber

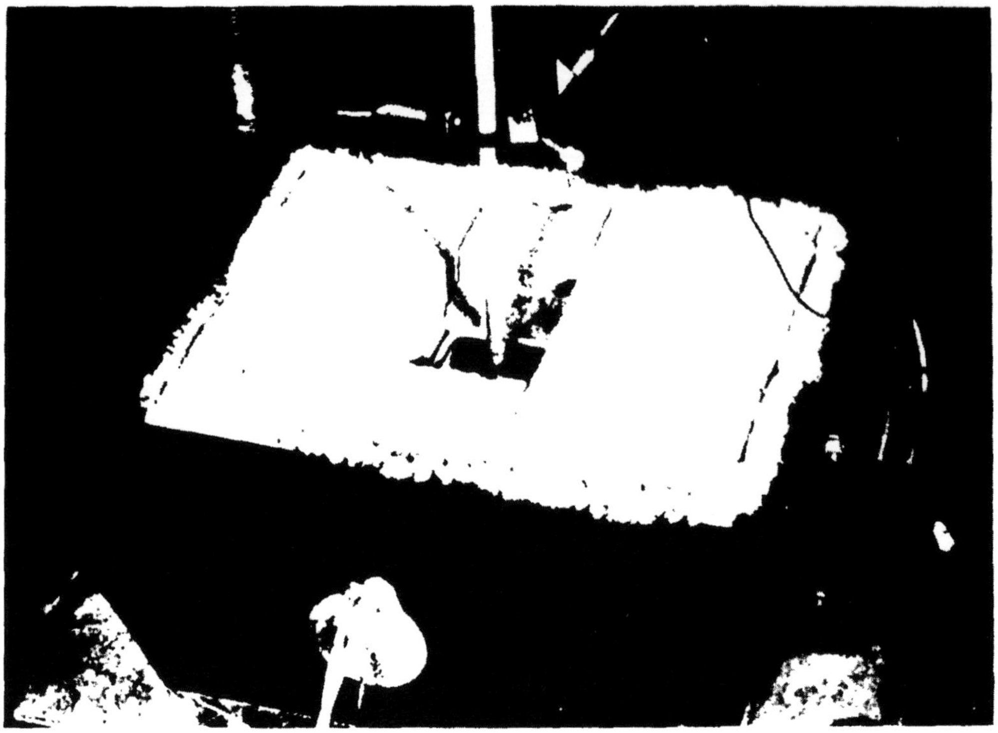

Fig. 3.2 – Brazing filler metal melting furnace with ingot, boat, and thermocouple in place

Fig. 3.3 – Brazing filler metal melting furnace with CO_2 gas outlet in place

(4) Weigh filter and place in collection chamber
(5) Start heating and fume collection
(6) Heat to test temperature and hold 3 minutes
(7) Turn off heat, continue fume collection for 10 minutes
(8) Weigh filter
(9) Weigh ingot before and after cleaning and drying

All weighings were made to ±0.1 mg. Times were recorded to the even minute. The relative humidity in the laboratory was recorded at least hourly. The minimum relative humidity recorded was 36 percent and the maximum, 54 percent; 16 of the 60 tests conducted were outside the range of 42 to 50 percent. Blank filters run for the length of an average sampling time showed small weight gains of approximately 0.12 percent which might be attributed to moisture pickup. Used filters were sealed in polyethylene bags and stored in a desiccator until transfer for analysis.

All tests were conducted at random with respect to test temperature and flux or gas cover. Each filler metal was evaluated in a separate group of tests. Test temperatures for BAg-1 were 1200, 1350, and 1500° F (649, 732, and 816° C). The low cadmium filler metal was tested at 1250, 1350, and 1500° F (677, 732, and 816° C).

Summaries of the laboratory data collected on the fume generation characteristics of the brazing filler metals tested are found in Appendix C. These data were summarized and are included in the report where their discussion occurs.

Section IIIA. Total Fume Generation Characteristics

The data collected on the amount of fume produced by each of the two filler metals studied at these different temperatures were averaged and are given in Tables 3.1 and 3.2. The data from these tables are shown graphically in Fig. 3.4 and 3.5. The average loss in weight of the ingots is also given along with other desirable data.[10] As mentioned previously, tests made with the glass fiber filters and the BAg-1 filler metals were excluded from consideration. This explains the fewer number of tests for BAg-1 than for the low cadmium filler metal.

10. Ingot weight losses are discussed later in this report.

Table 3.1
Brazing fume generation by filler metals covered with AWS Type 3A flux

	1200° F (649° C) BAg-1 filler metal, avg.	1250° F (677° C) Low cadmium filler metal, avg.
Number of tests	3	4
Time molten, min	6	7
Sampling time, min	25	27
Ingot loss, g	0.0422	0.0058
Fume weight, g	0.0684	0.0550
	1350° F (732° C)	
	BAg-1 filler metal, avg.	Low cadmium filler metal, avg.
Number of tests	3	4
Time molten, min	8	8
Sampling time, min	28	26
Ingot loss, g	0.0527	0.0252
Fume weight, g	0.0770	0.0602
	1500° F (816° C)	
	BAg-1 filler metal, avg.	Low cadmium filler metal, avg.
Number of tests	3	6
Time molten, min	12	16
Sampling time, min	30	31
Ingot loss, g	0.1095	0.1896
Fume weight, g	0.1670	0.0933

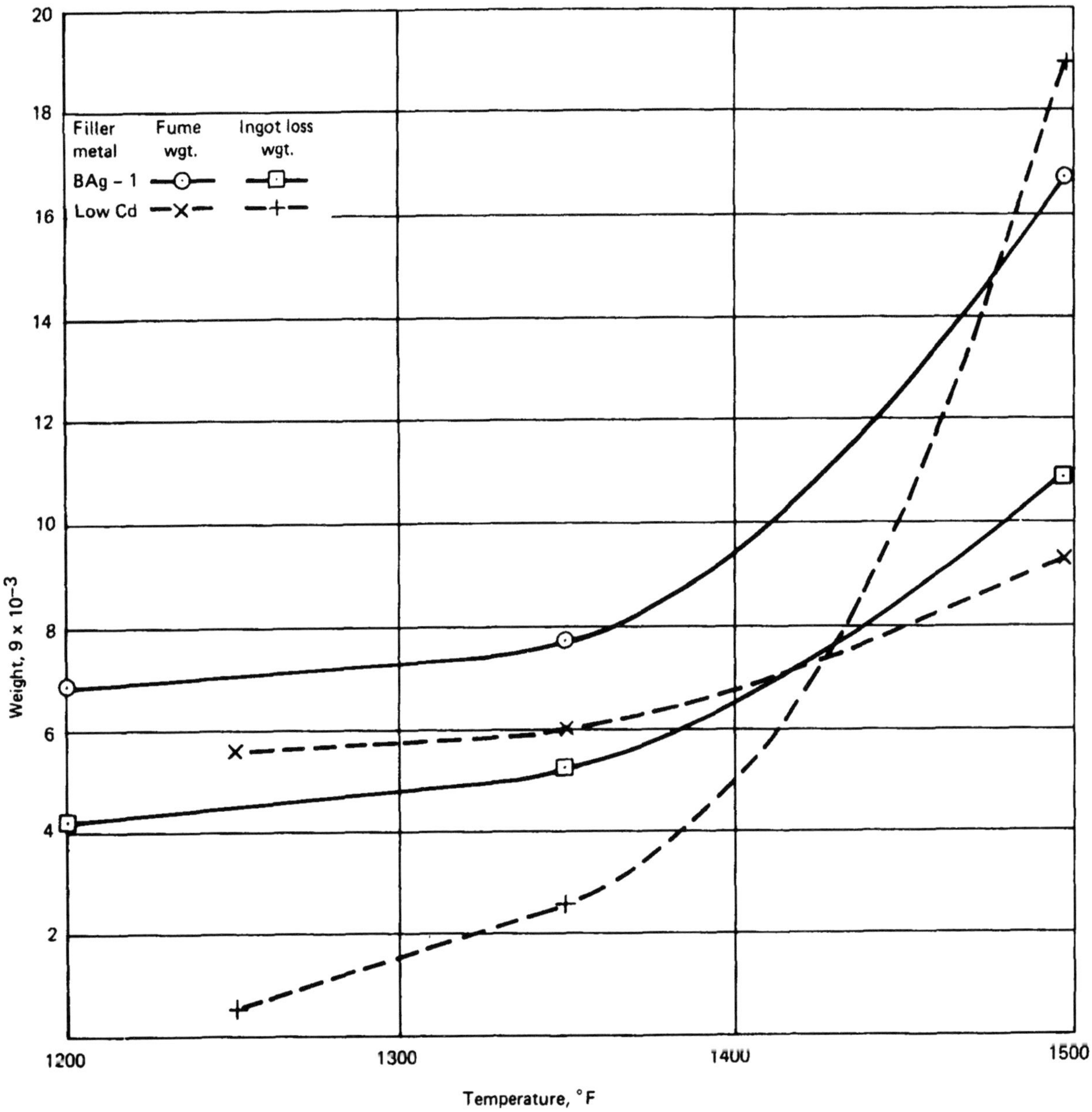

Fig. 3.4—Brazing fume generation and metal losses by flux covered filler metal

Flux Covered Melts

Examination of the fume weight data for both filler metals when melted under a flux (Table 3.1) shows the trend that would be expected: increased temperature results in increased fume weight. The actual weight of fume produced was smaller than expected. This may be attributed to the relationship between the filter location and the molten metal surface, although no fume could be wiped from the collection chamber walls even after several test runs. A different sampling system, such as a smaller, completely closed chamber, would be required to validate or refute the magnitude of the fume weights produced. Also, the influence of the presence of a flux and the composition of the fume must be considered.

The data show a large increase in weight of fume produced at 1500° F (816° C) for both filler metals when compared to that produced at lower temperatures. This would be expected because cadmium boils at 1409° F (765° C) and zinc has high vapor pressure (approx. 300 mm of Hg) at the same temperature. The rate of fume formation at the two lower temperatures was nearly the same, but this rate increased significantly at 1500° F whether or not the rates are calculated on the basis of time molten or total sampling time. For example, using the fume weights produced by BAg-1 filler metal and the average time molten or sampling time, the calculated rates of fume formation are as follows:

Temperature, °F	°C	Rate for time molten, g/min.	Rate for sampling time, g/min.
1200	649	0.0114	0.0027
1350	732	0.0096	0.0028
1500	816	0.0139	0.0056

The low cadmium filler metal appears to produce about one-third less fume than the BAg-1 filler metal.

Gas Covered Melts

The data in Table 3.2 summarize the fume formation attributes of the two brazing filler metals studied where they are "protected" by a flowing CO_2 gas blanket rather than a flux. The results parallel those obtained where a flux was used. Thus, the same comments on the general trends apply.

A significant difference is noted when the amounts of fume produced by the BAg-1 filler metal are compared to those from the low cadmium filler metal. In all cases more fume was produced by BAg-1 under CO_2 gas than under a flux; the opposite was true for the low cadmium filler metal. There appear to be two possible explanations for such an event: (a) the higher vapor pressure of cadmium at all temperatures and (b) an observed more tenacious and impervious dross that formed on the low cadmium filler metal.

The observations made while making fume collections on melts protected by a flowing CO_2 atmosphere indicate

Table 3.2
Brazing fume generation by CO_2 covered filler metals

	1200° F (649° C)	1250° F (677° C)
	BAg-1 filler metal, avg.	Low cadmium filler metal, avg.
Number of tests	3	4
Time molten, min	6	7
Sampling time, min	27	30
Ingot loss, g	0.0134	0.0079
Fume weight, g	0.1035	0.0435
	1350° F (732° C)	
	BAg-1 filler metal, avg.	Low cadmium filler metal, avg.
Number of tests	3	4
Time molten, min	8	9
Sampling time, min	27	32
Ingot loss, g	0.0446	0.0311
Fume weight, g	0.0924	0.0357
	1500° F (816° C)	
	BAg-1 filler metal, avg.	Low cadmium filler metal, avg.
Number of tests	4	4
Time molten, min	12	15
Sampling time, min	33	35
Ingot loss, g	0.0626	0.1144
Fume weight, g	0.2127	0.0635

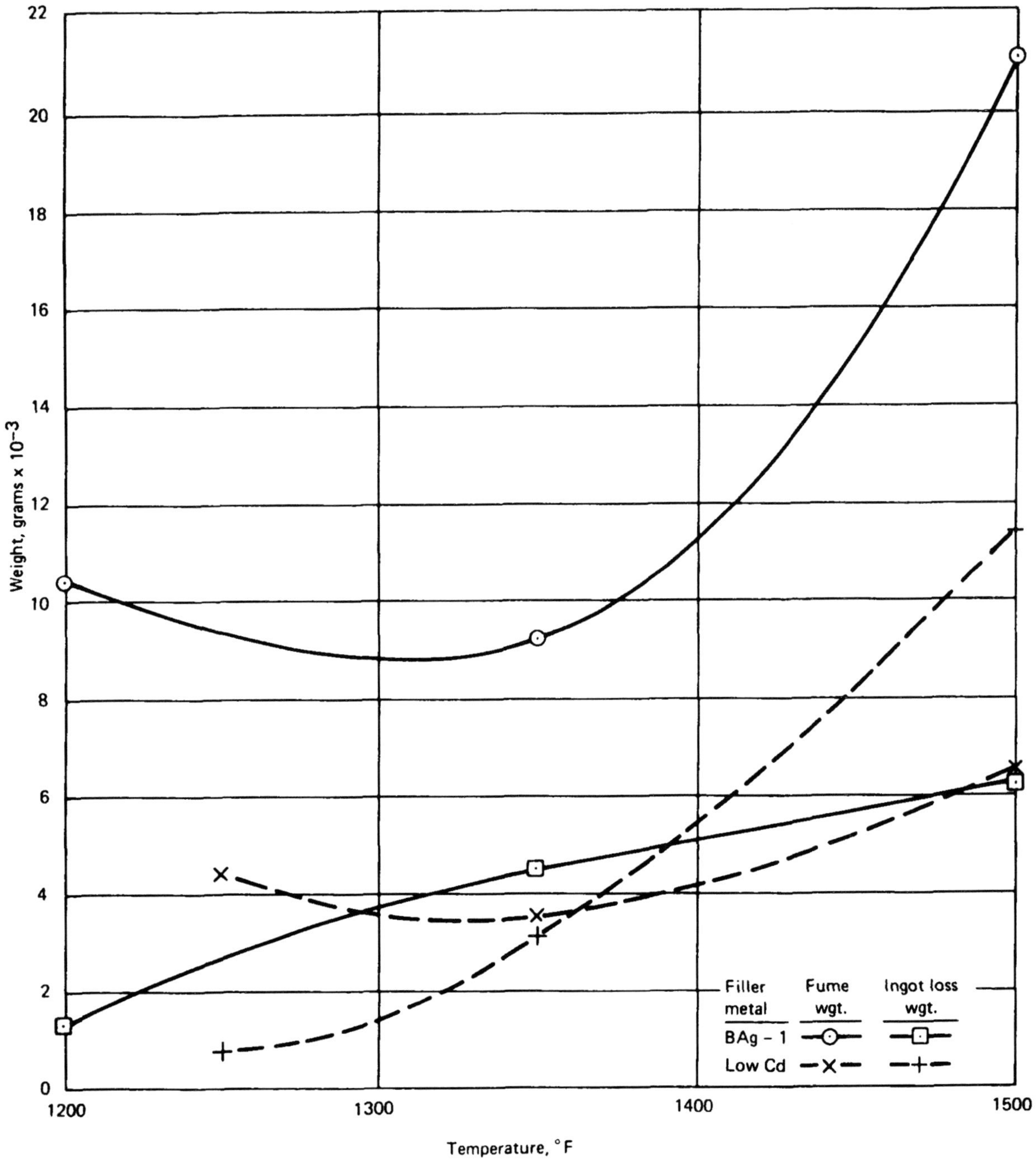

Fig. 3.5–Brazing fume generation and metal losses by CO_2 gas covered filler metal

that an improved technique should be used when making such studies. The use of a completely protective atmosphere, rather than a flowing blanket, is indicated to exclude air.

Ingot Weight Losses

The decrease in weight of the brazing filler metal melt (attributable to metal losses to fume) was considered a possible means of evaluating a particular filler metal's fume generation characteristics. Consequently, these losses were recorded during the study and are summarized in Tables 3.1 and 3.2.

In the case of flux covered melting, the losses recorded follow the expected trend: i.e., the greater losses occur at the higher temperatures. There is not a direct relationship between the ingot weight loss and the weight of fumes produced. In no case do the ingot metal losses calculated to oxides equal the fume weight. This should be true because no measure of the contribution made by the flux is evident. It is not reasonable to draw further conclusions from the weight loss data collected especially from that obtained from the higher temperature exposures. This is because of difficulties encountered in removing the fused flux from the ingot without also removing metal. The low cadmium filler metal was especially bad in this respect; it froze with a significant surface roughness which retained the flux residue.

The ingot weight loss data for tests made under a CO_2 blanket follow the proper trend as discussed for other situations. The magnitude of the losses recorded is debatable because of the adherent dross formed.

Because of the flux and dross removal difficulties, it is believed that a fume measurement procedure is a better way than metal weight losses to evaluate brazing filler metals for fume generation characteristics.

Section IIIB. Fume Composition

The composition of the fume produced by the brazing filler metals tested was appraised by analyzing the filter by the atomic absorption method. Analyses were made for cadmium, copper, zinc, and fluorine. The filters were chosen randomly from the tests made. Separate filters were used for the metal and for the fluorine analyses. The composition of the fume when the filler metals were melted under flux is given in Table 3.3. Table 3.4 contains the analysis of fume obtained when flowing CO_2 gas covered the melt. The data in Tables 3.3 and 3.4 are also shown graphically in Figs. 3.6 and 3.7.

Flux Covered Melts

Examination of the data in Table 3.3 shows that as the test temperature increases, the concentration of the more volatile constituents of the system increases. This was expected. In all cases the flux contributes most to the fume composition. Beyond these two facts that can be derived from the data, much concern must be shown for the tabulated results. A major concern is the low total composition shown by analysis. The highest total obtained on the low cadmium material at 1500° F (816° C) is 34.51 percent when the metals are calculated as oxides. Three possible explanations were immediately apparent: moisture pickup, the use of separate filters for metals and for fluorine, and the possibility that the flux constituent of the fume contains material in addition to fluorine.

A study of moisture pickup was made by running blank tests (all test conditions duplicated without brazing filler metal or flux) to determine if the filter weight increased. It was found that a weight gain attributable to moisture did occur. For seven blank tests made at 45 to 48 percent relative humidity, the average weight gain was 0.0070 grams or about 0.12 percent. The application of a correction factor for moisture based on the blank runs was not sufficient to overcome the fume weight deficit in the analysis. For the example given, 34.51 percent would become 38.06 percent.

The addition of data from two samples to get a single analysis for the fume produced can lead to errors in absolute values. It is believed that fume produced under like conditions should have essentially the same composition and, therefore, can be used to show relative values of composition. The error introduced by this technique cannot account for the loss recorded.

The use of a more thorough analysis for other fume constituents and the form they take in the fume should eliminate an important portion on the losses recorded. An indication of the possibility of other important fume constituents was obtained and is covered later in this section of the report.

Gas Covered Melts

The data recorded in Table 3.4 for the atomic absorption analysis of fume from brazing filler metals protected by a blanket of flowing CO_2 are questionable. The trends noted in the total fume and the flux covered melt analytic data are not evident in Table 3.4. The low cadmium filler metal appears to produce a higher cadmium fume than BAg-1. This is the opposite of other results. Also, the data show the greatest cadmium concentration in the fume from the filler metals tested at 1350°F (732°C). These observations and the very low recovery for the BAg-1, 1500°F (816°C) tests lead to a concern for the validity of all the data.

Discussion of Analytical Data

The experiences gained from the work covered by this report permit a number of suggestions to be made to those who may initiate future brazing fume studies, particularly studies whose objective is to compare characteristics of different filler metals.

Table 3.3
Composition of fumes from flux covered filler metals covered with AWS Type 3A flux

	1200° F (649° C) BAg-1 filler metal			1250° F (677° C) Low cadmium filler metal		
Element	Fume wgt., g	Element wgt., g	% as element	Fume wgt., g	Element wgt., g	% as element
Cadmium	0.0821	0.000320	0.39	0.1035	0.000110	0.11
Copper	0.0821	0.000008	0.01	0.1035	0.000010	0.01
Zinc	0.0821	0.000023	0.03	0.1035	0.000020	0.02
Fluorine-total	0.0453	0.000420	0.93	0.0198	0.000870	4.39
Fluorine-soluble	0.0453	0.000120	0.26	0.0198	0.000150	0.76
			1350° F (732° C)			
Cadmium	0.0802	0.000390	0.49	0.0332	0.000240	0.72
Copper	0.0802	0.000008	0.01	0.0332	0.000010	0.03
Zinc	0.0802	0.000019	0.02	0.0332	0.000050	0.15
Fluorine-total	0.0645	---	---*	0.0365	0.001800	4.93
Fluorine-soluble	0.0645	0.000330	0.51	0.0365	0.000250	0.69
			1500° F (816° C)			
Cadmium	0.2809	0.007500	2.67	0.1269	0.002100	1.65
Copper	0.2809	0.000011	0.01	0.1269	0.000010	0.01
Zinc	0.2809	0.005450	1.94	0.1269	0.005200	4.10
Fluorine-total	0.0831	0.006400	7.70	0.0687	0.012100	17.61
Fluorine-soluble	0.0831	0.003200	3.85	0.0687	0.006800	9.90

*Sample destroyed during analysis procedure.

Notes:

1. Copper in blank filter shows 0.000007 g equivalent to 0.01 percent in the table. Thus, copper must be at a very low level in the fume.
2. Cadmium and zinc are found in the blank filter at a slightly lower level than copper.

(1) The small quantities of fume produced and its apparent extremely small particle size present problems in filter choice, chamber design, sample weight-analysis balance, and sampling environments. To minimize these problems, it is advisable to use a fume collection chamber with the least volume possible which can be made part of a closed system. The chamber should contain an inert or readily controllable atmosphere. Filters or fume collection media must be capable of collecting very fine particles (<1μm) and should be small enough to avoid masking of actual fume weights by large blank filter weights. Continuous sampling through an atomic absorption type of analytical system appears worthy of consideration instead of filter sampling systems.

(2) Attention to the details of actual filler metal melting and fluxing practices are also required. Melt sizes, quantities of flux used, exposed melt surface areas, and heating and cooling rates need attention. Melt size, particularly with respect to exposed areas, may alter fume generation rate because of flux cover variation. Crucibles should be a metal which is wet by the filler metal to avoid flux collection around the melt. This implies combined weight determination that includes the crucible. Weighing exercises should include checks of flux losses in addition to filler metal losses. Ideally, all tests should begin in a cool chamber and heating furnace to minimize differences in heating and cooling rates. The closed chamber already suggested should eliminate any influence of humidity in the test area.

(3) Analytical procedures for fume analysis should be validated for the tests being conducted. An early phase of any analytical work should be directed toward determining the chemical character of the fume (metal, oxides, soluble, insoluble, etc.). Early analytical work should include complete analyses. Subsequent analysis requirements may be relaxed when all the data are collected.

Application of these suggestions and others that may

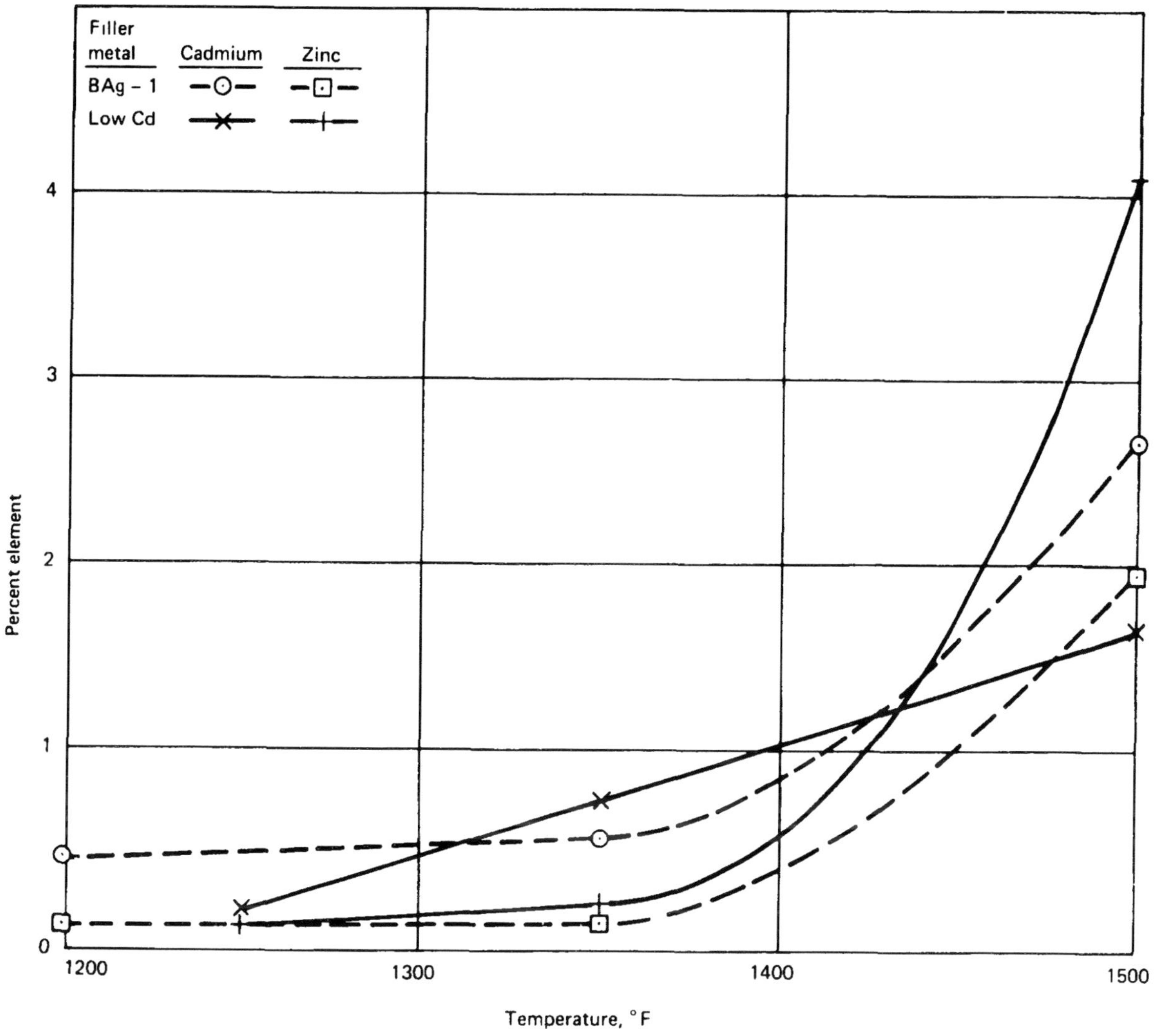

Fig. 3.6–Composition of fumes from flux covered brazing filler metals

occur to the researcher may produce more convincing data on the fumes produced by brazing filler metal usage than were obtained in this study.

Possible sources of error in the analytical data, moisture, and test result combining were discussed previously. In addtion, it was not feasible in the present program to identify the form taken by fluorine or the other elements in the flux when they became fume. Such information, if available, could add significantly to the recovery picture. No explanation of the poor recovery recorded on analyses of the fume from gas protected melts is apparent, except for the possibility that brazing fume particle size was too small to be collected on the membrane filter. No data were collected or are known on the particle sizes found in brazing fumes.

An effort was made to find possible causes of low analytical recovery by analyzing selected filters spectographically. In this way a complete elemental identification and a semiquantitative composition were found for fume generated at 1350°F (732°C) by each filler metal when melted with or without a flux cover. The results are shown in Table 3.5. The compositions are given in milligrams which constitute a quasi-quantitative (±50 percent or better) analysis of the fume and the filter. The data show the same discrepancy noted in the atomic absorption analyses. When the filter blank composition is subtracted, the remaining components fail to account for all fume that was measured gravimetrically.

A further analytical check was made for ammonium salts not detectable by atomic absorption or emission

Table 3.4
Composition of fumes from CO_2 covered brazing filler metal

Element	1200° F (649° C) BAg-1 filler metal			1250° F (677° C) Low cadmium filler metal		
	Fume wgt., g	Element wgt., g	% as element	Fume wgt., g	Element wgt., g	% as element
Cadmium	0.0639	0.000090	0.14	0.0483	0.000260	0.54
Copper	0.0639	0.000008	---	0.0483	0.000020	0.04
Zinc	0.0639	0.000013	0.02	0.0483	0.000020	0.04
			1350° F (732° C)			
Cadmium	0.0805	0.000230	0.29	0.0177	0.000840	4.75
Copper	0.0805	0.000009	0.01	0.0177	0.000010	0.06
Zinc	0.0805	0.000330	0.41	0.0177	0.000070	0.11
			1500° F (816° C)			
Cadmium	0.5127	0.000150	0.03	0.1154	0.002900	2.51
Copper	0.5127	0.000007	---	0.1154	0.000020	0.02
Zinc	0.5127	0.000020	---	0.1154	0.005200	4.50

spectroscopy. Filters were leached and analyzed for NH_3, Cl, SO_4, and NO_3. Although the results were not "finely tuned," they showed that such salts were not part of the fume.

Ingot Composition and Weight Changes

A record of ingot weight losses and the composition of the residual ingot was developed to evaluate the possibilities of rating brazing filler metal fuming characteristics without resorting to the more complex fume collection and analysis procedure. The ingot weight loss expressed in percent and the final composition of the ingots analyzed are given in Tables 3.6 and 3.7.

The ingot weight losses are quite small and appear nearly independent of the type of melt cover at either 1200, 1250, or 1350°F (649, 677, or 732°C). At 1500°F (816°C), the CO_2 cover seems more protective than the brazing flux. This is probably a manifestation of the formation of a protective dross, as mentioned in a previous section of this report. The ingot loss trends are the same as recorded for fume generation characteristics.

On the basis of the results recorded, the measurement of ingot metal loss would be a dubious procedure. Although the data show less metal loss for the low cadmium filler metal, the difference is not great enough for such an estimation except at 1500°F (816°C).

Comparative atomic absorption analysis for filler metal composition before and after melting is not feasible as a procedure for measuring metal in the fume. All the analytical data in Tables 3.6 and 3.7 are too close to the initial composition and within the limits of experimental error.

Discussion

A study was made to develop information on the fuming characteristics of two widely different silver-base brazing filler metals when used properly and improperly. The methods used were based on the best judgment of a committee of brazing experts, the experience of Battelle in similar fusion welding studies, and the availability of equipment used in welding studies.

The results of the study verify the trends that might have been derived intuitively from the materials and test conditions used:

(1) Increasing the temperature increases the amount of fume.

(2) Metal analysis of the fume depends upon the filler metal composition.

(3) Fluxes, when they are used, are a major contributor to the fume produced (see Table 3.3).

Beyond demonstrating trends, the data collected are of questionable value due to their erratic nature. More thorough analysis of the fume would be expected to provide some additional conclusions, especially in the case of tests made with fluxes.

In the following paragraphs, the procedures and meth-

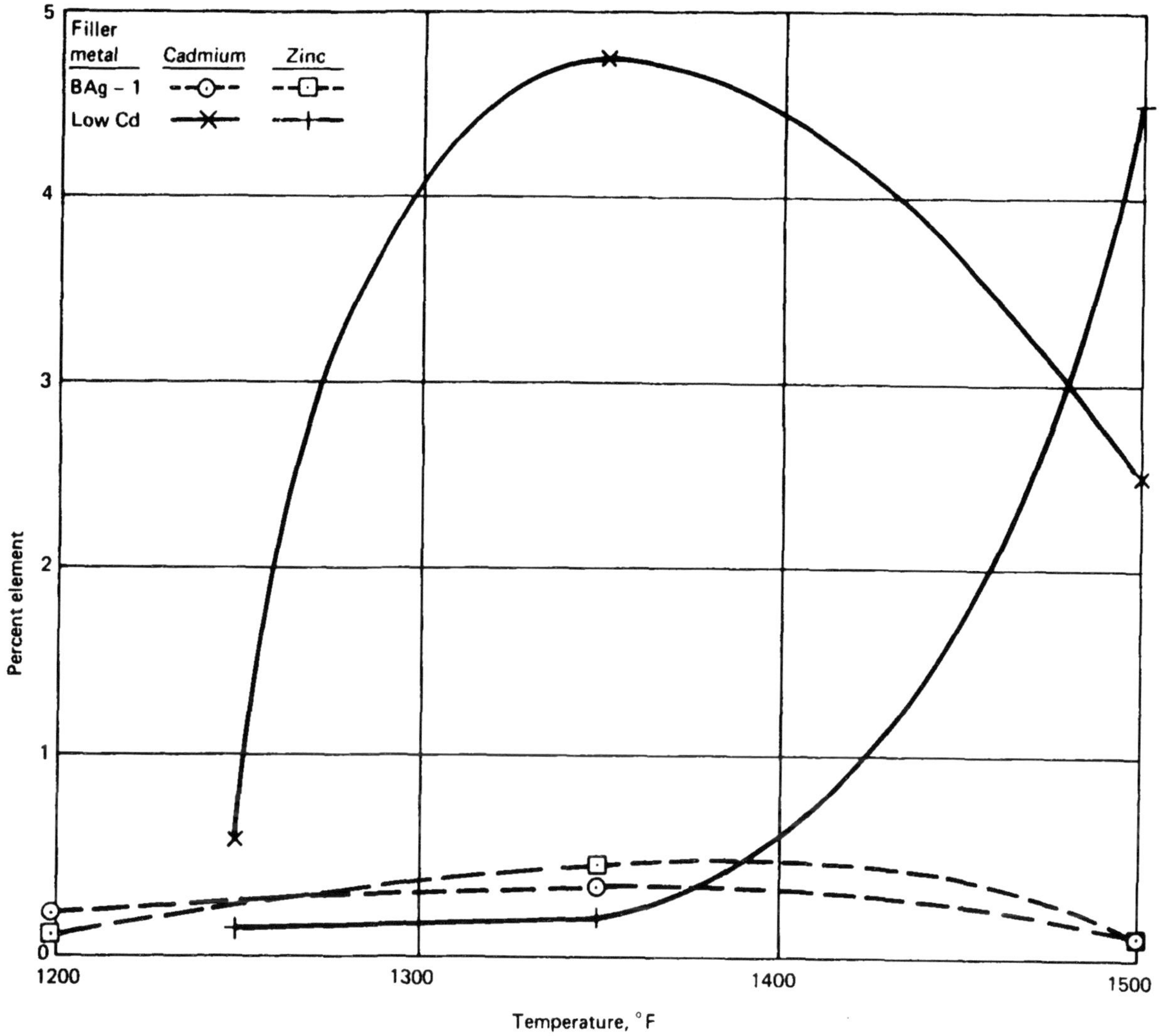

Fig. 3.7–Composition of fumes from CO_2 covered brazing filler metals

ods used in the study are discussed to indicate how they would be changed in any future work. The results would undoubtedly provide much more convincing information than is now on hand.

Filler Metal Handling and Surface Protection

The boats used were pure alumina and were not uniform in inside dimensions. This caused a variation in surface area which, it is believed, contributed to another problem. A controlled weight of flux was used. Because of the variable surface area and the fact that the filler metal did not wet the boat, this permitted the flux to almost completely envelop the melt. It was observed that the envelope was seldom complete, and a window of bare filler metal surface occurred on most melts. Its size varied considerably, permitting easy vaporization of some of the melt components in some tests and not in others. In addition, the surface area of flux was not constant, a condition that could lead to differing amounts of flux vapors in the fume.

Future studies should use metal boats of uniform size. The boat material should be wet by the filler metal, preventing inconstant envelopment by the flux of the filler metal melt. The flux quantity should be controlled so that a constant depth or a bare melt window of constant area is maintained.

A flowing "protective" gas blanket over the molten brazing filler metal surface does not provide adequate protection for quantitative measurement of fuming characteristics. The surface of the molten metal became covered with an "oxide" skin during the tests, which inconsistently inhibited the vaporization of melt constituents, producing inaccurate fume quantity measurements. If in new studies it is desirable to evaluate fuming

Table 3.5
Results of spectrographic analyses of fume samples collected at 1350° F (732° C)

Test no.	Filler metal	Melt cover	Fume weight, mg	All elements detected and calculated weight in the filter, mg																				
				Ca	Na	K	Al	Si	Mg	Fe	B	Ba	Zn	Cd	Ag	Cu	P	Mn	Pb	Cr	Sn	Ti	Ni	Sr
34	LC*	Flux	1057	400	80	200	40	40	20	12	40	2	4	40	1	10	4	2	1	4	2	2	1	1
28	BAg-1	Flux	865	400	80	600	15	40	20	12	40	2	12	80	1	10	4	2	1	4	2	2	1	1
29	BAg-1	CO_2	923	400	40	80	20	40	20	12	8	2	80	250	0.4	10	4	2	1	4	1	2	1	1
55	LC	CO_2	416	400	40	40	15	40	20	12	8	2	4	80	4	10	4	2	1	15	2	2	1	1
Blank filter				400	20	20	8	40	20	12	2	2	2	<1	<0.1	10	4	2	0.4	4	0.5	1	1	1

*Low cadmium filler metal.

Table 3.6
Composition and weight changes due to melting under a flux

Blank analysis	BAg-1 filler metal 25.4 Cd, 16.5 Zn	Low cadmium filler metal 7.02 Cd, 19.6 Zn
	1200° F (649° C)	1250° F (677° C)
Test number	7	48
Ingot weight loss, g	0.0495	0.0004
Percent weight loss	0.06	<0.01
Final cadmium content, %	25.9	6.93
Final zinc content, %	17.0	19.7
	1350° F (732° C)	
Test number	11	40
Ingot weight loss, g	0.0487	0.0413
Percent weight loss	0.06	0.05
Final cadmium content, %	25.6	6.94
Final zinc content, %	17.0	19.7
	1500° F (816° C)	
Test number	5	35
Ingot weight loss, g	0.1125	0.2275
Percent weight loss	0.13	0.28
Final cadmium content, %	25.2	6.95
Final zinc content, %	17.1	19.5

Table 3.7
Composition and weight changes due to melting under CO_2 cover

Initial blank analysis (atomic absorption)	BAg-1 filler metal 25.4 Cd, 16.5 Zn	Low cadmium filler metal 7.02 Cd, 19.6 Zn
	1200° F (649° C)	1250° F (677° C)
Test number	10	50
Ingot weight loss, g	0.0301	0.0119
Percent weight loss	0.04	0.01
Final cadmium content, %	25.7	7.03
Final zinc content, %	17.0	19.4
	1350° F (732° C)	
Test number	6	45
Ingot weight loss, g	0.0495	0.0243
Percent weight loss	0.06	0.03
Final cadmium content, %	25.6	7.00
Final zinc content, %	16.9	19.6
	1500° F (816° C)	
Test number	2	37
Ingot weight loss, g	0.0795	0.1083
Percent weight loss	0.09	0.13
Final cadmium content, %	25.6	6.94
Final zinc content, %	16.7	19.5

characteristics from the free surface of a brazing filler metal, a closed, inert atmosphere system should be used.

Filter System

In the program reported here, filters with a relatively large (33 in.^{2} or 21 290 mm^2) surface area and weight (5.9 to 6.4 grams) were used because they fit the available fume collection system. Also, the quantity of fume to be expected from the filler metals at 1500°F (816°C) was not known initially. In practice, the fume quantities were small at 1500°F (816°C) and even smaller at lower temperatures. Thus, the accuracy of the weights measured was minimized because the ratio of the filter weight to the fume weight was high (generally between 60 to 1 and 120 to 1). In addition, the influence of moisture and outside dust pickup during the test was maximized by lack of large fume quantities.

Since brazing filler metals and fluxes of the type studied produce small quantities of fume, especially when used properly, the clean filter weight should be small. This should also be used in small chambers to minimize the air-gas flow required to trap all the fume.

In this study the filter pore size was 1.2 μm; a smaller pore size filter would be more appropriate for collecting the amounts of fume evolved when brazing. Smaller pore sizes would assure greater confidence for collection of all the fume without concern for clogging the filter.

Test Chamber

The test chamber used for the brazing fume collection was large in comparison to the size of furnace and filler metal melt utilized. This is shown in Fig. 3.1. It was quite satisfactory for welding fume studies where large volumes of gases and fume are produced that require high flow rates through the filter.

A fume collection chamber for brazing fumes should be as small as permitted by the melting furnace used. The filter area and pumping rate should be commensurate with the chamber size.

Analytical Procedures

The atomic absorption procedures used in the work reported are accepted by industry and government alike for welding and brazing fume study. The previously discussed improvements in testing and sampling procedures should eliminate the analytical problems that were encountered. A broader analysis of the fume from tests made with fluxes would be required in any new study.

Summary of Part III

The data collected on the amount of fume produced by each of the two brazing filler metals studied at three different temperatures were averaged and are given in Tables 3.1 and 3.2. The data from these tables are shown graphically in Figs. 3.4 and 3.5. The average loss in weight of the ingots is also given along with other desirable data.

The major conclusions drawn from a study of the meager information in the literature were:

(1) The cadmium oxide fume hazard has received much attention without significant study of the basic causes of the problem.

(2) The possibility of harmful fluoride concentrations is mentioned briefly. As in the case of cadmium, there is a dearth of useful information.

(3) Almost no information exists which can be used to define the brazing environment or the safe practices to be followed.

In this study the actual weight of fume produced was smaller than expected.

The data show a large increase in weight of fume produced at 1500° F (816° C) for both filler metals when compared to that produced at lower temperatures. This would be expected because cadmium boils at 1409° F (765° C) and zinc has high vapor pressure (approx. 300 mm of Hg) at the same temperature. The rate of fume formation at the two lower temperatures was nearly the same, but this rate increased significantly at 1500° F. The low cadmium filler metal appears to produce about one-third less fume than the BAg-1 filler metal.

The data in Table 3.2 summarize the fume formation attributes of the two brazing filler metals studied where they are "protected" by a flowing CO_2 gas blanket rather than a flux. The results parallel those obtained where a flux was used.

A significant difference is noted when the amounts of fume produced by the BAg-1 filler metal are compared to those from the low cadmium filler metal. In all cases, more fume was produced by BAg-1 under CO_2 gas than under a flux; the opposite was true for the low cadmium filler metal.

The ingot weight loss measurements follow the expected trend: that is, the greater losses occur at the higher temperatures. There is no direct relationship between the ingot weight loss and the weight of fumes produced.

The fume produced by the brazing filler metals tested was analyzed for cadmium, copper, zinc, and fluorine. The composition of the fume when the filler metals were melted under flux is given in Table 3.3. Table 3.4 contains the analysis of fume obtained when flowing CO_2 gas covered the melt. The data in Tables 3.3 and 3.4 are also shown graphically in Figs. 3.6 and 3.7.

The results of the study indicate these trends:

(1) Increasing the temperature increases the amount of fume.

(2) Metal analysis of the fume depends on the filler metal composition.

(3) Fluxes, when used, are a major contributor to the fume produced (see Table 3.3).

Part IV. Thermal Spraying Fumes and Gases

Introduction

Oxyacetylene gas combustion spraying is not only the oldest of the thermal spraying processes but also the most widely used because equipment costs are minimal and spraying skills can be acquired readily. This process can be used to spray metals with melting temperatures below those produced by oxyacetylene flame (about 5000° F or 2760° C). The more recently developed arc spraying process is suitable for spraying all metals that are available in wire form. While equipment costs for arc spraying are higher than those for flame spraying, important savings can be realized through higher deposition rates.

The fumes produced during the thermal spraying of metal wires occurs as a result of the process in which the wire is melted and the molten droplets are atomized by a blast of compressed air to form minute particles that are propelled against the substrate. The substrate neither contributes to the production of fumes nor affects the fume composition, because the sprayed coating is primarily bonded to the substrate by mechanical linkages; there is little or no alloying between the coating and the substrate. This contrasts with metal surfacing, a process that is somewhat similar to thermal spraying. Surfacing more nearly resembles welding because the substrate is melted and the coating is metallurgically bonded to it. In this instance, the effect of the substrate on fume characteristics must not be overlooked. Thermally sprayed coatings are further characterized by inhomogeneities, and the presence of oxides in the coating indicates that oxidation of the metallic particles occurs during their flight to the substrate.

The potential problems associated with the fumes produced by thermal spraying operations have been recognized by the American Welding Society, and appropriate precautionary measures have been included in publications concerned with such processes. However, the literature on health and safety contains few references to thermal spraying and even less quantitative data on fume levels in spraying areas. Thus, data obtained in the course of this research should provide guidelines for estimating the magnitude of problems presented by specific thermal spraying applications.

Scope

The thermal spraying processes investigated in this part of the AWS research program were gas combustion spraying and electric arc spraying. The materials sprayed were wires 1/8 in. in diameter for combustion spraying and 1/16 in. in diameter for arc spraying.

Limitations

The materials sprayed were aluminum, aluminum bronze, zinc, and a proprietary ausenitic steel. The fume collection studies were limited to tests conducted at recommended spraying conditions.

Plasma arc spraying was not investigated, nor was detonation spraying. Detonation spraying is a propietary process and little is known publicly about its characteristics.

The composition of the fumes was determined only for the stainless steel experiments. The aluminum, aluminum bronze, and zinc fumes were presumed to resemble the wire compositions as metal particles or oxides and nitrides. Only limited examination of the fume particle sizes was conducted. Although large particles would present a health problem of lesser magnitude than small ones, further research would be needed to determine the distribution of particles in a fume sample by size. Examination of particles with the scanning electron microscope would more fully characterize them.

Program

The metals being tested were sprayed within the Battelle fume collection chamber onto a carbon steel sheet. Fume samples were collected in triplicate on glass fiber filters that were weighed before and immediately after each test to determine the weight of fumes collected. The wire consumed and the sprayed deposit weight were also measured.

In addition to determining fume generation rates, average fume generation characteristics were computed as the weight of fumes per weight of metal sprayed (readily convertible to a percentage of the wire weight consumed), and the weight of fumes per weight of metal deposited (convertible to a percentage of deposit weight).

The composition of the stainless steel fume was determined by optical emission spectroscopy. The fume particles themselves (for each of the materials) were studied with the light microscope.

Equipment, Materials, Procedures

Thermal Spraying Equipment

The following thermal spraying units were used in this investigation:

(1) A Metco 10E wire spraying gun was used for gas combustion spraying. This is an oxyacetylene gun with controls to regulate the rate at which wire is consumed. The pressure and flow rates of oxygen, acetylene, and compressed air were monitored and regulated by auxiliary control units.

(2) Studies of the electric arc thermal spraying process were conducted with a TAFA Arcspray 375 unit. This is a cart-mounted unit that includes the power supply with control console, twin wire reels, and a portable gun that can be either hand-held or mounted on a tool post.

Fume Collection Chamber

Fumes produced by thermal spraying operations were collected in the Battelle fume collection chamber. The features of this chamber and the manner in which its performance was evaluated are presented fully in Part II. In brief, the chamber is a sheet metal structure with a volume of about 33 cu ft (0.93 m^3). The upper part of the chamber is tapered to form an area in which the filters and holder are located; the fumes are removed from the chamber by an exhaust blower attached to the exit side of the filter holder. The front of the chamber has a hinged opening to provide access for equipment installation and adjustment; the opening can be sealed for fume collection experiments. In studying fumes produced by spraying, however, the access opening was not closed during sampling because of the size of the spraying unit and because of the need to ignite and adjust the flame of the gas combustion equipment. The suction rate of the blower was high enough to preclude any appreciable loss in fumes during the sampling interval.

Fume samples for gravimetric analysis were collected on glass fiber absolute filters. Cellulose membrane filters were used to collect samples for chemical analysis.

Materials

The following materials were selected for the thermal spraying studies with the concurrence of the AWS Project Committee on Spraying and Surfacing:

(1) For gas combustion spraying—aluminum, zinc, and austenitic stainless steel (Type 316)

(2) For electric arc spraying—aluminum, aluminum-bronze (90Cu-9Al-0.5Fe), zinc, and austenitic stainless steel (18Cr-5Ni proprietary alloy)

The wires for gas combustion spraying were 0.125 in. (3.2 mm) in diameter; those for electric arc spraying had a diameter of 0.0625 in. (1.6 mm).

Fume Sampling Procedures

Fume samples were collected in accordance with the following procedures:

(1) The glass fiber absolute filter was weighed to 0.1 mg on an analytical balance.

(2) The filter was placed in the holder located at the apex of the fume collection chamber, the sealing gaskets were installed, and the holder was clamped in place.

(3) The spraying equipment was adjusted to produce the desired spraying conditions.

(4) The blower that exhausted the chamber was turned on. Immediately thereafter, metal was sprayed on a grit-blasted steel substrate for a period of one-half minute.

(5) With the blower functioning, fumes were collected until the chamber atmosphere was clear.

(6) The filter was weighed immediately after its removal from the filter holder to determine the weight of fumes collected.

Samples of fumes associated with specific spraying operations were collected in triplicate to average the inconsistencies of sampling.

The following were recorded during these studies: (a) filter weight before and after sampling, (b) spraying conditions, (c) wire characteristics (diameter, length consumed, and weight consumed), and (d) sprayed deposit weight.

During the fume studies, the manufacturers recommended conditions were used for spraying the respective base metals. These conditions are listed in Tables 4.1 and 4.4.

The substrate used during the thermal spraying studies was 1/16 in. thick (1.6 mm thick) carbon steel sheet. The steel sheet was grit-blasted to remove surface contaminants and to provide a surface to which the sprayed metal would adhere well.

Gravimetric Analysis. Fume samples were collected to determine the rate at which fumes were produced when various metals were deposited on a substrate by gas combustion and electric arc spraying processes. With these data, it was also possible to calculate other values of interest to the user of such processes, particularly the amount of fumes per unit of weight of metal consumed and the amount of fumes per unit weight of metal deposited.

Section IVA. Gas Combustion Spraying

The conditions chosen for the gas combustion spraying tests are shown in Table 4.1.

Average fume generation characteristics of three thermally sprayed metals are shown in Table 4.2; complete data on the results of the experimental studies are presented in Table D1, Appendix D. The fume composition of the type 316 stainless steel wire is shown in Table 4.3.

Table 4.1
Gas combustion spraying conditions

Metal	Gas	Flow rate		Pressure	
		ft^3/hr	liters/min	psi	kPa
Aluminum	Oxygen	44	21	30	206.8
	Acetylene	40	19	15	103.4
	Air	53	25	65	448.2
Zinc	Oxygen	45	21	32	220.6
	Acetylene	42	20	15	103.4
	Air	53	25	65	448.2
Stainless steel	Oxygen	46	22	28	193.1
	Acetylene	33	16	14	96.6
	Air	51	24	55	379.2

Table 4.2
Average fume generation characteristics of gas combustion sprayed metals

	Metal wire sprayed		
	Aluminum	Zinc	Stainless steel
Fume generation rate, g/min	0.28	2.31	0.45
Melting rate, kg/h (lb/h)	3.22 (7.09)	7.79 (17.14)	1.64 (3.61)
Deposition rate, kg/h (lb/h)	2.81 (6.18)	4.54 (9.99)	1.18 (2.60)
Weight of fumes/weight of metal sprayed, g/kg*	5.34	17.94	16.60
Weight of fumes/weight of metal deposited, g/kg*	6.43	31.14	23.04

*g/kg ÷ 10 = %

Table 4.3
Composition of fumes produced during gas combustion and electric arc spraying of stainless steels by optical emission spectroscopy

Element	Concentration according to spraying method, weight %	
	Gas combustion[1, 3]	Electric arc[2, 3]
Ba	<0.01	<0.01
Si	0.8	0.8
Mn	2-4	8-15
Fe	30-50	30-50
Mg	<0.01	<0.01
Cr	8-15	10-20
Sn	0.02	0.02
Ni	5-10	4-6
Ga	<0.01	<0.01
Al	0.03	0.5
Mo	2-4	0.1
Ca	0.01	0.03
V	0.08	0.05
Cu	0.5	0.2
Zn	1-3	1-3
Ce	0.5	0.05
Zr	0.01	0.01
K	<0.1	<0.1

1. Type 316 stainless steel deposited by gas combustion spraying. Nominal composition 0.08C - 2Mn - 1Si - (16 - 18Cr) - (10 - 14Ni) - (2 - 3Mo) - Fe (bal.)
2. Proprietary stainless steel deposited by electric arc spraying. Nominal composition 0.06C - 0.03Si - (8 - 9Mn) - (17 - 18Cr) - (5 - 6Ni) - Fe (bal.)
3. Fume sample weights 0.164 g for gas combustion sprayed wire and 0.227 g for electric arc sprayed wire.

Discussion

Under the conditions used for gas combustion spraying, fumes were produced at the highest rate during the spraying of zinc. This appeared to be a reflection of the high vapor pressure of zinc. However, when the rates at which fumes were produced are considered as a function of the metal and the amount of metal being sprayed, there was little difference in the fume quantities associated with the spraying of zinc and stainless steel.

The manganese content of the fumes produced during gas combustion spraying was high in comparison to the manganese content of the stainless steel wire. This result is in agreement with data obtained during the Part II studies on the composition of fumes from welding electrodes. High manganese contents may be a function of the oxidation potential of the gas used to propel particles toward the substrate.

Zinc was detected in the stainless steel fume sample. While the source for zinc is not known, it may have been present as a contaminant in the fume collection chamber or on the rolls of the spraying equipment from prior spraying operations.

Nickel and chromium contents of the fumes associated with gas combustion spraying tended to be lower than the contents of these elements in the wire being sprayed.

The fume particles produced by gas combustion spraying were consistently larger than those produced by arc welding, to a significant number.

Section IVB. Electric Arc Spraying

The conditions chosen for the electric arc spraying tests are shown in Table 4.4.

Table 4.4
Electric arc spraying conditions

Metal	Current, A	Voltage, V
Aluminum	230-250	13-15
Zinc	200-225	12-15
Stainless steel	220-230	13
Aluminum-bronze	240-260	13

The average fume generation characteristics of four thermally sprayed wires are shown in Table 4.5; complete data on the results of the experimental studies are presented in Table D2, Appendix D. The fume composition for the proprietary stainless steel wire used for arc spraying is shown in Table 4.3, column 3.

Discussion

In electric arc spraying, the rates at which fumes were produced during the spraying of zinc, stainless steel, and aluminum bronze were quite similar; the fume generation rate for aluminum was somewhat lower. However, aluminum produced the most fumes as a function of the weight of metal sprayed, because the volume of aluminum sprayed was greater than that of the other metals, due to its low density.

Using the ratio of weight of fumes to weight of metal sprayed as a criterion, the fume formation propensity of aluminum was greatest when this metal was sprayed by the electric arc process; the spraying process had little effect on the propensity of the other metals to produce fumes.

The fumes from electric arc spraying had a chromium content lower than the chromium content of the wire; the nickel content of the fumes and wire was essentially the same. Lower than expected contents of nickel and chromium in the fumes from welding electrodes containing these elements were commonly observed in previous studies.

Again note that zinc was detected in the stainless steel fume sample. Its source is not known, but it may have been present as a contaminant in the fume collection chamber or on the feed rolls of the spraying equipment.

It should be noted that the above comments are based on the results obtained during this investigation. Different results may be produced with other diameters of wire or other spraying conditions.

The characteristics of fume particles produced during thermal spraying operations were studied with the light microscope. Limited examinations of fume samples revealed the presence of innumerable minute spherical particles of metal that had solidified from the molten state after the atomization process. While particles with diameters as large as 0.002 in. (50 μm) were observed, most were much smaller in size (0.0002 in. [5 μm] or less). On balance, the fume particles produced by electric arc spraying were smaller than those produced by gas combustion spraying. In contrast, particle size distribution studies have shown that most welding fume particles are smaller than 0.00004 in. (1.0 μm).

The composition of the fumes produced during the gas combustion and electric arc spraying of stainless

Table 4.5
Average fume generation characteristics of electric arc sprayed wires

	Aluminum	Zinc	Stainless steel	Aluminum-bronze
Fume generation rate, g/min	0.92	1.44	1.51	1.34
Melting rate, kg/h (lb/h)	1.32 (2.90)	4.10 (9.06)	4.66 (10.25)	3.89 (8.57)
Deposition rate, kg/h (lb/h)	0.60 (1.32)	2.00 (4.40)	2.35 (5.16)	1.79 (3.94)
Weight of fumes/weight of metal sprayed, g/kg*	41.72	21.05	19.75	20.85
Weight of fumes/weight of metal sprayed, g/kg*	90.84	43.24	38.67	45.49

*g/kg ÷ 10 = %

steel alloys was determined by optical emission spectroscopy (Table 4.3). This semiquantitative analytical method provides a quick overview of a wide spectrum of the elements present in fume samples; however, it is more accurate in detecting the presence of small elemental concentrations than large ones. If more accurate compositional data are needed, optical emission spectroscopy should be supplemented by other analytical techniques (e.g., atomic absorption analysis or wet chemistry) for selected elements. In Table 4.3, the sum of the fume constituents approaches 100 percent, if it is assumed that all elements are converted to oxides.

Depending on the metal being sprayed, the fume generation rates for gas combustion and electric arc spraying ranged from 0.28 to 2.31 and 0.92 to 1.51 g/min, respectively. Although precise comparisons are difficult to make because of differences in materials, processes, and process parameters, these fume generation rates are similar in magnitude to those encountered in many arc welding operations that were examined during the laboratory investigation of arc welding fumes (Part II). This implies that the potential health hazards presented by thermal spraying and arc welding are also similar. Such may well be the case, but additional work over a range of operating conditions would be needed to substantiate this broad premise, because welding fume generation rates were subject to dramatic changes depending on the welding process, electrode, and welding conditions. The rate at which fumes are produced during thermal spraying operations may be subject to similar variations. As in the case of arc welding, research should be directed toward defining significant relationships between the fume generation rate and the variables that control the spraying process. Although fume particles with larger diameters than those associated with arc welding fumes were observed during spraying, more work would be needed to determine if larger particles are present in amounts sufficient to reduce potential fume hazards.

Of the characteristics investigated, fume generation rates are most important because they provide a direct measure of the amount of fumes that will be produced under specified spraying conditions. For example, assume that fumes are being produced at the rate of 2.00 g/min during gas combustion spraying of zinc in an unventilated room with a volume of 283 m^3 (10,000 ft^3). Assuming further that the fumes diffuse uniformly throughout the room (a simplistic assumption, since fume particles are subject to gravitational, electrical, and other forces), the fumes concentration in the room will exceed the threshold limit value (TLV) for zinc oxide of 5 mg/m^3 (as defined by the American Conference of Governmental Industrial Hygienists) in less than 1 minute. To alleviate this situation, ventilation sufficient to effect an air change in the room each minute would be needed to maintain a fume level below current TLV's.

If the spraying efficiency is known, the ratio of weight of fumes to weight of metal sprayed provides an indication of the amount of fumes that will be produced in depositing a sprayed coating of a given area and thickness.

There are numerous methods available to industry to insure that thermal spraying operations are safe. These range from portable exhaust systems that can be positioned to remove fumes from the spraying area to dry or wet spraying booths with facilities to trap and retain fumes. The manufacturers of such equipment should be consulted for assistance in selecting a unit whose performance matches the requirements of the spraying facility.

Summary of Part IV

A limited research program was undertaken to examine the fume formation characteristics of thermally sprayed metal wires. In this work, fumes were sampled during the gas combustion and electric arc spraying of aluminum, zinc, stainless steel, and aluminum-bronze (arc sprayed only). The wire diameters were 0.125 in. (3.2 mm) for the gas combustion sprayed metals and 0.0625 in. (1.6 mm) for those that were arc sprayed. The resulting data were used to calculate the rate at which fumes were produced and other related parameters.

The average fume generation characteristics are summarized in Tables 4.2 and 4.5.

Part V. Fumes and Gases from Oxygen Cutting

Introduction

Oxygen cutting encompasses several processes that are used to cut, groove, scarf, and otherwise shape ferrous metals by means of the chemical reaction between iron and a confined stream of high-purity oxygen. Carbon and many alloy steels can be cut readily in this manner; highly alloyed and heat-resistant steels require flux or iron powder additions in the oxygen stream to facilitate the chemical reaction. Oxygen cutting processes are differentiated by the gas used as a fuel for the preheating flames. Among the fuel gases used by industry are acetylene, natural gas, propane, stabilized methylacetylene, and others. Acetylene is the most commonly used fuel gas because of its ready availability, high flame temperature, and familiar flame characteristics. However, other fuel gases (e.g., natural gas or propane) may be used, depending on availability and the economics of the cutting applications.

The fumes associated with oxygen cutting operations occur as the result of the chemical reaction between iron and high-purity oxygen and the process through which molten oxides and unoxidized metal are removed from the kerf by the cutting jet. Fume quantities are dependent upon the cutting process parameters, base metal thickness, base metal composition, surface contaminants, and other factors. Fume compositions are governed by the composition of the metals being cut and any coating that may be on the surface of the metal.

Although the potential problems presented by oxygen cutting fumes have been noted by the American Welding Society (see ANSI Z49.1-1973, Safety in Welding and Cutting), there are few references to this subject in the technical literature on safety and health. Fume samples have been collected in various industrial environments and analyzed to determine concentrations of specific elements and compounds in cutting areas (Refs. 5.1 and 5.2). However, little has been done to determine the rate at which fumes are produced during oxygen cutting operations. Such information is of interest to the welding engineer, the industrial hygienist, and the ventilation specialist. The data obtained in the course of this investigation partially fill this void in the technical knowledge and provide the impetus for further research.

Scope

Research was directed toward a study of the fumes produced during oxyacetylene and oxynatural gas (methane) cutting of low carbon steel plate in thicknesses of 1/2, 1, and 2 in. (12.7, 25.4, and 50.8 mm).

Cutting was done under the normal conditions recommended by the cutting torch manufacturer, using various gas pressures and various travel speeds. Total fume samples were collected to accumulate data for use in determining fume generation rates and quantities relating fume generation rates to important process variables.

Limitations

Fume compositions were analyzed only by optical emission spectroscopy (except for iron, which was confirmed by wet chemical analysis). Concentrations of major elements cannot be determined accurately by optical emission spectroscopy because of difficulty in interpreting the spectrographic record and because of the lack of suitable standards. As a result, concentrations of such elements are expressed in a range of weight percentages instead of discrete weight percentages. If more accurate compositional data are needed, optical emission spectroscopy should be supplemented by other analytical techniques such as atomic absorption analyses or wet chemistry methods.

Particle sizes of fume deposits were examined briefly with the light microscope. Additional research would be needed to more fully characterize cutting fumes. Such research should include studies to determine the distribution of fume particles by size and examination of fume deposits with the scanning electron microscope.

Acetylene and methane were the gases chosen for this study. Methane was chosen as a cutting fuel gas instead of natural gas because it was readily available in cylinders.

No study of arc cutting or plasma arc cutting was made, for lack of readily accessible equipment.

Program

Cutting experiments were conducted within the Battelle fume collection chamber. Fume samples for gravimetric analyses were collected on glass fiber absolute filters. Cellulose membrane filters were used to collect fumes for chemical analysis and were weighed before and immediately after each run. Fume generation data were computed as both the rate of generation in grams per minute and the weight generated per unit length of cut. The composition of the fume during cutting of carbon steel plate was determined by optical emission spectroscopy and wet chemistry (iron only). The fume particles were studied with the light microscope.

Equipment, Materials, and Procedures

Equipment and Facilities

Cutting Equipment: The equipment used during this investigation is described below:

(1) An Oxweld W-22 torch was used for oxyacetylene cutting. As indicated below, it was equipped with tips appropriate to the thickness of plate being cut.

Plate thickness, in.	Plate thickness, mm	Tip number	Cutting orifice diameter, in.	Cutting orifice diameter, mm
1/2	12.7	4	0.04	1.0
1	25.4	6	0.06	1.5
2	50.8	8	0.08	2.0

(2) A Harris 63-2 torch equipped with the following tips was used for oxymethane cutting.

Plate thickness, in.	Plate thickness, mm	Tip number	Cutting orifice diameter, in.	Cutting orifice diameter, mm
1/2	12.7	OOVVC	0.036	0.9
1	25.4	0-1/2VVC	0.40	1.0
2	50.8	1VVC	0.046	1.2

Fume Collection Chamber. Samples of oxygen cutting fumes were collected in the Battelle fume collection chamber. The features of this chamber and the manner in which its performance was evaluated are discussed fully in Part II. Briefly, the chamber was a sheet metal structure with a volume of about 33 ft^3 (0.93 m^3). The upper third of the chamber tapered inward to form an area in which the filters and holder were located; fumes were removed from the chamber by an exhaust blower attached to the exit side of the filter holder. The front of the chamber had a hinged opening to provide access for equipment installation and adjustment; the opening could be sealed for fume collection experiments. In this study, the access opening was not closed during sampling because of the size of the cutting torch and the need to ignite it and adjust the flame before cutting ensued. The suction rate of the blower was sufficiently high to prevent any significant loss in fumes during the sampling.

The chamber was equipped with a turntable and a variable speed motor to permit cutting to be done at selected rates. A steady-rest was provided for supporting the cutting torch.

Materials

The cutting studies were conducted using hot-rolled low carbon steel plate with thicknesses of 1/2, 1, and 2 in. (12.7, 25.4, and 50.8 mm); the plate was grit-blasted to remove rust and other surface contaminants. The oxygen was welding grade, supplied from cylinders; the acetylene and methane were commercial grade, in cylinders.

Fume Sampling Procedures

The plate to be cut was positioned on the turntable. Then the speed of the turntable was adjusted to produce the selected cutting speed, and fume samples were collected in accordance with the following procedures:

(1) The glass fiber absolute filter was weighed to 0.1 mg on an analytical balance.

(2) The filter was placed in the holder located in the apex of the fume collection chamber, the sealing gaskets were installed, and the holder was clamped in place.

(3) The oxygen and fuel gas gages were set to produce the desired cutting conditions. To avoid introducing soot into the fume chamber, the cutting torch was ignited, and the flame was adjusted before the torch was placed in the chamber.[11]

11. To ignite the cutting torch, the fuel gas valve was opened slightly and the torch was lit with a sparklighter. Then the oxygen valve was opened gradually until a neutral flame was obtained. Successive adjustments of the oxygen and fuel gas flow rates were made to produce a flame with the desired intensity. While the fuel gas flow rate was increased during these adjustments, the valve was not opened completely.

(4) The blower that exhausted the chamber was started. Immediately thereafter, cutting began and continued for a period of one-half minute.

(5) With the blower functioning, fumes were collected during the cutting period and afterwards until the chamber atmosphere cleared (about 5 minutes).

(6) The filter was weighed immediately after its removal from the filter holder to determine the weight of fumes collected.

While sóme fumes were generated below the workpiece, the workpiece did not appear to interfere with their collection since there was little evidence of settling of the fumes.

Samples of fumes produced under specific cutting conditions were collected in triplicate and the following data were recorded: (a) filter weight before and after sampling, (b) plate thickness, and (c) cutting conditions (oxygen and fuel gas pressures, cutting speed, etc.).

Gravimetric analyses of fume samples were conducted to determine the rate at which fumes were produced during the cutting of various thicknesses of carbon steel plate with the oxyacetylene and oxymethane cutting processes. Ideally, plates with thicknesses of 1/2, 1, and 2 in. (12.7, 25.4, and 50.8 mm) were cut using conditions recommended by the manufacturer of the cutting equipment to produce a drop cut. Cutting was also done at other oxygen and fuel gas pressures and other cutting speeds to determine the effect of these variables on the rate at which fumes were produced. For each cutting condition, the fume generation rate and the amount of fumes per unit length of plate cut were determined.

Section VA. Oxyacetylene Cutting

The results of the oxyacetylene cutting studies are summarized in Table 5.1; the data upon which this table is based are included in Appendix E (Tables E1, E2, and E3). The following comments are based on an examination of the experimental data.

Cutting under Normal Conditions

The fume generation rate increased as a function of plate thickness when plates with thicknesses of 1/2 to 2 in. (12.7 to 50 mm) were cut using recommended gas pressures and cutting speeds (Fig. 5.1). This trend is reasonable, since the rate at which fumes are produced is dependent upon the volume of metal that is oxidized and removed by the cutting jet to form the kerf, and this volume increases with increasing plate thickness. However, the fume generation rate did not increase in direct proportion to the plate thickness. Extrapolation down to zero plate thickness would indicate fume generation at 0.35 g/min. with no material being cut. The fume generation curve is presumed to drop off as sketched in Fig. 5.1.

The generation rate for 2 in. (50 mm) plate seems low, since the wider kerf of those cuts removed more than twice as much metal as that removed from the 1 in. (25.4 mm) plate; the oxygen orifice diameters were 0.08 in. (2.0 mm) and 0.06 in. (1.5 mm) respectively (Tables E3, Appendix E). Although actual kerf widths were not recorded, the kerfs were observed to be wider in the thicker plate.

Figure 5.1 also shows the variation in the ratio of fume weight to the length of cut as a function of plate thickness (dashed curve). This factor appears to be closely proportional to plate thickness.

Cutting at Various Gas Pressures

Deviations from recommended oxygen and acetylene pressures have a generally adverse effect on the quality of cut produced during oxygen cutting because the oxygen-acetylene ratio is unbalanced. As indicated in Table 5.1, such deviations also affected fume generation rates and ratios of fume weight to length of cut. For any given plate thickness, the experimental data showed a consistent trend toward higher fume generation rates with increased oxygen and acetylene pressures and lower rates with decreased gas pressures (Fig. 5.1A). Weights of fumes per unit length of plate cut varied in the same manner (Fig. 5.1B). The data plotted are for cutting done at recommended speeds; other results would be obtained by using other cutting speeds. For example, half-inch plate cut at 9.5 in./min (4 mm/s) produced 50% more fumes per minute than the 13 in./min (5.5 mm/s) cut.

Extrapolation of the curves for 2 in. and 1 in. plates to lower oxygen pressure would suggest no fume generation below 20 to 25 psi (138 to 172 kPa). The data for 1/2 in. plate extrapolate to zero pressure. The same relationship is true for the ratios of fume weights to lengths of cut.

These results should not be interpreted to mean that potential hazards presented by cutting fumes could be minimized by the use of lower-than-recommended pressures, because the quality of the cut deteriorated when reduced gas pressure was used.

Cutting at Various Speeds

Variations in cutting speed appeared to have less effect on the fume generation rates than variations in gas pressures. For example, when 1/2 and 1 in. thick plates were cut at speeds higher and lower than those recommended with normal gas pressures, fume generation rates trended upward slightly with increasing cutting rates (Fig. 5.1C). The upward trend was more pronounced when 2 in. thick plate was cut. Since more metal is being melted and oxidized at high cutting speeds, fume generation rates tend to increase.

A stronger trend was evident in ratios of fume weight to unit length of cut (Fig. 5.1D) This ratio appears to decrease hyperbolically as the speed of cutting increases. The high fume ratio at low cutting speeds correlates with the poor quality of cut obtained at low speed.

Table 5.1
Effect of variations in oxyacetylene cutting conditions on average fume generation characteristics

Operating conditions[1]	Oxygen pressure		Acetylene pressure		Plate thickness		Cutting speed		Fume generation rate, g/min	Weight of fumes/length of plate cut	
	kPa	psi	kPa	psi	mm	in.	mm/s	in./min		g/m	g/in.
Normal	241.3	35	34.5	5	12.7	0.50	5.50	13	0.42	1.26	0.032
High gas pressures	275.8	40	48.3	7	12.7	0.50	5.50	13	0.46	1.39	0.035
Low gas pressures	206.8	30	20.7	3	12.7	0.50	5.50	13	0.34	1.01	0.026
High cutting speed	241.3	35	34.5	5	12.7	0.50	6.77	16	0.49	1.21	0.031
Low cutting speed	241.3	35	34.5	5	12.7	0.50	4.23	10	0.45	1.77	0.045
Normal	275.8	40	34.5	5	25.4	1.00	4.66	11	0.54	1.93	0.049
High gas pressures	310.3	45	48.3	7	25.4	1.00	4.66	11	0.64	2.27	0.058
Low gas pressures	241.3	35	20.7	3	25.4	1.00	4.66	11	0.35	1.25	0.032
High cutting speed	275.8	40	34.5	5	25.4	1.00	5.93	14	0.57	1.60	0.041
Low cutting speed	275.8	40	34.5	5	25.4	1.00	3.39	8	0.54	2.68	0.068
Normal	275.8	40	41.4	6	50.8	2.00	2.75	6.5	0.67	4.05	0.103
High gas pressures	310.3	45	55.2	8	50.8	2.00	2.75	6.5	0.81	4.89	0.124
Low gas pressures	241.3	35	27.6	4	50.8	2.00	2.75	6.5	0.48	2.86	0.073
High cutting speed	275.8	40	41.4	6	50.8	2.00	4.02	9.5	0.72	2.98	0.076
Low cutting speed	275.8	40	41.4	6	50.8	2.00	1.48	3.5	0.53	5.96	0.151

1. Tip number (cutting orifice diameter): 1/2 in. (12.7 mm) thick plate: No. 4 (0.04 in. [1.02 mm]); 1 in. (25.4 mm) thick plate: No. 6 (0.06 in. [1.5 mm]); 2 in. (50.8 mm) thick plate: No. 8 (0.08 in. [2.0 mm]).

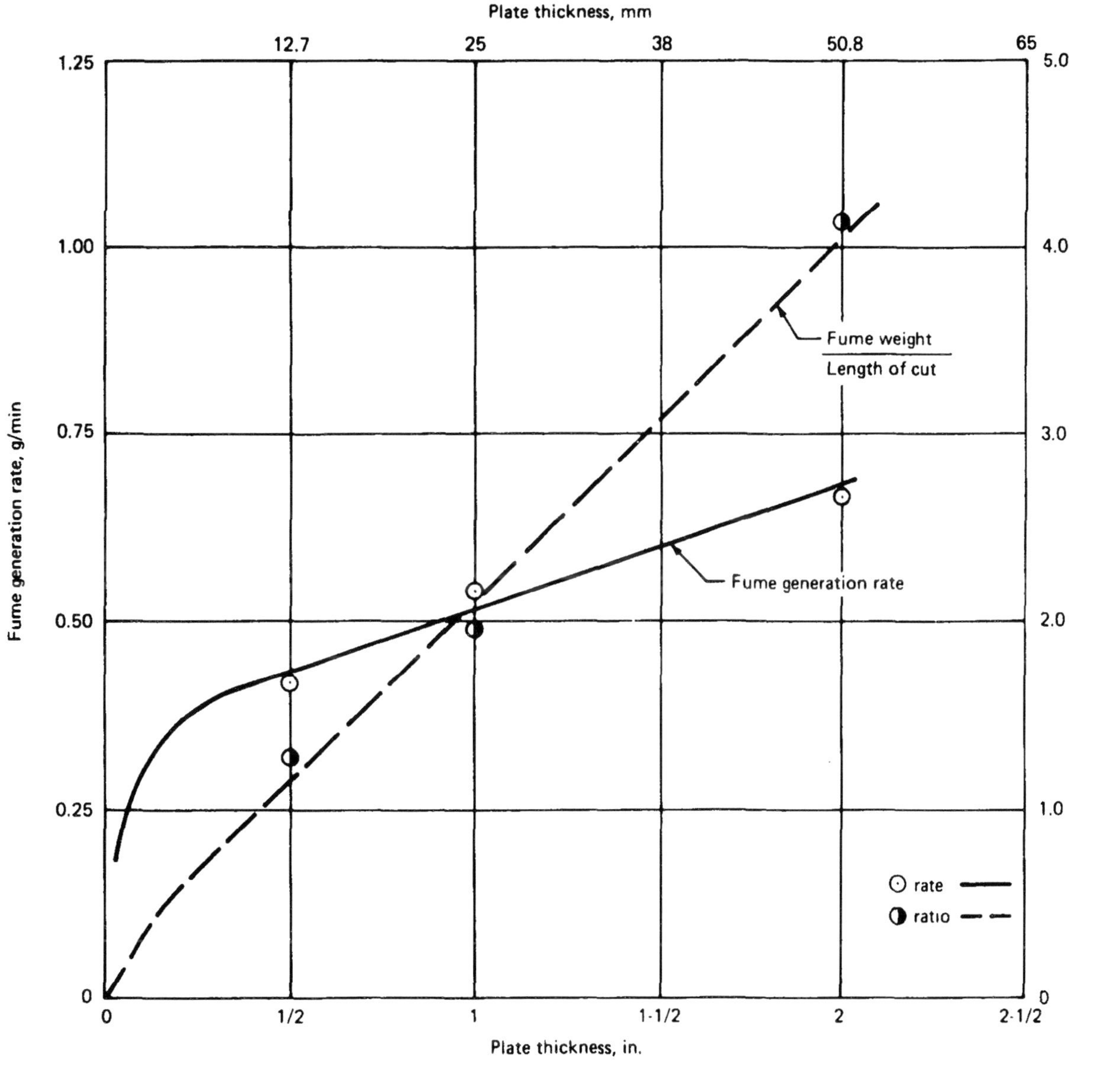

Fig. 5.1–Fume generation rate and weight of fumes per length of plate cut as a function of plate thickness during oxyacetylene cutting at recommended oxygen and acetylene pressures

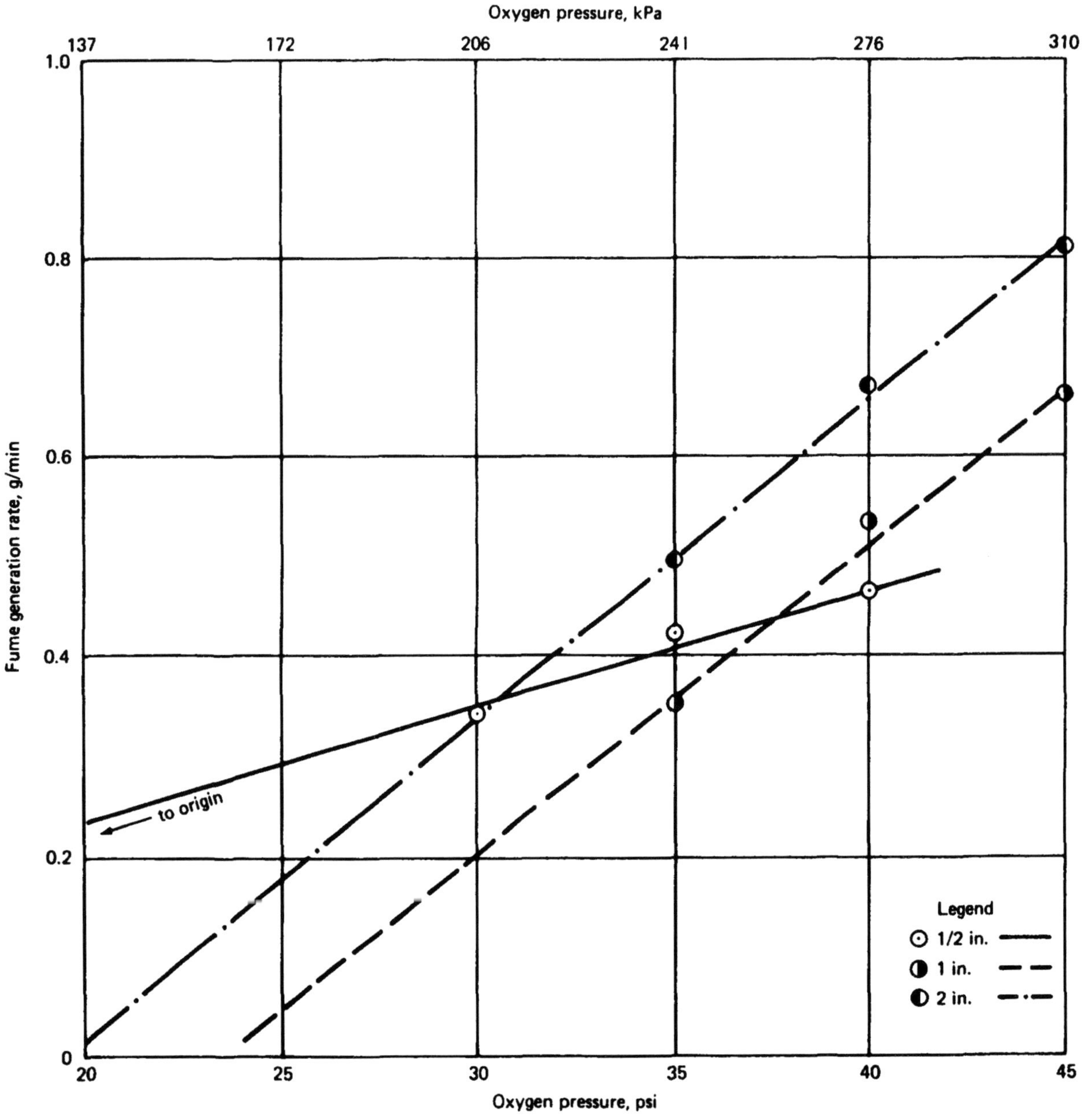

Fig. 5.1A–Fume generation rate as a function of gas pressure during oxyacetylene cutting at recommended speeds

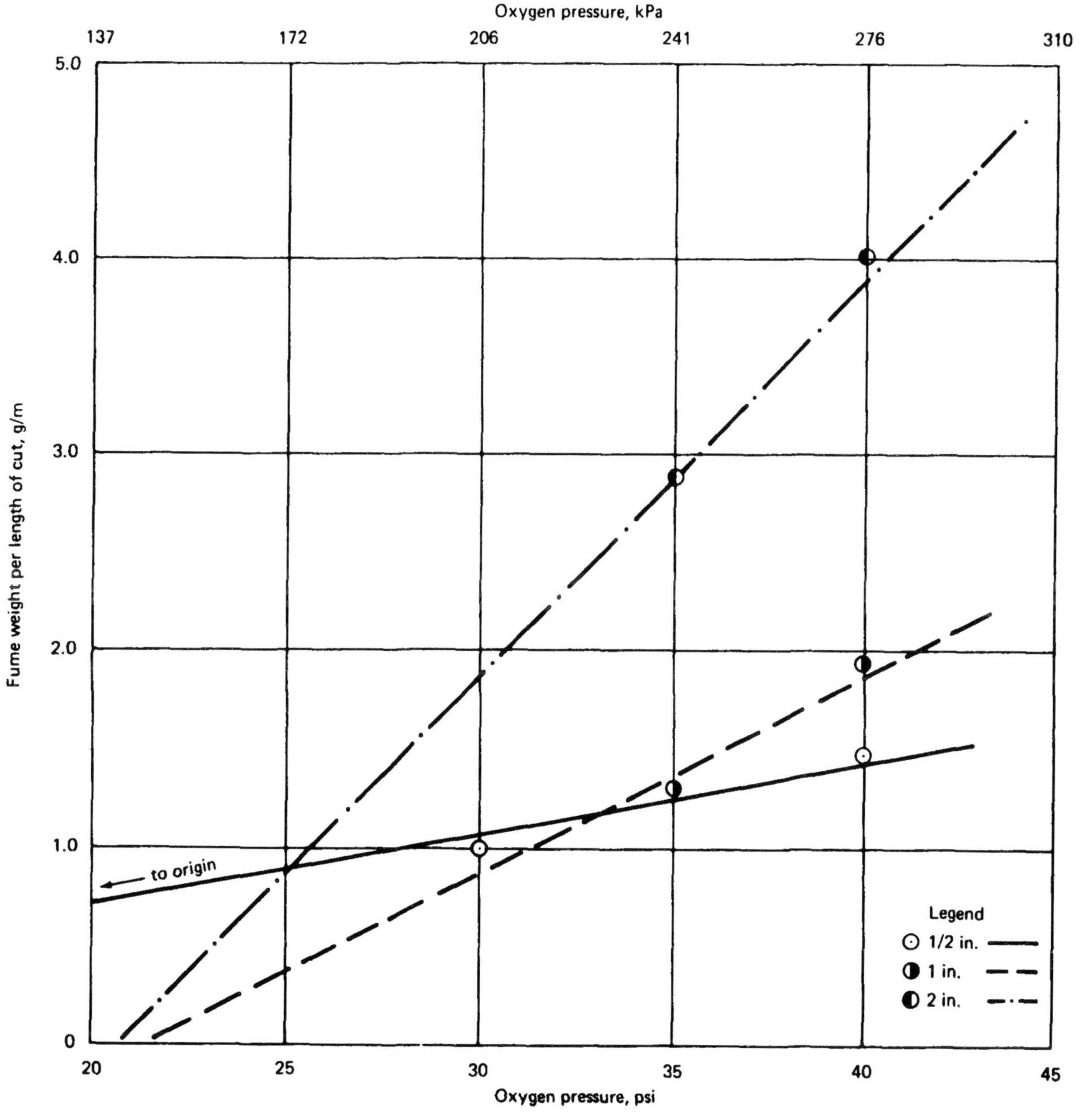

Fig. 5.1B–Weight of fumes per length of plate cut as a function of oxygen pressure during oxyacetylene cutting at recommended speeds.

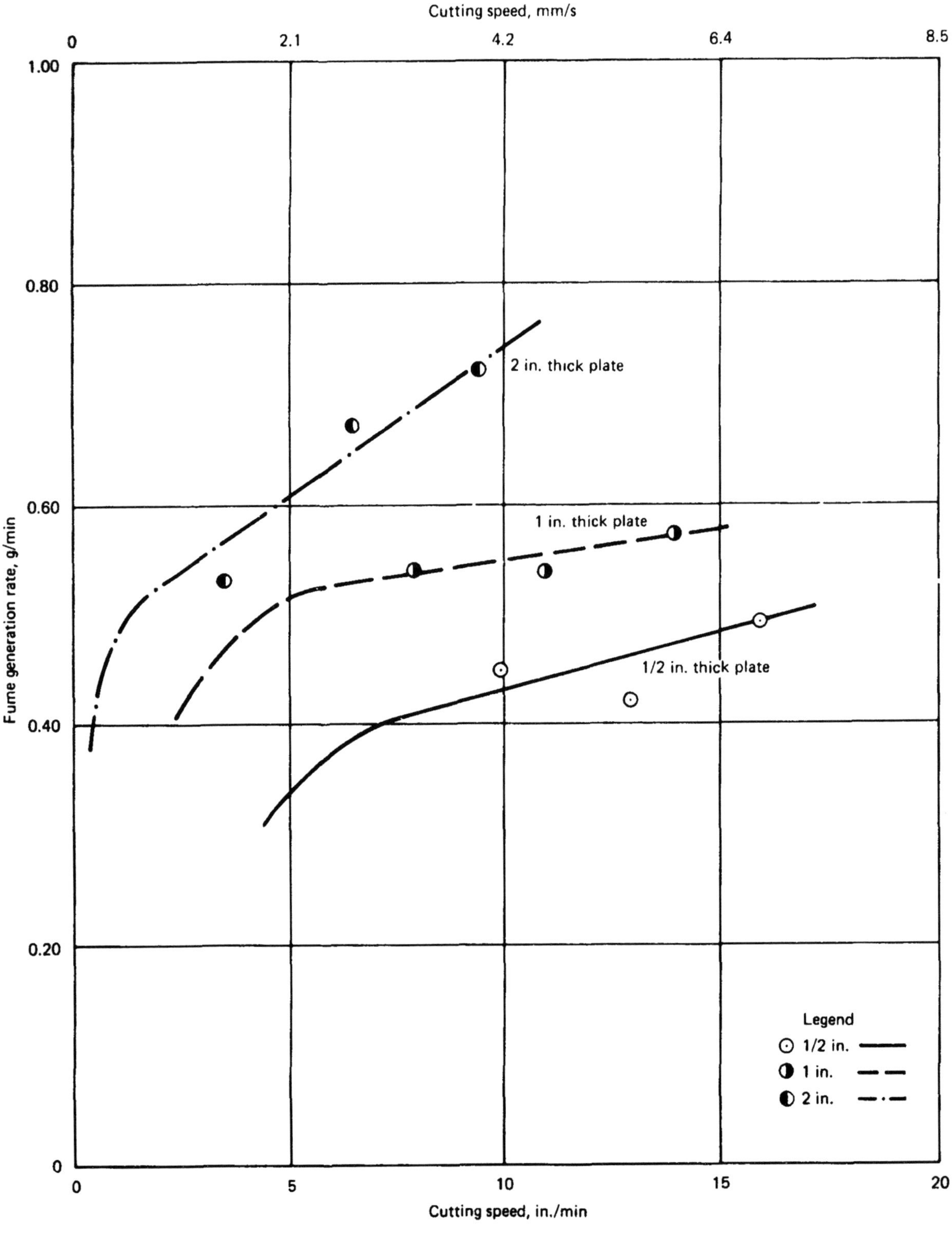

Fig. 5.1C—Fume generation as a function of cutting speed during oxyacetylene cutting of carbon steel plate at recommended oxygen and acetylene pressures

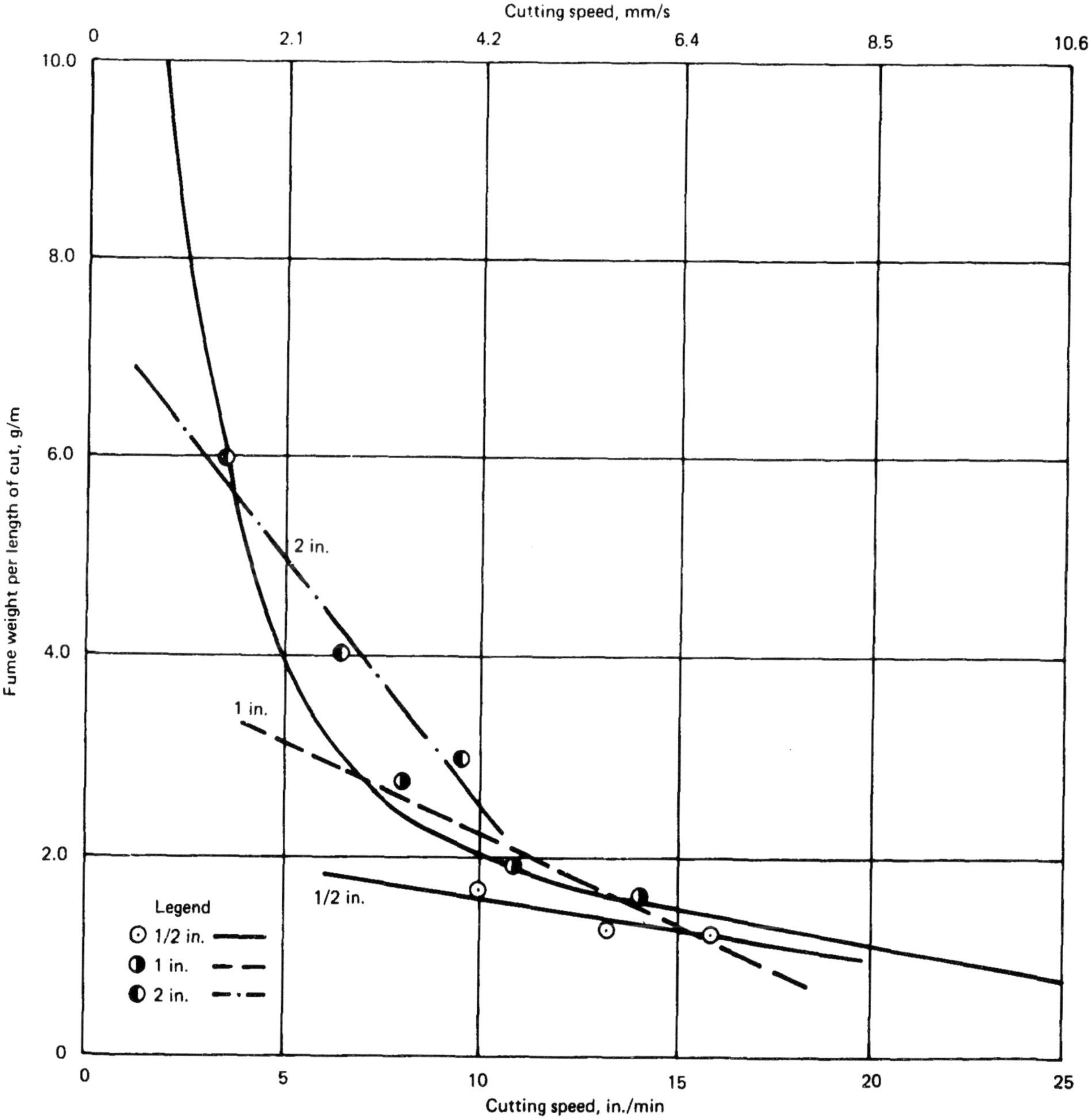

Fig. 5.1D – Weight of fumes per length of plate cut as a function of speed during oxyacetylene cutting at recommended oxygen and acetylene pressures

Table 5.2
Effect of variation in oxymethane cutting conditions on average fume generation characteristics

Operating conditions[1]	Oxygen pressure		Methane pressure		Plate thickness		Cutting speed		Fume generation rate, g/min	Weight of fumes/ length of plate cut	
	kPa	psi	kPa	psi	mm	in.	mm/s	in./min		g/m	g/in.
Normal	517.1	75	6.9	1	12.7	0.50	7.79	18.4	0.60	1.28	0.033
High gas pressures	689.4	100	6.9	1	12.7	0.50	7.86	18.6	0.68	1.47	0.037
Low gas pressures	344.7	50	6.9	1	12.7	0.50	6.95	16.4	0.30	0.71	0.018
High cutting speed	689.4	100	6.9	1	12.7	0.50	8.93	21.1	0.27	0.50	0.013
Low cutting speed	689.4	100	6.9	1	12.7	0.50	6.90	16.3	0.59	1.42	0.037
Normal	689.7	100	6.9	1	25.4	1.00	6.01	14.2	0.53	1.47	0.037
Low gas pressures	517.1	75	20.7	3	25.4	1.00	4.23	10.0	0.61	2.38	0.061
High cutting speed	689.4	100	6.9	1	25.4	1.00	8.66	20.5	0.35	0.67	0.017
Low cutting speed	689.4	100	6.9	1	25.4	1.00	2.89	6.8	0.36	2.08	0.056
Normal	723.9	105	6.9	1	50.8	2.00	5.28	12.5	0.54	1.70	0.043
Low gas pressures	517.1	35	27.6	4	50.8	2.00	1.48	3.5	0.14	1.57	0.040
Low cutting speed	723.9	105	6.9	1	50.8	2.00	4.78	11.3	0.41	1.43	0.036

1. Tip number (cutting orifice diameter): 1/2 in. (12.7 mm) thick plate: No. 00VVC (0.036 in. [0.9 mm]); 1 in. (25.4 mm) thick plate: No. 0-1/2 VVC (0.040 in. [1.0 mm]); 2 in. (50.8 mm) thick plate: No. 1 VVC (0.046 in. [1.2 mm]).

Section VB. Oxymethane Cutting

The results of studies to investigate the fumes produced during oxymethane cutting are summarized in Table 5.2; the data upon which this table is based are shown in Tables E4, E5, and E6, Appendix E. This process was examined in somewhat less detail than the oxyacetylene process. For example, the cutting of 1 and 2 in. thick plate at high oxygen pressures was not done, because it was decided that use of oxygen at substantially higher-than-recommended levels (100 psi for 1 in. thick plate and 105 psi for 2 in. thick plate) would not be relevant to the research objectives.

The following comments are based on an examination of the data presented in Table 5.2.

Cutting Under Normal Conditions

The rates at which fumes were produced during the oxymethane cutting of carbon steel with thicknesses of 1/2 to 2 in. ranged from 0.53 to 0.60 g/min. (Fig. 5.2). These rates are similar in magnitude to those observed during the oxyacetylene cutting studies. However, in oxyacetylene cutting the fume generation rates tended to increase with increasing plate thickness; in oxymethane cutting, these rates were almost constant. Extrapolation to zero plate thickness would again show fume production with no metal being cut. Additional research would be needed to substantiate these results.

When the thicker plates were being cut, fume weights as a function of unit length of cut (dashed curve) were considerably less than those determined for oxyacetylene cutting (Fig. 5.1).

Cutting at Various Pressures

Limited testing was done. All results have been plotted in Figs. 5.2A and 5.2B. When 1/2 in. thick plate was cut at the recommended speed using higher-than-recommended oxygen pressure, the fume generation rate increased. At that higher pressure, slightly higher speed dramatically reduced the fume rate, while slightly lower speed reproduced the rate obtained at the recommended speed and pressure. At lower oxygen pressure, the lower speed also reduced the fume rate. The curves suggested in Fig. 5.2A are drawn in the manner of the oxyacetylene curves of Fig. 5.1A.

The data for 1 in. thick plate show greater dependency on speed than on oxygen pressure. Judging from the effect of pressure on cuts in 1/2 in. plate, the one low pressure reading for 1 in. plate seems questionably high.

The three data points for 2 in. plate seem to fit a line drawn through the origin (see Fig. 5.2A).

The ratio of fume weight to length of cut for oxymethane cutting (Fig. 5.2B) presents a picture similar to the rate data. Low oxygen pressure again produced a lower ratio of fumes to length except for 1 in. thick plate.

Cutting at Various Speeds

Contrary to the case of oxyacetylene cutting, half of the fume generation rates obtained for oxymethane cutting at various speeds (Fig. 5.2C) fell in a scatter-band directly proportional to speed. The exceptions were two experiments at high speed (21 in./min), two experiments at low oxygen pressure (50 psi in 1/2 in. thick plate and 75 psi in 1 in. plate), and the "normal" data point for 1/2 in. plate. If the accuracy of those points is not questioned, curves must be drawn showing a maximum rate of fume generation at intermediate speeds to explain the variations found at the same oxygen pressures.

The ratio graph (Fig. 5.2D) showing fume weight per length of cut as a function of speed shows an overriding benefit of high speed cutting. In two of the three pairs of experiments comparing low and high oxygen pressures at the same cutting speed, the use of low pressure was beneficial in reducing the fume ratio. Nevertheless, these results should be confirmed by further studies before recommendations are made.

Discussion

Fume Composition

To determine the composition of the fumes produced during oxyacetylene and oxymethane cutting operations, fume samples were collected on cellulose membrane filters and analyzed by optical spectroscopy. This semiquantitative analytical technique provides a means for obtaining a quick overview of a wide spectrum of the elements in fume samples; however, it is more accurate in detecting the presence of elements in small amounts than in large amounts. If more accurate compositional data are needed, optical emission spectroscopy should be supplemented by other analytical techniques such as atomic absorption analyses or wet chemistry methods.

In this instance, the content of iron in oxygen cutting fumes was lower than expected when the fume samples were analyzed by optical emission spectroscopy. Subsequent analysis for total iron by wet chemistry produced iron concentrations in the expected range.[12]

In Table 5.3, the sum of the major constituents (iron, manganese, and silicon) approaches 100 percent, if it is assumed that these elements are converted to oxides.

An examination of the data presented in Table 5.3 indicates that there was little or no difference in the elemental composition of the fumes associated with oxyacetylene and oxymethane. These results were anticipated since both cutting studies were conducted with the same type of steel plates.

12. Wet chemistry analyses for total iron were performed by a program sponsor

Table 5.3
Composition of fumes produced during oxygen cutting of carbon steel plate with acetylene and methane fuel gases by optical emission spectroscopy

Element	Concentration according to cutting method, weight percent	
	Oxyacetylene cutting	Oxymethane cutting
Ba	0.01	0.01
Si	0.1	0.05
Mn	0.3	0.3
Fe	69.28*	67.22*
Mg	<0.01	<0.01
Cr	0.02	0.02
Sn	0.01	0.01
Ga	0.01	0.01
Ni	0.05	0.05
Al	0.02	0.02
Mo	0.03	0.04
Ca	0.02	0.02
V	<0.01	<0.01
Cu	0.1	0.1
Zn	<0.1	<0.1
Co	0.02	0.02
Zr	0.02	0.02
K	<0.1	<0.1

*Total iron was determined by wet chemical analysis.

Fume Particles

Limited examinations of fume deposits were made with the light microscope to determine if cutting fume particles were significantly larger than those produced during arc welding operations. A few spherical particles of metal that had solidified from the molten state were observed; particles with diameters up to 0.00063 in. (16 μm) were noted. However, the bulk of the deposit was in the form of agglomerated particles, as in the case of welding fumes.

Importance of Fume Generation Rates

Fume generation rates are important because they provide a measure of the amount of fumes that will be produced when plate is cut under specified conditions. They also provide an indication of the fume concentrations that will exist in an area after an interval of cutting. For example, assume that iron oxide fumes are produced at the rate of 0.60 g/min during oxygen cutting of carbon steel plate in an unventilated room with a volume of 10,000 ft^3. (283 m^3). Assuming further that the fumes diffuse uniformly throughout the room, the fume concentration in the room will exceed the threshold limit value (TLV) for iron oxide of 5 mg/m^3 [13] (as defined by the American Conference of Governmental Industrial Hygienists) in about 2.5 minutes. Ventilation provided at the rate of about 120 m^3/min should alleviate this problem.

As in the case of arc welding, there are many available methods for minimizing the potential hazards of oxygen cutting. Hoods and down-draft tables provide effective means to remove cutting fumes from the atmosphere. Water tables are particularly useful with automatic cutting equipment that incorporates a multiplicity of torches.

13. Note that the value accepted by OSHA for iron oxide is 10 mg/m^3. See General Industry Safety and Health Regulations, Title 29 CFR, Part 1910.1000

Summary of Part V

Limited research was conducted to investigate the fumes associated with oxygen cutting operations. In this work, fumes were sampled during the oxyacetylene and oxymethane cutting of carbon steel plates with thicknesses of 1/2, 1, and 2 in. (12.7, 25.4, and 50.8 mm). Cutting was done at recommended conditions to obtain data that were used to calculate base-line fume generation rates and amounts of fume as a function of the length of the cut. Cutting was also done using other-than-recommended oxygen and fuel gas pressures and cutting speeds to study the effect of these process variables on fume generation characteristics.

Depending on the thickness of metal being cut, the fume generation rates for oxyacetylene and oxymethane cutting were similar in magnitude when recommended cutting conditions were used; they ranged from 0.42 to 0.67 g/min for oxyacetylene cutting and 0.53 to 0.60 g/min for oxymethane cutting. Such rates are comparable to those encountered in many arc welding operations studied in the laboratory investigation of welding fumes.

Fume generation rates increased somewhat over base-line levels when cutting was done at higher-than-recommended gas pressures or speeds. However, this may not constitute an additional problem, since cutting conditions that deviate substantially from recommended levels frequently result in unacceptable cuts.

Ratios of fume weight to unit length of cut were determined for various cutting applications during this program. If the amount of cutting to be done is known, this factor can be used to estimate the quantity of fumes that will be produced in completing specific tasks.

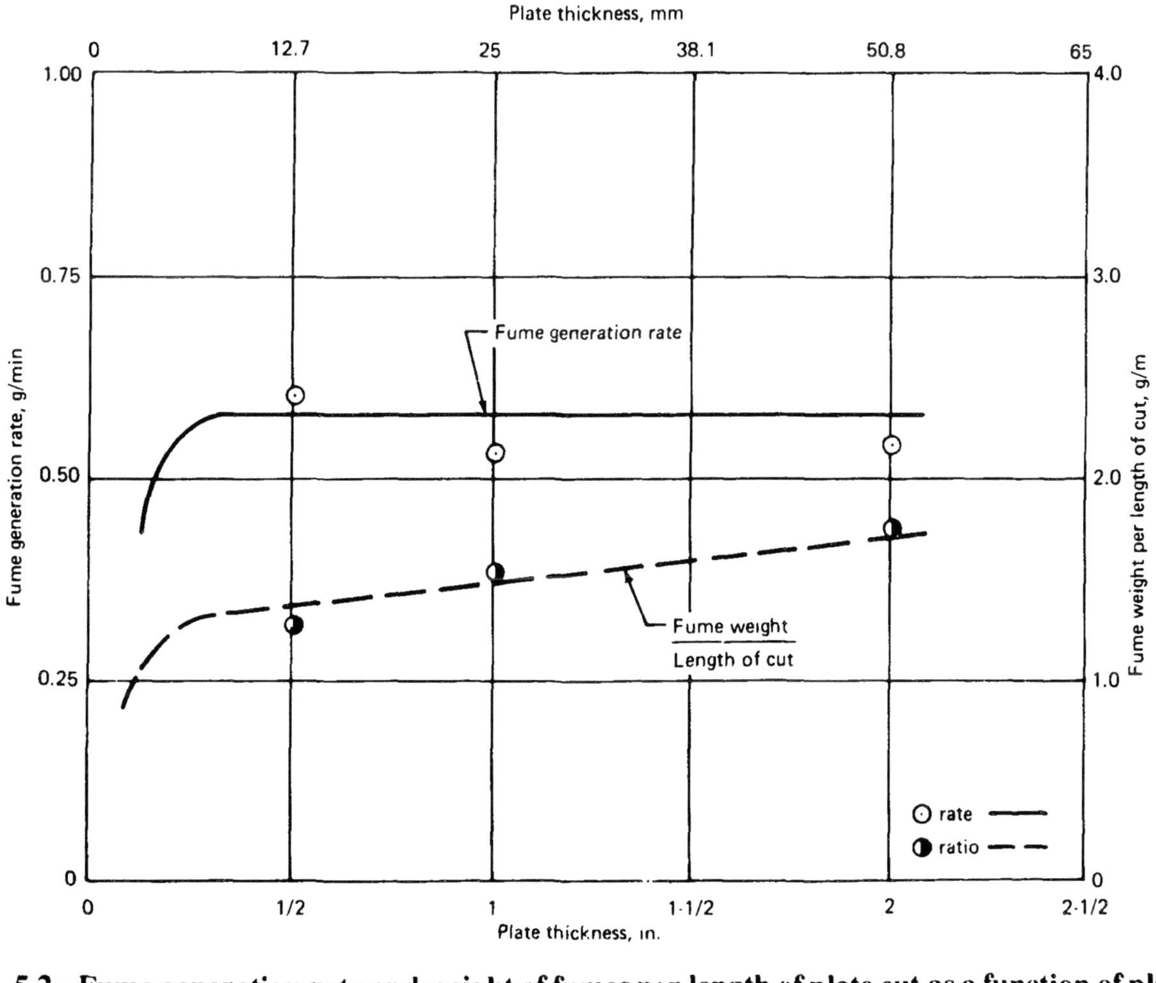

Fig. 5.2—Fume generation rate and weight of fumes per length of plate cut as a function of plate thickness during oxymethane cutting

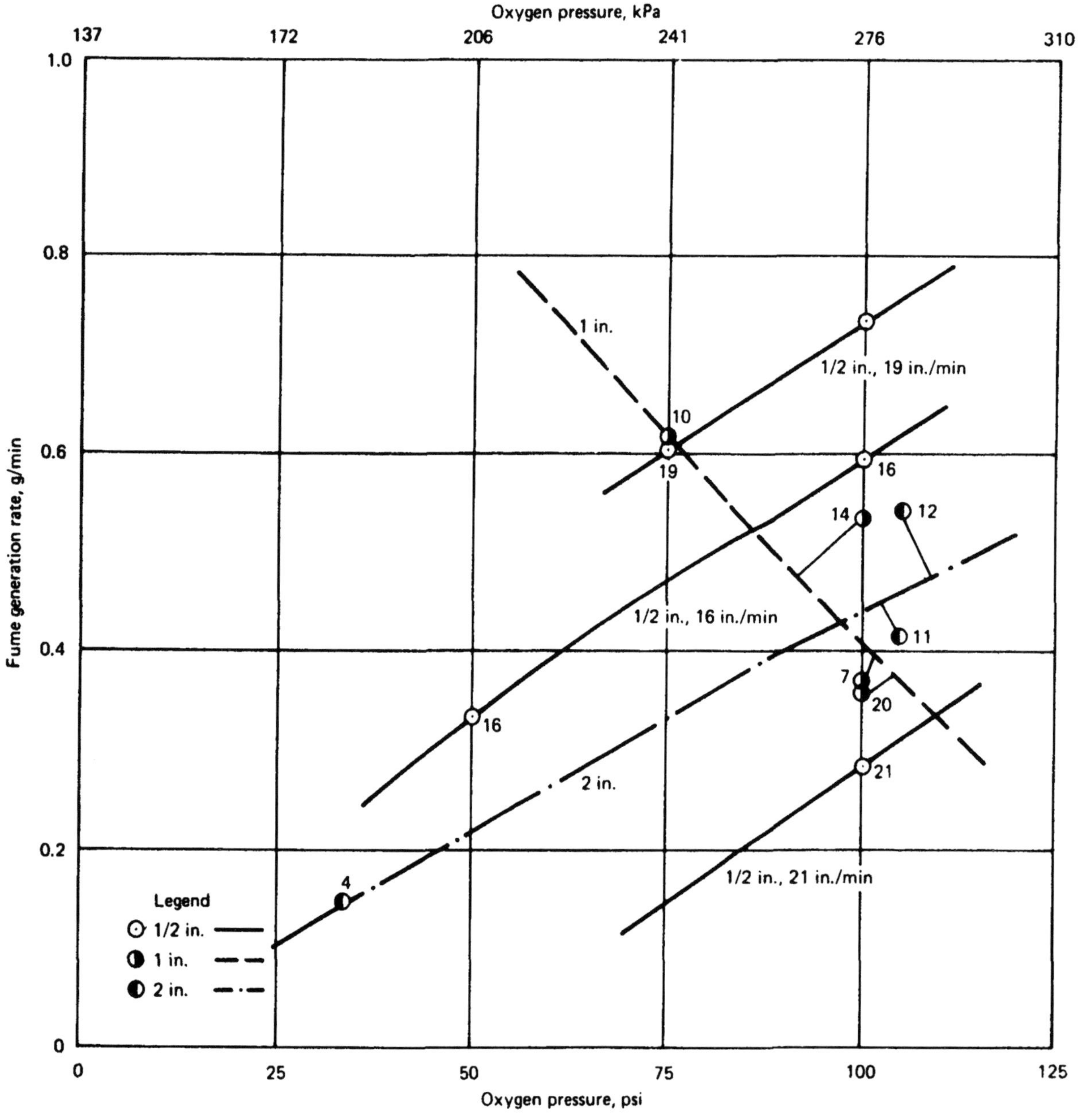

Fig. 5.2A—Fume generation rate as a function of gas pressure during oxymethane cutting at various speeds (in./min, beside data points)

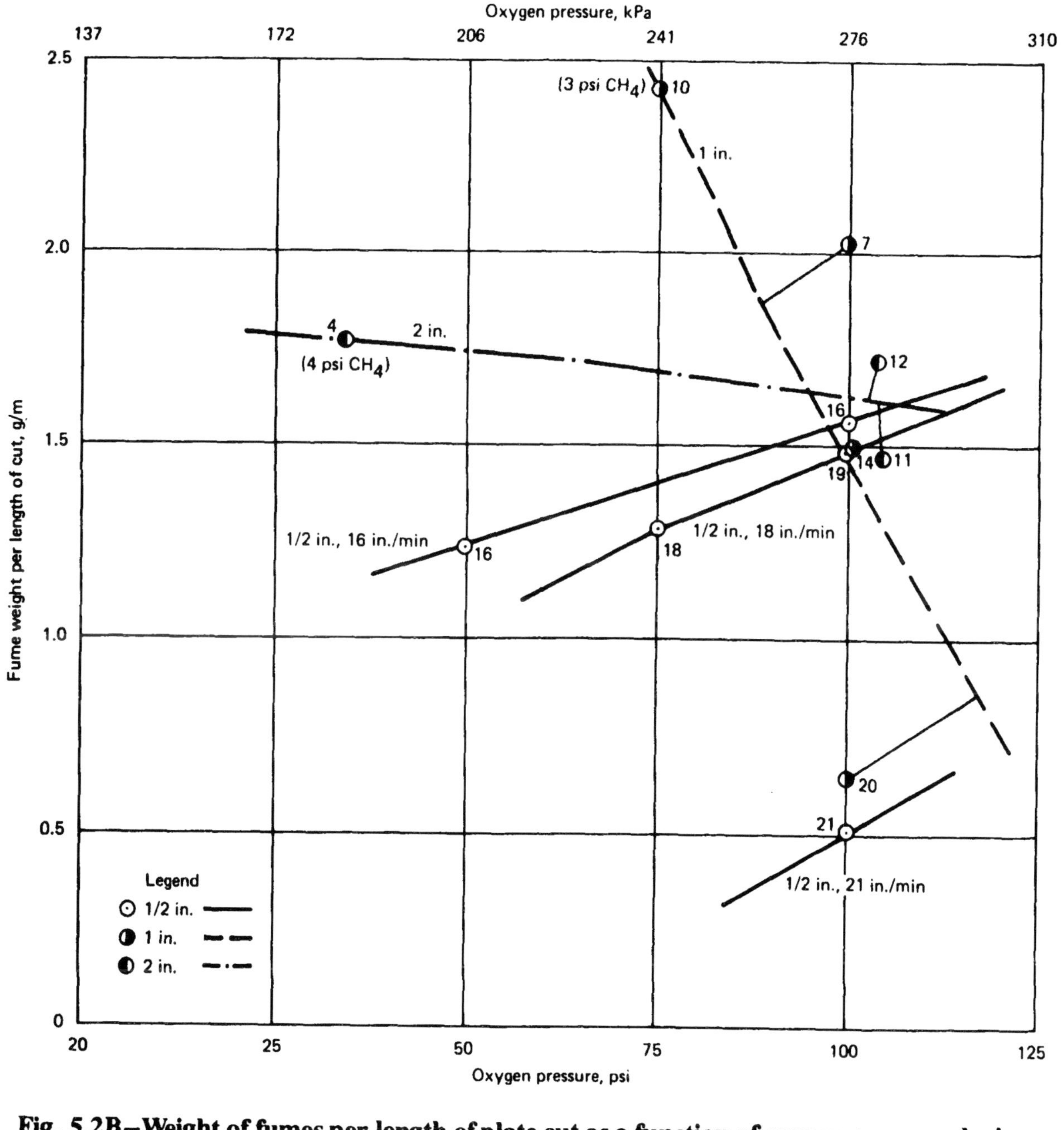

Fig. 5.2B–Weight of fumes per length of plate cut as a function of oxygen pressure during oxymethane cutting at various speeds (methane pressure 1 psi [6.9 kPa] except as noted; in./min, beside data points)

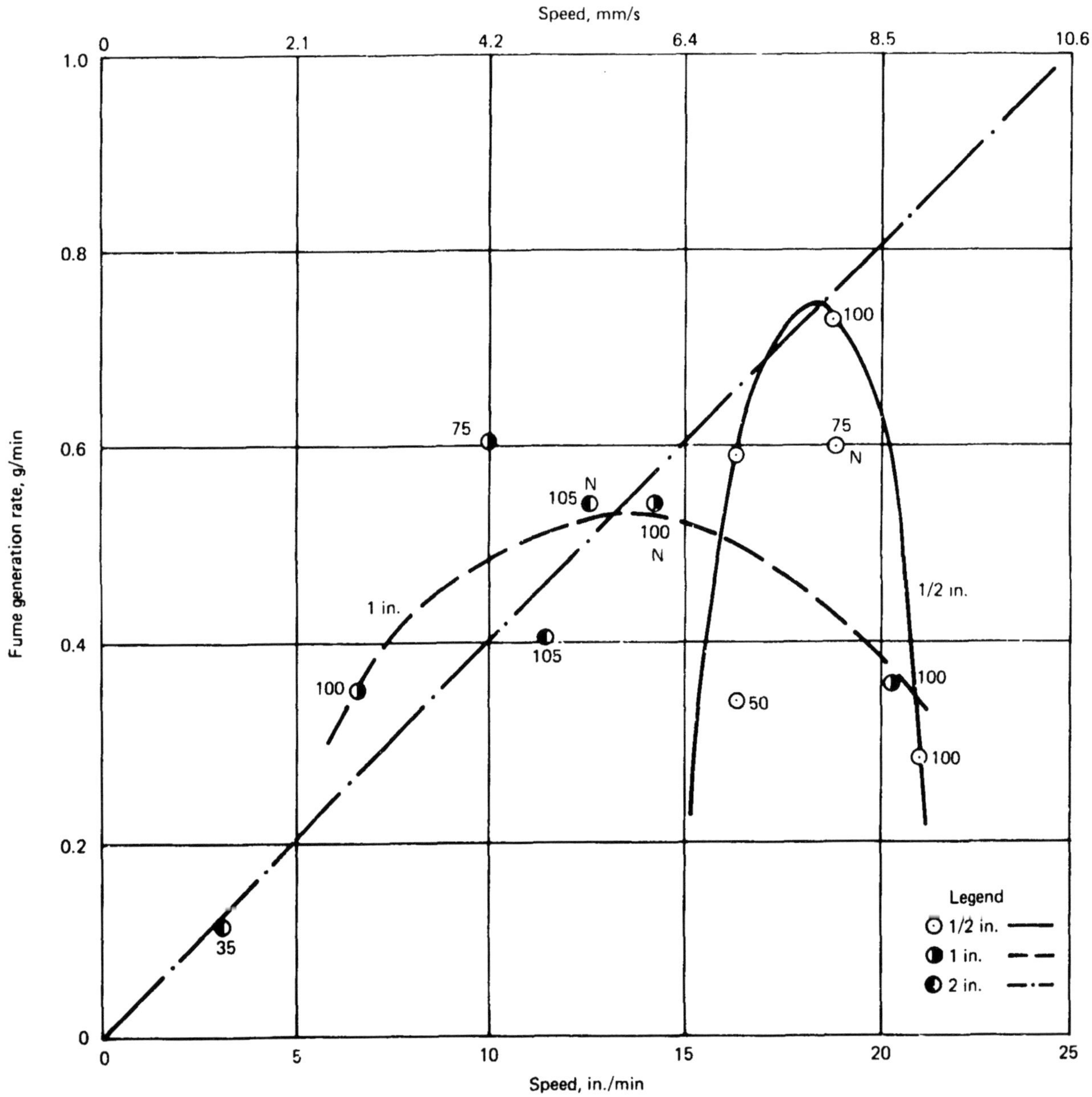

Fig. 5.2C–Fume generation rate as a function of speed during oxymethane cutting at various oxygen pressures (psi, beside data points)

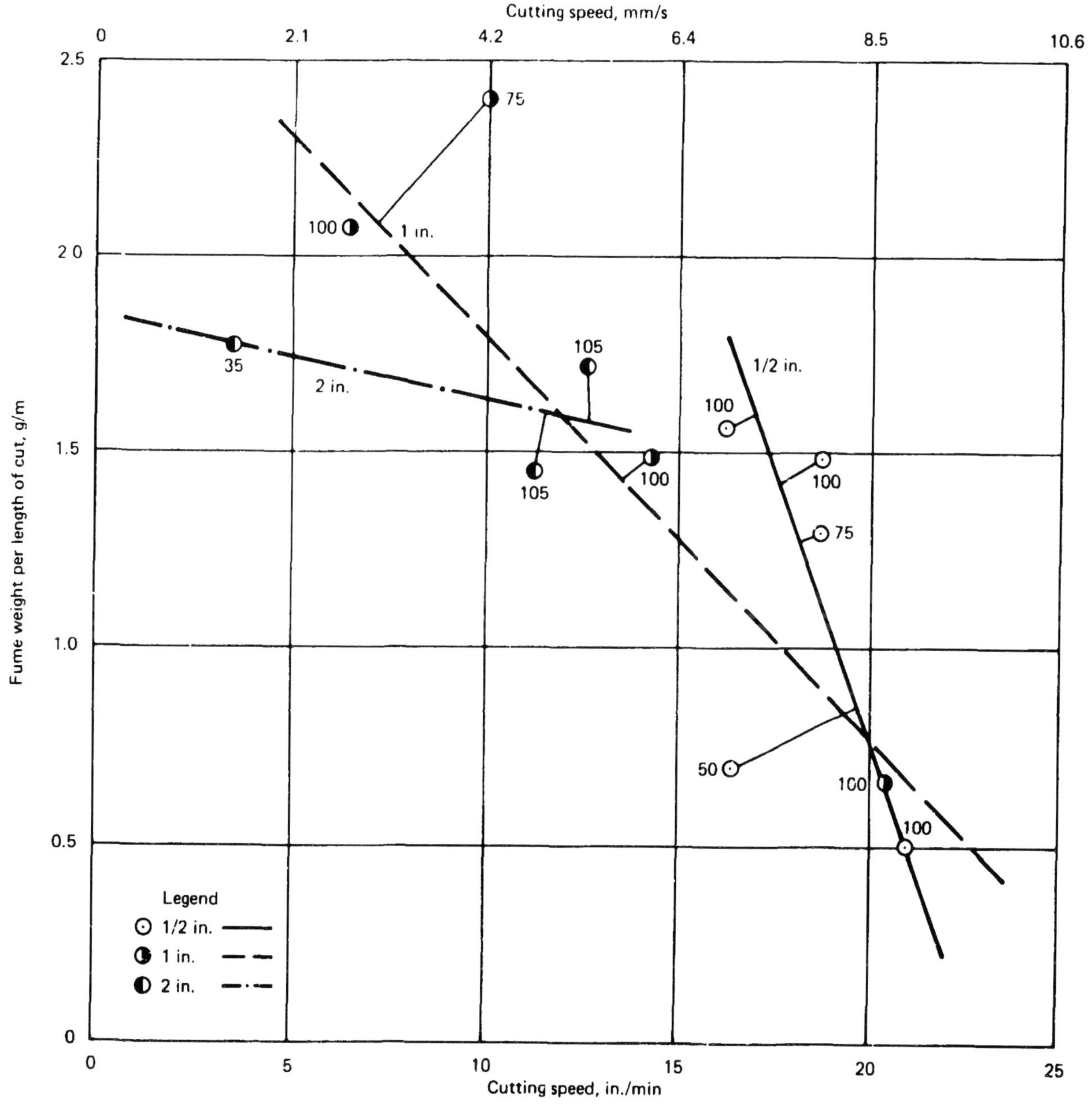

Fig. 5.2D—Weight of fumes per length of plate cut as a function of cutting speed during oxymethane cutting at various oxygen pressures (psi, beside data points)

Appendix A: Table A1
(Electrodes Used in Part II)

Table A1
List of electrodes investigated during program

Electrode classification	Electrode no.[1]	Electrode type	Electrode diam., mm	Electrode diam., in.	Welding current, A (nominal)	Purpose	Table
					Covered electrodes		
E6010 (IP)	1	Carbon steel	4	5/32	145-150	Fume generation characteristics; current density, diameter, arc length, and electrode angle effects	B1, B33, B36, and B39
E6010 (IP)	2	Carbon steel	4	5/32	145-150	Fume generation characteristics	B1
E6010 (IP)	11	Carbon steel	4	5/32	155-160	Fume generation characteristics, current and voltage effects	B1, B27, and B35
E6010 (IP)	1	Carbon steel	2.4	3/32	55	Current density and diameter effects	B33
E6010 (IP)	1	Carbon steel	3.2	1/8	85-90	Current density and diameter effects	B33
E6010 (IP)	1	Carbon steel	4.8	3/16	180-190	Current density and diameter effects	B33
E6013	3	Carbon steel	4	5/32	160	Fume generation characteristics; arc length, and electrode angle effects	B2, B36, and B39
E6013	4	Carbon steel	4	5/32	150-155	Fume generation characteristics	B2
E6013	6	Carbon steel	4	5/32	145-150	Fume generation characteristics	B2
E6013	4	Carbon steel	3.2	1/8	155	Current density, polarity, and diameter effects	B30 and B31
E6013	4	Carbon steel	4.8	3/16	140-150	Current density, polarity, and diameter effects	B30 and B31
E6013	4	Carbon steel	2.4	3/32	50-53	Current density and diameter effects	B34
E6013	4	Carbon steel	3.2	1/8	83	Current density and diameter effects	B34
E6013	4	Carbon steel	4	5/32	135-150	Current density and diameter effects	B34
E6013	4	Carbon steel	4.8	3/16	200-205	Current density and diameter effects	B34
E6013	34	Carbon steel	4	5/32	160-170	Fume generation characteristics – iron powder additions	B37
E7014	30	Carbon steel	4	5/32	170-180	Fume generation characteristics – iron powder additions	B37
E7016	29	Carbon steel	4	5/32	170-175	Fume generation characteristics – iron powder additions	B38
E7018	5	Carbon steel	4	5/32	175-180	Fume generation characteristics	B3
E7018	10	Carbon steel	4	5/32	165-175	Fume generation characteristics; arc length and electrode angle	B3, B36, and B39
E7018	12	Carbon steel	4	5/32	170-175	Fume generation characteristics	B3
E7018	10	Carbon steel	4	5/32	130-210	Current effects	B26
E7018	10	Carbon steel	2.4	3/32	60-65	Current density and diameter effects	B32
E7018	10	Carbon steel	3.2	1/8	100	Current density and diameter effects	B32
E7018	10	Carbon steel	4	5/32	155-160	Current density and diameter effects	B32
E7018	10	Carbon steel	4.8	3/16	220-230	Current density and diameter effects	B32
E7018	36	Carbon steel	4	5/32	175-180	Fume generation characteristics – iron powder effects	B38
E7018	72	Carbon steel	6.4	1/4	390-400	Fume generation characteristics at high current	B40

Table A1 (cont.)
List of electrodes investigated during program

Electrode classification	Electrode no.[1]	Electrode type	Electrode diam., mm	Electrode diam., in.	Welding current, A (nominal)	Purpose	Table
E7024	7	Carbon steel	4	5/32	200-220	Fume generation characteristics	B4
E7024	8	Carbon steel	4	5/32	205-230	Fume generation characteristics	B4
E7024	9	Carbon steel	4	5/32	205-215	Fume generation characteristics	B4
E7024	35	Carbon steel	4	5/32	205-220	Fume generation characteristics – iron powder additions	B37
E7028	32	Carbon steel	4	5/32	225-240	Fume generation characteristics – iron powder additions	B38
E7028	73	Carbon steel	6.4	1/4	385-395	Fume generation characteristics at high currents	B40
E8018 C3	13	Low alloy steel	4	5/32	155-160	Fume generation characteristics	B5
E8018 C3	14	Low alloy steel	4	5/32	175-180	Fume generation characteristics	B5
E8018 C3	15	Low alloy steel	4	5/32	165	Fume generation characteristics	B5
E9018 B3	18	Low alloy steel	4	5/32	165	Fume generation characteristics	B6
E9018 B3	17	Low alloy steel	4	5/32	175-180	Fume generation characteristics	B6
E9018 B3	16	Low alloy steel	4	5/32	160-165	Fume generation characteristics	B6
E316-15	22	Stainless steel	4	5/32	148	Fume generation characteristics	B7
E316-15	20	Stainless steel	4	5/32	155	Fume generation characteristics	B7
E316-15	19	Stainless steel	4	5/32	150	Fume generation characteristics	B7
E316-16	26	Stainless steel	4	5/32		Fume generation characteristics	B8
E316-16	20	Stainless steel	4	5/32		Fume generation characteristics	B8
E316-16	25	Stainless steel	4	5/32		Fume generation characteristics	B8
E410-16	21	Stainless steel	4	5/32	145	Fume generation characteristics	B9
E410-16	23	Stainless steel	4	5/32	160	Fume generation characteristics	B9
E410-16	24	Stainless steel	4	5/32	160	Fume generation characteristics	B9
ENi-CI	28	Nickel-base	4	5/32	135	Fume generation characteristics	B10
ENiCU-2	33	Nickel-base	4	5/32	145	Fume generation characteristics	B10
Inconel 625	62	Nickel-base	4	5/32	140-155	Fume generation characteristics	B10
Haynes C-276	69	Nickel-base	4	5/32	130-135	Fume generation characteristics	B10
Haynes 25	70	Cobalt-base	4	5/32	135-140	Fume generation characteristics	B10

Table A1 (cont.)
List of electrodes investigated during program

Electrode classification	Electrode no.[1]	Electrode type	Electrode diam., mm	Electrode diam., in.	Welding current, A (nominal)	Purpose	Table
					Flux cored electrodes		
E70T-1	40	Carbon steel	2.4	3/32	470-485	Fume generation characteristics; current density and diameter effects	B11 and B41
E70T-1	42	Carbon steel	2.4	3/32	435-445	Fume generation characteristics	B11
E70T-1	41	Carbon steel	2.4	3/32	455-465	Fume generation characteristics	B11
E70T-1	46	Carbon steel	2.4	3/32	435-445	Fume generation characteristics	B11
E70T-1	47	Carbon steel	2.4	3/32	435	Fume generation characteristics	B11
E70T-1	41	Carbon steel	2.4	3/32	345-600	Current and voltage effects	B28
E70T-1	45	Carbon steel	1.6	1/16	325-345	Current density and diameter effects	B41
E70T-1	40	Carbon steel	2.4	3/32	445-460	Shielding gas effects	B25
E70T-1	42	Carbon steel	2.4	3/32	445	Shielding gas effects	B25
E70T-4	48	Carbon steel	2.4	3/32	385-390	Fume generation characteristics	B12
E70T-4	56	Carbon steel	2.4	3/32	370-390	Fume generation characteristics	B12
E70T-4	49	Carbon steel	2.4	3/32	400-410	Fume generation characteristics	B12
E70T-5	44	Carbon steel	2.4	3/32	425-445	Fume generation characteristics	B13
E70T-5	50	Carbon steel	2.4	3/32	440-450	Fume generation characteristics	B13
81-C3	51	Low alloy steel	2.4	3/32	440-445	Fume generation characteristics	B14
91-B3	52	Low alloy steel	2.4	3/32	450	Fume generation characteristics	B14
308L	53	Stainless steel	2.4	3/32	380-400	Fume generation characteristics	B15
316L	55	Stainless steel	2.4	3/32	340-390	Fume generation characteristics	B15
316L	67	Stainless steel	2.4	3/32	390-405	Fume generation characteristics	B15

Table A1 (cont.)
List of electrodes investigated during program

Electrode classification	Electrode no.[1]	Electrode type	Electrode diam., mm	Electrode diam., in.	Welding current, A (nominal)	Purpose	Table
					Solid electrodes		
E70S-3	54	Carbon steel	1.1	0.045	Note 2	Fume generation characteristics – shielding gas and metal transfer effects	B16
E70S-3	57	Carbon steel	1.1	0.045	Note 2	Fume generation characteristics – shielding gas and metal transfer effects	B17
E70S-3	58	Carbon steel	1.1	0.045	Note 2	Fume generation characteristics – shielding gas, metal transfer, and current effects	B18 and B29
E70S-5	59	Carbon steel	1.1	0.045	Note 2	Fume generation characteristics – shielding gas and metal transfer effects	B19
ER316	65	Stainless steel	0.9	0.035	165-175	Fume generation characteristics	B20
Inconel 625	63	Nickel-base	1.1	0.045	190-195	Fume generation characteristics	B21
ERNiCu-7	64	Nickel-base	1.1	0.045	250-260	Fume generation characteristics	B21
Haynes C-276	68	Nickel-base	0.9	0.035	165	Fume generation characteristics	B21
Haynes 25	71	Cobalt-base	1.1	0.045	200-205	Fume generation characteristics	B21
ER4043	61	Aluminum-base	1.2	3/64	160-165	Fume generation characteristics	B22
ER5356	60	Aluminum-base	1.2	3/64	150-165	Fume generation characteristics	B23
ERCu	76	Copper-base	1.1	0.045	205-210	Fume generation characteristics	B24
ERCu Al-A2	75	Copper-base	1.1	0.045	210-215	Fume generation characteristics	B24

1. The electrode number identifies the specific electrode and its manufacturer. One manufacturer may be represented by more than one electrode number.
2. The welding current is dependent upon the shielding gas and metal transfer mode (see appropriate tables for current levels).

Appendix B: Tables B1 to B43 (Tabular Data on Fume Generation Characteristics of Electrodes Used in Part II)

Table B1
Fume generation characteristics of 3.97 mm (5/32 in.) diameter E6010 (IP) covered electrodes

Electrode	Code	Voltage, V avg	Current, A avg	Fume generation rate: Measured g/min	Fume generation rate: Normalized g/min/I^{2}x10^{-5}	Wgt. of fume/wgt. of electrode used x 100	Melting rate kg/h	Melting rate lb/h	Deposition rate kg/h	Deposition rate lb/h	Wgt. of fume/wgt. of deposited metal g/kg*	Weight of fume/length of electrode used g/m	Weight of fume/length of electrode used g/in.
E6010 IP	1	31	145	1.08	4.82	3.74	1.74	3.83	---	---	---	4.14	0.11
E6010 IP	1	32	150	1.07	4.77	3.65	1.77	3.89	---	---	---	4.10	0.10
E6010 IP	1	31	150	1.09	5.13	3.70	1.77	3.90	---	---	---	4.14	0.11
E6010 IP	1	30	150	1.08	4.82	3.51	1.85	4.08	1.38	3.04	47.05	3.88	0.10
Avg				1.08	4.89	3.65	1.78	3.93				4.07	0.11
SD				>0.01	0.17	0.10	0.05	0.11				0.12	0.01
E6010 IP	2	28	150	0.59	2.62	2.13	1.66	3.65	---	---	---	2.76	0.07
E6010 IP	2	29	148	0.58	2.63	2.19	1.58	3.45	---	---	---	2.36	0.06
E6010 IP	2	29	150	0.57	2.47	2.13	1.61	3.55	---	---	---	2.36	0.06
E6010 IP	2	28	150	0.58	2.59	2.23	1.86	4.09	1.47	3.24	23.77	2.06	0.05
Avg				0.58	2.58	2.19	1.68	3.69				2.39	0.06
SD				>0.01	0.07	0.04	0.13	0.28				0.29	0.01
E6010 IP	11	27	160	1.19	4.97	3.65	1.96	4.32	---	---	---	4.17	0.11
E6010 IP	11	28	155	1.20	4.70	3.68	1.96	4.32	---	---	---	4.17	0.11
E6010 IP	11	28	155	1.20	4.68	3.67	1.96	4.32	---	---	---	4.17	0.11
E6010 IP	11	28	160	1.20	4.99	5.79	1.90	4.18	1.32	2.91	54.36	4.17	0.11
Avg				1.20	4.84	3.70	1.95	4.29				4.17	0.11
SD				>0.01	0.17	0.06	0.03	0.07				0.00	0.00
E6010	1	27	145	0.45	2.15	1.60	1.70	3.73	1.29	2.84	20.08	1.72	0.04
E6010	1	27	145	0.46	2.20	1.64	1.70	3.73	1.30	2.85	21.44	1.77	0.04
E6010	1	26	145	0.43	2.06	1.46	1.78	3.92	1.50	3.30	17.32	1.63	0.04
E6010	1	26	145	0.44	2.09	1.54	1.72	3.79	1.52	3.34	17.39	1.66	0.04
E6010	1	26	145	0.44	2.09	1.48	1.79	3.94	1.51	3.32	17.49	1.63	0.04
Avg				0.45	2.12	1.54	1.74	3.82	1.42	3.13	18.92	1.68	0.04
SD				0.01	0.06	0.08	0.05	0.10	0.12	0.26	2.09	0.06	<0.01
E6010 IP	11	23.5	150	0.83	3.69	2.74	1.83	4.00	1.43	3.15	34.78	3.17	0.08
E6010 IP	11	23.0	150	0.83	3.69	2.70	1.84	4.06	1.42	3.13	35.00	3.13	0.08
E6010 IP	11	23.0	150	0.83	3.69	2.68	1.86	4.10	1.43	3.15	34.78	3.09	0.08
Avg				0.83	3.69	2.71	1.84	4.06	1.43	3.14	35.85	3.13	0.08
SD				>0.01	>0.01	0.03	0.02	4.06	>0.01	0.01	0.13	0.04	<0.01

*g/kg $\times$ 10 = weight of fume expressed as a percentage of the deposit weight

Notes

Electrode polarity: positive

Welding speed: Code 1, 2, and 11 electrodes. 4.02 mm/s (9.5 in./min)

Electrode weight: Code 1 electrode: 111.02 g/m (2.82 g/in.)
Code 2 electrode: 108.27 g/m (2.75 g/in.)
Code 11 electrode: 114.57 g/m (2.91 g/in.)

Metal deposition rate determined for only 1 test per electrode.

SD: standard deviation (sigma)

Table B2
Fume generation characteristics of 3.97 mm (5/32 in.) diameter E6013 covered electrodes

Electrode	Code	Voltage, V avg	Current, A avg		Fume generation rate: Measured g/min	Fume generation rate: Normalized g min⁻¹ I⁻² ×10⁻⁵	Wgt of fume/wgt. of electrode used ×100	Melting rate kg/h	Melting rate lb/h	Deposition rate kg/h	Deposition rate lb/h	Wgt of fume/wgt. of deposited metal g/kg*	Weight of fume/length of electrode used g/m	Weight of fume/length of electrode used g/in.
E6013	3	24	160		0.58	2.26	1.81	1.92	4.22	1.32	2.91	26.33	2.17	0.06
E6013	3	25	160		0.58	2.27	1.81	1.92	4.23	1.37	3.01	25.44	2.20	0.06
E6013	3	25	160		0.58	2.27	1.77	1.97	4.33	1.37	3.01	25.48	2.18	0.06
				Avg	0.58	2.27	1.80	1.94	4.26	1.35	2.98	25.75	2.19	0.06
				SD	<0.01	<0.01	0.02	0.03	0.06	0.03	0.06	0.51	0.02	0.00
E6013	4	20	150		0.46	2.03	1.63	1.68	3.69	1.18	2.59	23.27	1.94	0.05
E6013	4	23	150		0.49	2.16	1.70	1.71	3.77	1.22	2.68	23.89	2.01	0.05
E6013	4	20	155		0.49	2.06	1.63	1.83	4.02	1.32	2.91	22.48	1.93	0.05
				Avg	0.48	2.08	1.65	1.74	3.83	1.24	2.73	23.21	1.96	0.05
				SD	0.02	0.07	0.04	0.08	0.17	0.07	0.16	0.71	0.05	0.00
E6013	6	---	152		0.32	1.39	1.03	1.87	4.12	1.36	2.99	14.25	1.34	0.03
E6013	6	22	147		0.31	1.41	0.99	1.86	4.10	1.34	2.95	13.77	1.26	0.03
E6013	6	24	150		0.31	1.40	1.01	1.86	4.10	1.30	2.87	14.45	1.31	0.03
				Avg	0.31	1.40	1.01	1.86	4.10	1.33	2.93	14.16	1.31	0.03
				SD	>0.01	<0.01	0.02	0.01	0.02	0.03	0.06	0.35	0.04	0.00

*g/kg − 1% = weight of fume expressed as a percentage of the deposit weight

Notes:

Electrode polarity: positive

Welding speed: Code 3, 4, and 6 electrodes: 3.81 mm/s (9 in./min)

Electrode weight: Code 3 electrode: 122.05 g/m (3.10 g/in.)
Code 4 electrode: 117.32 g/m (2.98 g/in.)
Code 6 electrode: 133.86 g/m (3.40 g/in.

Metal deposition rate determined for only 1 test per electrode

SD: standard deviation (sigma)

Table B3
Fume generation characteristics of 3.97 mm (5/32 in.) diameter E7018 covered electrodes

Electrode	Code	Voltage, V avg	Current, A avg		Fume generation rate: Measured g/min	Fume generation rate: Normalized g/min/I^{2}x10^{-5}		Wgt. of fume/wgt. of electrode used x100		Melting rate kg/h	Melting rate lb/h	Deposition rate kg/h	Deposition rate lb/h	Wgt. of fume/wgt of deposited metal g/kg*		Weight of fume/length of electrode used g/m	Weight of fume/length of electrode used g/in.
E7018	5	24	180		0.57	1.78		1.64		2.12	4.66	---	---	---		2.43	0.06
E7018	5	23	180		0.56	1.74		1.51		2.25	4.95	---	---	---		2.39	0.06
E7018	5	24	175		0.58	1.90		1.64		2.12	4.67	---	---	---		2.45	0.06
E7018	5	23	180		0.58	1.78		1.60		2.16	4.78	1.68	3.70	20.59		2.44	0.06
				Avg	0.57	1.80	Avg	1.60	Avg	2.17	4.77				Avg	2.43	0.06
				SD	>0.01	0.07	SD	0.06	SD	0.06	0.13				SD	0.03	0.00
E7018	10	23	165		0.59	2.16		1.57		2.25	4.96	---	---	---		2.31	0.06
E7018	10	23	170		0.58	2.00		1.57		2.21	4.87	---	---	---		2.30	0.06
E7018	10	22	175		0.58	1.90		1.53		2.28	5.02	---	---	---		2.29	0.06
E7018	10	23	175		0.58	1.90		1.49		2.35	5.16	1.71	3.77	20.35		2.25	0.06
				Avg	0.58	1.99	Avg	1.54	Avg	2.27	5.00				Avg	2.28	0.06
				SD	>0.01	0.12	SD	0.04	SD	0.06	0.12				SD	0.03	0.00
E7018	12	23	170		0.60	2.07		1.63		2.21	4.85	---	---	---		2.46	0.06
E7018	12	24	170		0.60	2.09		1.64		2.21	4.85	---	---	---		2.49	0.06
E7018	12	23	175		0.61	1.98		1.65		2.21	4.87	---	---	---		2.48	0.06
E7018	12	23	170		0.60	2.07		1.62		2.21	4.87	1.65	3.62	21.83		2.42	0.06
				Avg	0.60	2.05	Avg	1.64	Avg	2.21	4.86				Avg	2.46	0.06
				SD	>0.01	0.05	SD	0.01	SD	0.00	0.01				SD	0.03	0.00

*g/kg ÷ 10 = weight of fume expressed as a percentage of the deposit weight

Notes

Electrode polarity: positive

Welding speed: Code 5, 10, and 12 electrodes: 3.92 mm/s (9.25 in./min)

Electrode weight: Code 5 electrode: 148.42 g/m (3.77 g/in.)
Code 10 electrode: 148.42 g/m (3.77 g/in.)
Code 12 electrode: 145.57 g/m (3.70 g/in.)

Metal deposition rate determined for only 1 test per electrode.

SD: standard deviation (sigma)

Table B4-
Fume generation characteristics of 3.97 mm (5/32 in.) diameter E7024 covered electrodes

Electrode	Code	Voltage, V avg	Current, A avg	Fume generation rate: Measured g/min	Fume generation rate: Normalized $g/min/I^2 \times 10^{-5}$	Wgt. of fume/wgt of electrode used x 100	Melting rate kg/h	Melting rate lb/h	Deposition rate kg/h	Deposition rate lb/h	Wgt. of fume/wgt of deposited metal g/kg*	Weight of fume/length of electrode used g/m	Weight of fume/length of electrode used g/in
E7024	7	39	200	0.55	1.38	0.67	4.97	10.93	---	---	---	1.72	0.04
E7024	7	40	220	0.52	1.08	0.62	5.05	11.10	---	---	---	1.61	0.04
E7024	7	41	220	0.54	1.11	0.64	5.08	11.17	---	---	---	1.65	0.04
E7024	7	38	210	0.55	1.25	0.77	4.29	9.43	2.97	6.54	11.11	1.63	0.04
Avg				0.54	1.21	0.67	4.84	10.66				1.65	0.04
SD				0.01	0.13	0.07	0.37	0.82				0.05	0.00
E7024	8	38	225	0.54	1.06	0.63	5.08	11.18	---	---	---	2.05	0.05
E7024	8	38	225	0.54	1.07	0.65	5.02	11.04	---	---	---	2.05	0.05
E7024	8	35	230	0.56	1.05	0.68	4.92	10.82	---	---	---	2.13	0.05
E7024	8	41	205	0.57	1.35	0.72	4.78	10.51	3.27	7.20	10.43	2.17	0.05
Avg				0.55	1.13	0.67	4.95	10.89				2.10	0.05
SD				0.01	0.15	0.04	0.13	0.29				0.06	0.00
E7024	9	39	210	0.42	0.96	0.63	4.02	8.85	---	---	---	1.67	0.04
E7024	9	36	215	0.41	0.89	0.61	4.02	8.84	---	---	---	1.61	0.04
E7024	9	37	210	0.42	0.96	0.62	4.08	8.97	---	---	---	1.66	0.04
E7024	9	41	205	0.47	1.11	0.64	4.39	9.66	3.14	6.90	8.92	1.80	0.04
Avg				0.43	0.98	0.63	4.13	9.08				1.69	0.04
SD				0.02	0.09	0.01	0.18	0.39				0.08	0.00

*g/kg ÷ 10 = weight of fume expressed as a percentage of the deposit weight

Notes:

Electrode polarity: positive

Welding speed: Code 7 electrode: 466 mm/s (11 in./min)
Code 8 and 9 electrodes: 3.60 mm/s (8.5 in./min)

Electrode weight: Code 7 electrode: 214.96 g/m (5.46 g/in.)
Code 8 electrode: 215.75 g/m (5.48 g/in.)
Code 9 electrode: 205.51 g/m (5.22 g/in.)

Metal deposition rate determined for only 1 test per electrode

SD: standard deviation (sigma)

Table B5
Fume generation characteristics of 3.97 mm (5/32 in.) diameter E8018C3 covered electrodes

Electrode	Code	Voltage, V avg	Current, A avg		Fume generation rate: Measured g/min	Fume generation rate: Normalized $g/min/I^2 \times 10^{-5}$		Wgt. of fume/wgt. of electrode used x 100		Melting rate kg/h	Melting rate lb/h		Deposition rate kg/h	Deposition rate lb/h		Wgt. of fume/wgt. of deposited metal g/kg*		Weight of fume/length of electrode used g/m	Weight of fume/length of electrode used g/in.
E8018C3	13	22	160		0.45	1.74		1.21		2.22	4.88		1.51	3.32		17.74		1.81	0.05
E8018C3	13	24	158		0.44	1.77		1.22		2.15	4.73		1.47	3.24		17.88		1.81	0.05
E8018C3	13	23	160		0.45	1.74		1.23		2.18	4.79		1.50	3.30		17.83		1.81	0.05
				Avg	0.44	1.75	Avg	1.22	Avg	2.18	4.80	Avg	1.49	3.29	Avg	17.82	Avg	1.81	0.05
				SD	>0.01	0.01	SD	0.01	SD	0.03	0.07	SD	0.02	0.04	SD	0.07	SD	0.00	>0.01
E8018C3	14	23	175		0.46	1.52		1.28		2.18	4.79		- - -	- - -		- - -		1.91	0.05
E8018C3	14	23	175		0.46	1.51		1.27		2.19	4.81		- - -	- - -		- - -		1.89	0.05
E8018C3	14	23	175		0.49	1.59		1.33		2.19	4.82		- - -	- - -		- - -		2.00	0.05
E8018C3	14	23	178		0.47	1.54		1.31		2.16	4.76		1.52	3.34		18.69		1.95	0.05
				Avg	0.47	1.54	Avg	1.30	Avg	2.18	4.80						Avg	1.94	0.05
				SD	0.01	0.03	SD	0.03	SD	0.01	0.03						SD	0.05	>0.01
E8018C3	15	23	165		0.42	1.55		1.11		2.22	4.89		1.56	3.44		16.19		1.70	0.04
E8018C3	15	23	165		0.44	1.61		1.19		2.19	4.82		1.62	3.57		16.08		1.80	0.05
E8018C3	15	23	165		0.43	1.60		1.19		2.19	4.83		1.68	3.70		15.50		1.77	0.04
				Avg	0.43	1.58	Avg	1.16	Avg	2.20	4.85	Avg	1.62	3.57	Avg	15.92	Avg	1.76	0.04
				SD	0.01	0.03	SD	0.04	SD	0.02	0.04	SD	0.06	0.13	SD	0.37	SD	0.05	>0.01

*g/kg ÷ 10 = weight of fume expressed as a percentage of the deposit weight

Notes:

Electrode polarity: positive

Welding speed:
Code 13 electrode: 3.39 mm/s (8 in./min)
Code 14 electrode: 3.81 mm/s (9 in./min)
Code 15 electrode: 4.28 mm/s (10.1 in./min)

Electrode weight/unit length:
Code 13 electrode: 147.71 g/m (3.75 g/in.)
Code 14 electrode: 150.50 g/m (3.82 g/in.)
Code 15 electrode: 151.30 g/m (3.84 g/in.)

The deposition rate for the Code 14 electrode was determined in only one experimental test.

SD: standard deviation (sigma)

Table B6
Fume generation characteristics of 3.97 mm (5/32 in.) diameter E9018B3 covered electrodes

Electrode	Code	Voltage, V avg	Current, A avg	Fume generation rate: Measured g/min	Fume generation rate: Normalized g/min/I^2x10^{-5}	Wgt. of fume/wgt. of electrode used x 100	Melting rate kg/h	Melting rate lb/h	Deposition rate kg/h	Deposition rate lb/h	Wgt. of fume/wgt. of deposited metal g/kg*	Weight of fume/length of electrode used g/m	Weight of fume/length of electrode used g/in.
E9018B3	18	23	165	0.36	1.33	0.85	2.55	5.61	1.96	4.35	11.02	1.33	0.03
E9018B3	18	23	165	0.36	1.32	0.85	2.54	5.58	1.90	4.23	11.21	1.32	0.03
E9018B3	18	23	165	0.36	1.33	0.86	2.53	5.58	1.90	4.22	11.33	1.34	0.03
Avg				0.36	1.33	0.85	2.54	5.59	1.92	4.27	11.19	1.33	0.03
SD				>0.01	0.01	>0.01	0.01	0.02	0.03	0.07	0.16	0.01	>0.01
E9018B3	17	24	175	0.46	1.46	1.22	2.28	5.01	1.85	4.11	14.89	1.72	0.04
E9018B3	17	24	180	0.46	1.45	1.24	2.22	4.89	1.80	4.00	15.18	1.76	0.04
E9018B3	17	24	180	0.47	1.47	1.19	2.35	5.17	1.88	4.18	14.75	1.69	0.04
Avg				0.46	1.46	1.22	2.28	5.03	1.84	4.10	14.94	1.73	0.04
SD				>0.01	0.01	0.02	0.06	0.14	0.04	0.09	0.22	0.03	>0.01
E9018B3	16	22	160	0.39	1.53	1.08	2.19	4.81	1.74	3.83	13.51	1.55	0.04
E9018B3	16	21	163	0.40	1.51	1.09	2.31	5.08	1.74	3.83	13.71	1.49	0.04
E9018B3	16	22	165	0.40	1.51	1.07	2.23	4.91	1.77	3.89	13.55	1.55	0.04
E9018B3	16	22	160	0.40	1.56	1.09	2.21	4.86	1.77	3.86	13.54	1.55	0.04
Avg				0.40	1.53	1.07	2.23	4.91	1.76	3.86	13.58	1.54	0.04
SD				>0.01	0.02	0.02	0.05	0.12	0.02	0.03	0.09	0.03	>0.01

*g/kg ÷ 10 = weight of fume expressed as a percentage of the deposit weight

Notes:

Electrode polarity: positive

Welding speed: Code 18 electrode: 4.23 mm/s (10. in./min)
Code 17 electrode: 3.81 mm/s (9 in./min)
Code 16 electrode: 3.39 mm/s (8 in./min)

Electrode weight/unit length: Code 18 electrode: 149.74 g/m (3.80 g/in.)
Code 17 electrode: 143.31 g/m (3.64 g/in.)
Code 16 electrode: 144.88 g/m (3.68 g/in.)

Metal deposition rate determined for only 1 test per electrode

SD: standard deviation (sigma)

Table B7
Fume generation characteristics of 3.97 mm (5/32 in.) diameter E316-15 covered electrodes

Electrode	Code	Voltage, V avg	Current, A avg		Fume generation rate: Measured g/min	Fume generation rate: Normalized $g/min/I^2 x10^{-5}$		Wgt. of fume/wgt. of electrode used x 100		Melting rate kg/h	Melting rate lb/h		Deposition rate kg/h	Deposition rate lb/h		Wgt. of fume/wgt. of deposited metal g/kg*		Weight of fume/length of electrode used g/m	Weight of fume/length of electrode used g/in.
E316-15	22	25	148		0.31	1.43		0.71		2.62	5.77		1.81	4.03		10.19		0.94	0.02
E316-15	22	25	148		0.32	1.47		0.73		2.62	5.77		1.88	4.18		10.14		0.97	0.02
E316-15	22	25	148		0.32	1.47		0.73		2.62	5.77		1.79	3.98		10.65		0.97	0.02
				Avg	0.32	1.46	Avg	0.73	Avg	2.62	5.77	Avg	1.83	4.06	Avg	10.32	Avg	0.96	0.02
				SD	0.01	0.03	SD	0.01	SD	0.00	>0.01	SD	0.05	0.11	SD	0.28	SD	0.02	>0.01
E316-15	27	29	155		0.37	1.55		0.77		2.92	6.42		1.99	4.49		11.13		1.08	0.03
E316-15	27	29	155		0.37	1.55		0.76		2.92	6.42		1.98	4.41		11.18		1.08	0.03
E316-15	27	29	155		0.38	1.53		0.78		2.92	6.42		2.07	4.59		10.93		1.10	0.03
				Avg	0.38	1.54	Avg	0.77	Avg	0.00	>0.01	Avg	2.02	4.59	Avg	11.08	Avg	1.09	0.03
				SD	0.01	0.01	SD	0.01	SD	0.00	>0.01	SD	0.05	0.10	SD	0.13	SD	0.01	>0.01
E316-15	19	25	150		0.28	1.22		0.64		2.57	5.66		2.04	4.49		8.10		0.82	0.02
E316-15	19	25	150		0.27	1.24		0.63		2.60	5.72		2.10	4.61		7.79		0.80	0.02
E316-15	19	25	150		0.28	1.24		0.65		2.57	5.66		2.05	4.54		8.17		0.82	0.02
				Avg	0.28	1.22	Avg	0.64	Avg	2.58	5.68	Avg	2.06	4.54	Avg	8.02	Avg	0.81	0.02
				SD	>0.01	0.01	SD	0.01	SD	0.02	0.04	SD	0.03	0.07	SD	0.20	SD	0.01	>0.01

*g/kg ÷ 10 = weight of fume expressed as a percentage of the deposit weight

Notes:

Direct current: electrode positive

Welding speed: Code 19 electrode: 4.23 mm/s (10 in./min)
Code 22 electrode: 5.08 mm/s (12 in./min)
Code 27 electrode: 3.81 mm/s (9 in./min)

Electrode weight/unit length: Code 19 electrode: 145.26 g/m (3.69 g/in.)
Code 22 electrode: 132.28 g/m (3.36 g/in.)
Code 27 electrode: 141.73 g/m (3.60 g/in.)

Metal deposition rate determined for only 1 test per electrode

SD: standard deviation (sigma)

Table B8
Fume generation characteristics of 3.97 mm (5/32 in.) diameter E316-16 covered electrodes

Electrode	Code	Voltage, V avg	Current, A avg	Fume generation rate: Measured g/min	Fume generation rate: Normalized g/min/I ^{2}x10^{-5}	Wgt. of fume/wgt. of electrode used x 100	Melting rate kg/h	Melting rate lb/h	Deposition rate kg/h	Deposition rate lb/h	Wgt. of fume/wgt. of deposited metal g/kg*	Weight of fume/length of electrode used g/m	Weight of fume/length of electrode used g/in.
E316-16	26	22	150	0.20	0.89	0.46	2.59	5.72	1.89	4.17	6.33	0.59	0.01
E316-16	26	23	150	0.22	0.97	0.51	2.60	5.71	1.90	4.18	6.92	0.67	0.02
E316-16	26	23	150	0.20	0.91	0.47	2.60	5.72	1.90	4.19	6.43	0.63	0.02
				Avg 0.21	0.92	Avg 0.48	Avg 2.60	5.72	Avg 1.90	4.18	Avg 6.56	Avg 0.63	0.02
				SD 0.01	0.04	SD 0.02	SD 0.01	0.01	SD 0.01	0.01	SD 0.32	Avg 0.04	>0.01
E316-16	20	24	150	0.31	1.39	0.74	2.53	5.57	1.57	3.46	11.98	0.95	0.02
E316-16	20	25	150	0.30	1.34	0.72	2.53	5.57	1.59	3.49	11.44	0.91	0.02
E316-16	20	24	150	0.31	1.38	0.74	2.52	5.55	1.50	3.31	12.34	0.94	0.02
				Avg 0.31	1.37	Avg 0.73	Avg 2.53	5.56	Avg 1.55	3.42	Avg 11.92	Avg 0.93	0.02
				SD 0.01	0.02	SD 0.01	SD 0.01	0.01	SD 0.05	0.10	SD 0.45	SD 0.02	0.00
E316-16	25	23	147	0.24	1.14	0.59	2.53	5.56	1.70	3.74	8.70	0.85	0.02
E316-16	25	24	147	0.25	1.16	0.50	2.52	5.55	1.74	3.83	8.50	0.85	0.02
E316-16	25	23	147	0.25	1.15	0.59	2.53	5.56	1.80	3.96	8.32	0.86	0.02
				Avg 0.25	1.15	Avg 0.59	Avg 2.53	5.56	Avg 1.75	3.84	Avg 8.50	Avg 0.85	0.02
				SD 0.01	0.01	SD 0.01	SD 0.01	0.01	SD 0.05	0.11	SD 0.19	SD 0.01	0.00

*g/kg ÷ 10 = weight of fume expressed as a percentage of the deposit weight

Notes

Electrode polarity: positive

Welding speed: Code 26 electrode: 4.44 mm/s (10.5 in./min)
Code 20 and 25 electrodes: 4.23 mm/s (10 in./min)

Electrode weight: Code 26 electrode: 132.68 g/m (3.37 g/in.)
Code 20 electrode: 127.07 g/m (3.23 g/in.)
Code 25 electrode: 145.26 g/m (3.68 g/in.)

Metal deposition rate determined for only 1 test per electrode

SD: standard deviation (sigma)

Table B9
Fume generation characteristics of 3.97 mm (5/32 in.) diameter E410-16 covered electrodes

Electrode	Code	Voltage, V avg	Current, A avg	Fume generation rate: Measured g/min	Fume generation rate: Normalized g/min/I^{2}x10^{-5}	Wgt. of fume/wgt of electrode used x 100	Melting rate kg/h	Melting rate lb/h	Deposition rate kg/h	Deposition rate lb/h	Wgt. of fume/wgt of deposited metal g/kg*	Weight of fume/length of electrode used g/m	Weight of fume/length of electrode used g/in
E410-16	21	23	145	0.34	1.61	0.98	2.06	4.53	1.42	3.14	14.17	1.33	0.03
E410-16	21	25	145	0.32	1.54	0.94	2.06	4.53	1.42	3.13	13.58	1.27	0.03
E410-16	21	25	145	0.33	1.58	0.97	2.06	4.53	1.41	3.11	14.15	1.31	0.03
Avg				0.33	1.58	0.96	2.06	4.53	1.42	3.13	13.97	1.30	0.03
SD				0.01	0.03	0.03	0.02	>0.01	0.01	0.02	0.33	0.03	0.01
E410-16	23	22	158	0.29	1.17	0.82	2.12	4.66	1.47	3.24	11.80	1.11	0.03
E410-16	23	22	158	0.28	1.13	0.79	2.12	4.66	1.44	3.17	11.66	1.06	0.03
E410-16	23	22	158	0.26	1.15	0.81	2.11	4.64	1.45	3.20	11.78	1.08	0.03
Avg				0.28	1.15	0.81	2.12	4.65	1.46	3.20	11.75	1.08	0.03
SD				>0.01	0.02	0.01	0.01	>0.01	0.02	0.03	0.08	0.03	>0.01
E410-16	24	21	155	0.34	1.42	0.91	2.23	4.94	1.40	3.29	13.68	1.26	0.03
E410-16	24	22	158	0.33	1.35	0.90	2.24	4.93	1.47	3.22	13.71	1.28	0.03
E410-16	24	22	158	0.34	1.37	0.91	2.22	4.89	1.45	3.18	14.04	1.33	0.03
Avg				0.34	1.37	0.91	2.24	4.92	1.47	3.23	13.81	1.29	0.03
SD				>0.01	0.37	0.01	0.01	0.03	0.02	0.05	0.20	0.04	>0.01

*g/kg ÷ 10 = weight of fume expressed as a percentage of the deposit weight

Notes

Direct current electrode positive

Welding speed Code 21 electrode: 4.02 mm/s (9.5 in./min)
Code 23 electrode: 4.23 mm/s (10 in./min)
Code 24 electrode: 3.39 mm/s (8 in./min)

Electrode weight/unit length. Code 21 electrode: 134.10 g/m (3.41 g/in.)
Code 23 electrode: 135.10 g/m (3.43 g/in.)
Code 24 electrode: 137.82 g/m (3.50 g/in.)

Metal deposition rate determined for only 1 test per electrode

SD. standard deviation (sigma)

Table B10
Fume generation characteristics of 3.97 mm (5/32 in.) diameter high alloy covered electrodes

Electrode	Code	Voltage, V avg	Current, A avg		Fume generation rate: Measured g/min	Fume generation rate: Normalized $g/min/I^2 \times 10^{-5}$		Wgt. of fume/wgt of electrode used x 100		Melting rate kg/h	Melting rate lb/h		Deposition rate kg/h	Deposition rate lb/h		Wgt. of fume/wgt. of deposited metal g/kg*		Wgt. of fume/length of electrode used g/m	Wgt. of fume/length of electrode used g/in
ENi-CI	28	21	135		0.36	1.97		1.18		1.83	4.02		1.68	3.69		12.78		1.57	0.04
ENi-CI	28	21	135		0.36	1.98		1.20		1.80	3.96		1.70	3.74		12.74		1.57	0.04
ENi-CI	28	21	135		0.38	2.06		1.26		1.79	3.93		1.71	3.77		13.18		1.63	0.04
				Avg	0.37	2.01	Avg	1.22	Avg	1.81	3.97	Avg	1.70	3.73	Avg	12.90	Avg	1.51	0.04
				SD	0.01	0.05	SD	0.04	SD	0.02	0.05	SD	0.02	0.04	SD	0.25	SD	0.07	>0.01
ENiCu-2	33	25	143		0.31	1.52		0.74		2.51	5.52		1.80	3.96		10.29		1.02	0.03
ENiCu-2	33	27	143		0.31	1.51		0.73		2.51	5.52		1.85	4.07		9.94		1.02	0.03
ENiCu-2	33	27	143		0.31	1.52		0.74		2.51	5.52		1.85	4.07		10.01		1.02	0.03
				Avg	0.31	1.52	Avg	0.74	Avg	2.51	5.52	Avg	1.83	4.04	Avg	10.08	Avg	1.02	0.03
				SD	>0.01	0.01	SD	0.01	SD	0.00	>0.01	SD	0.02	0.06	SD	0.18	SD	0.00	>0.01
Inconel 625	62	24	155		0.32	1.35		0.71		2.75	6.06		2.07	4.56		9.36		0.97	0.02
Inconel 625	62	26	140		0.31	1.61		0.68		2.78	6.13		2.10	4.61		9.00		0.93	0.02
Inconel 625	62	26	145		0.32	1.54		0.70		2.78	6.11		2.08	4.59		9.34		0.96	0.02
				Avg	0.32	1.50	Avg	0.70	Avg	2.77	6.10	Avg	2.08	4.59	Avg	9.24	Avg	0.96	0.02
				SD	0.01	0.14	SD	0.01	SD	0.02	0.04	SD	0.01	0.03	SD	0.20	SD	0.02	>0.01
Haynes C-276	69	24	130		0.37	0.06		0.98		2.27	5.00		1.56	3.44		14.20		1.51	0.04
Haynes C-276	69	22	135		0.37	0.07		0.98		2.26	4.97		1.56	3.44		14.20		1.40	0.04
Haynes C-276	69	23	135		0.37	0.07		0.98		2.26	4.98		1.56	3.44		14.20		1.50	0.04
				Avg	0.37	0.07	Avg	0.98	Avg	2.26	4.98	Avg	1.56	3.44	Avg	14.20	Avg	1.47	0.04
				SD	0.00	0.01	SD	0.00	SD	0.01	0.02	SD	0.00	0.00	SD	0.00	SD	0.06	>0.01
Haynes 25	70	19	140		0.26	0.05		0.66		2.37	5.22		1.74	3.82		8.98		0.98	0.02
Haynes 25	70	19	140		0.26	0.05		0.64		2.44	5.37		1.75	3.86		8.89		0.97	0.02
Haynes 25	70	19.5	135		0.26	0.05		0.66		2.37	5.21		1.74	3.83		8.96		0.97	0.02
				Avg	0.26	0.05	Avg	0.65	Avg	2.40	5.27	Avg	1.74	3.84	Avg	8.94	Avg	0.97	0.02
				SD	0.00	0.00	SD	0.01	SD	0.04	0.09	SD	0.01	0.02	SD	0.05	SD	0.01	>0.01

*g/kg – 10 = weight of fume expressed as a percentage of the deposit weight

Notes:

Direct current: electrode positive

Welding speed: Code 28 electrode: 3.81 mm/s (9 in./min)
Code 33 electrode: 5.93 mm/s (14 in./min)
Inconel 625 electrode: 4.23 mm/s (10 in./min)
Code 69 electrode: 4.02 mm/s (9.5 in./min)
Code 70 electrode: 4.45 mm/s (10.5 in./min)

Electrode weight/unit length: Code 28 electrode: 131.42 g/m (3.34 g/in.)
Code 33 electrode: 138.68 g/m (3.52 g/in.)
Inconel 625 electrode: 138.97 g/m (3.53 g/in.)
Code 69 electrode: 153.76 g/m (3.90 g/in.)
Code 70 electrode: 150.10 g/m (3.81 g/in.)

Compositions: ENi-CI: 2C-4Si-1Mn-8Fe-2.5Cu-85Ni
ENiCu-2: 6Mn-1.5Si-2.5Fe-60 to 68 Ni-1Al-1Ti-3(Cb+Ta)-Bal Cu
Inconel 625: 61Ni-0.25Mn-0.25Si-2.5Fe-21.5Cr-0.2Al-0.2Ti-9Mo-3.65(Cb+Ta)
Haynes C-276: 2.5Co-16Cr-16Mo-4W-5Fe-0.08Si-1Mn-2Cu-2(Cb+Ta)-Bal.Ni
Haynes 25: 10Ni-20Cr-15W-3Fe-1Si-1Mn-Bal.Co

Metal deposition rate determined for only 1 test per electrode

SD: standard deviation (sigma)

Table B11

Fume generation characteristics of 2.38 mm (3/32 in.) diameter E70T-1 flux cored electrodes (CO_2 shielding)

Code	Voltage, V avg	Current, A avg	Wire feed rate mm/s	Wire feed rate in./min	Fume generation rate Measured g/min	Fume generation rate Normalized g/min/I ^{2}x10^{-5}	Wgt. of fume/wgt. of electrode used x 100	Melting rate kg/h	Melting rate lb/h	Deposition rate kg/h	Deposition rate lb/h	Wgt. of fume/wgt. of deposited metal g/kg*	Wgt. of fume/length of electrode used g/m	Wgt. of fume/length of electrode used g/in.
40	30	485	87.63	207	1.36	0.58	0.91	8.95	19.70	7.72	17.15	10.44	0.26	0.01
40	31	485	87.63	207	1.34	0.57	0.90	8.95	19.70	7.70	17.10	10.37	0.26	0.01
40	31	470	87.63	207	1.36	0.61	0.91	8.95	19.70	7.77	17.27	10.38	0.26	0.01
					Avg 1.36	0.59	Avg 0.91			Avg 7.73	17.17	Avg 10.40		
					SD 0.01	0.02	SD 0.01			SD 0.04	0.09	SD 0.04		
42	30	445	87.00	205.5	2.28	1.15	1.56	8.80	19.36	7.84	17.42	17.31	0.44	0.01
42	31	435	87.00	205.5	2.25	1.19	1.54	8.80	19.36	7.60	16.89	17.60	0.44	0.01
42	31	440	87.00	205.5	2.28	1.18	1.56	8.80	19.36	7.68	17.07	17.63	0.44	0.01
					Avg 2.27	1.17	Avg 1.55			Avg 7.71	17.13	Avg 17.51		
					SD 0.02	0.02	SD 0.01			SD 0.12	0.27	SD 0.17		
41	31	460	87.00	205.5	0.96	0.45	0.62	9.29	20.43	8.53	18.95	6.68	0.19	>0.01
41	30	465	87.00	205.5	0.96	0.44	0.62	9.29	20.43	8.67	19.27	6.66	0.19	>0.01
41	30	455	87.00	205.5	0.96	0.46	0.62	9.29	20.43	8.61	19.14	6.16	0.19	>0.01
					Avg 0.96	0.45	Avg 0.62			Avg 8.60	19.12	Avg 6.65		
					SD >0.01	0.01	SD >0.01			SD 0.07	0.16	SD 0.04		
46	30	435	88.90	210	1.33	0.70	0.90	8.84	19.45	7.92	17.60	9.95	0.25	0.01
46	31	435	88.90	210	1.33	0.70	0.90	8.84	19.45	7.95	17.66	9.95	0.25	0.01
46	31	445	88.90	210	1.33	0.67	0.90	8.84	19.45	7.88	17.50	10.02	0.25	0.01
					Avg 1.33	0.69	Avg 0.90			Avg 7.67	17.59	Avg 9.97		
					SD >0.01	0.02	SD >0.01			SD 0.04	0.08	SD 0.04		
47	30	435	88.90	210	1.83	0.97	1.21	9.10	20.03	7.62	16.94	14.25	0.34	0.01
47	31	435	88.90	210	1.83	0.97	1.21	9.10	20.03	7.68	17.07	14.15	0.34	0.01
47	30	435	88.90	210	1.83	0.97	1.21	9.10	20.03	7.72	17.15	14.11	0.34	0.01
					Avg 1.83	0.97	Avg 1.21			Avg 7.67	17.05	Avg 14.17		
					SD >0.01	>0.01	SD >0.01			SD 0.05	0.10	SD 0.07		

*g/kg ÷ 10 = weight of fume expressed as a percentage of the deposit weight

Notes:

Electrode polarity: positive

Electrode contact tube-to-work distance: 38 mm (1-1/2 in.)

Welding time: 30 s

Shielding gas flow rate: 21-24 liters/min (45-50 ft^3/h)

Welding speed: 6.77 mm/s (16 in./min)

Electrode weight/unit length:
Code No. 40: 28.35 g/m (0.72 g/in.)
Code No. 41: 29.53 g/m (0.75 g/in.)
Code No. 42: 27.95 g/m (0.71 g/in.)
Code No. 46: 27.56 g/m (0.70 g/in.)
Code No. 47. 28.35 g/m (0.72 g/in.)

Metal deposition rate determined for only 1 test per electrode

SD: standard deviation (sigma)

Table B12
Fume generation characteristics of 2.38 mm (3/32 in.) diameter E70T-4 self-shielded flux cored electrodes

Code	Voltage, V avg	Current, A avg	Wire feed rate mm/s	Wire feed rate in./min		Fume generation rate Measured g/min	Fume generation rate Normalized $g/min/I^2 \times 10^{-5}$		Wgt. of fume/wgt of electrode used x 100	Melting rate kg/h	Melting rate lb/h		Metal deposition rate kg/h	Metal deposition rate lb/h		Wgt. of fume/wgt of deposited metal g/kg*		Wgt of fume/length of electrode used g/m	Wgt of fume/length of electrode used g/in
48	31	390	93.98	222		2.96	1.95		1.91	9.30	20.47		7.70	17.10		22.85		0.52	0.01
48	32	385	93.98	222		2.99	2.02		1.93	9.30	20.47		7.86	17.47		22.59		0.53	0.01
48	32	385	93.98	222		2.98	2.01		1.92	9.30	20.47		7.80	17.34		22.66		0.53	0.01
					Avg	2.98	1.99	Avg	1.92			Avg	7.79	17.30	Avg	22.70			
					SD	0.02	0.04	SD	0.01			SD	0.08	0.19	SD	0.13			
56	29	370	91.86	217		1.76	1.29		1.13	9.37	20.62		8.41	18.50		12.56		0.32	0.01
56	30	390	96.94	229		1.88	1.23		1.14	9.89	21.76		8.80	19.35		12.83		0.32	0.01
56	30	390	96.94	229		1.89	1.24		1.15	9.89	21.76		8.89	19.56		12.78		0.33	0.01
56	31	390	96.94	229		1.90	1.25		1.15	9.89	21.76		8.85	19.48		12.87		0.33	0.01
					Avg	1.86	1.25	Avg	1.14			Avg	8.74	19.22	Avg	12.76	Avg	0.33	0.01
					SD	0.07	0.03	SD	0.01			SD	0.22	0.49	SD	0.13	SD	0.01	0.00
49	31	400	113.03	267		2.00	1.25		1.17	10.29	22.63		8.93	19.85		13.28		0.29	0.01
49	32	400	113.03	267		2.18	1.36		1.27	10.29	22.63		9.02	20.04		14.35		0.32	0.01
49	31	410	113.03	267		2.10	1.25		1.23	10.29	22.63		9.00	20.01		13.85		0.31	0.01
					Avg	2.09	1.29	Avg	1.22			Avg	8.99	19.97	Avg	13.83			
					SD	0.09	0.07	SD	0.05			SD	0.05	0.10	SD	0.53			

*g/kg ÷ 10 = weight of fume expressed as a percentage of the deposit weight

Notes

Electrode polarity positive

Electrode contact tube-to-work distance: 70 mm (2-3/4 in.)

Welding time: 30 seconds

Welding speed 6.77 mm/s (16 in./min)

Electrode weight/unit length: Code No. 48: 27.56 g/m (0.70 g/in.)
Code No. 49 25.20 g/m (0.64 g/in.)
Code No. 56: 27.95 g/m (0.71 g/in.)

Metal deposition rate determined for only 1 test per electrode.

SD. standard deviation (sigma)

Table B13

Fume generation characteristics of 2.38 mm (3/32 in.) diameter E70T-5 flux cored electrodes (CO_2 shielding)

Manu- Code	Voltage, V avg	Current, A avg	Wire feed rate mm/s	Wire feed rate in./min	Fume generation rate Measured g/min	Fume generation rate Normalized g/min/I ^{2}x10^{-5}	Wgt. of fume/ wgt. of electrode used x 100	Melting rate kg/h	Melting rate lb/h	Metal deposition rate kg/h	Metal deposition rate lb/h	Wgt of fume/ wgt. of deposited metal g/kg*	Wgt of fume/ length of electrode used g/m	Wgt of fume/ length of electrode used g/in.
44	30	425	87.63	207	2.27	1.26	1.55	8.95	19.70	7.59	16.86	17.77	0.43	0.01
44	31	445	87.63	207	2.26	1.14	1.54	8.95	19.70	7.47	16.60	17.95	0.43	0.01
44	31	445	87.63	207	2.26	1.14	1.54	8.95	19.70	7.51	16.68	17.88	0.43	0.01
					Avg 2.26	1.18	Avg 1.54			Avg 7.52	16.71	Avg 17.87		
					SD 0.01	0.07	SD >0.01			SD 0.06	0.13	SD 0.09		
50	31	440	93.98	222	3.26	1.68	2.08	9.40	20.67	8.19	18.21	23.63	0.58	0.01
50	30	440	93.98	222	3.25	1.68	2.07	9.40	20.67	8.16	18.13	23.66	0.58	0.01
50	31	450	93.98	222	3.25	1.61	2.08	9.40	20.67	8.19	18.21	23.59	0.58	0.01
					Avg 3.25	1.66	Avg 2.08			Avg 8.18	18.18	Avg 23.63		
					SD >0.01	0.04	SD >0.01			SD 0.02	0.05	SD 0.04		

*g/kg – 10 = weight of fume expressed as a percentage of the deposit weight

Notes:

Electrode polarity: positive

Electrode contact tube-to-work distance: 38 mm (1-1/2 in.)

Welding time: 30 seconds

Shielding gas flow rate: 21-24 liters/min (45-50 ft^3/h)

Welding speed: 6.77 mm/s (16 in./min)

Electrode weight/unit length: Code No. 44: 27.95 g/m (0.71 g/in.)
Code No. 50: 27.56 g/m (0.70 g/in.)

Metal deposition rate determined for only 1 test per electrode

SD: standard deviation (sigma)

Table B14
Fume generation characteristics of 2.38 mm (3/32 in.) diameter alloy steel flux cored electrodes (CO_2 shielding)

Electrode	Code	Voltage, V avg	Current, A avg	Wire feed rate mm/s	Wire feed rate in./min	Fume generation rate Measured g/min	Fume generation rate Normalized g/min/I ^{2}x10^{-5}	Wgt. of fume/ wgt. of electrode used x 100	Melting rate kg/h	Melting rate lb/h	Metal deposition rate kg/h	Metal deposition rate lb/h	Wgt. of fume/ wgt. of deposited metal g/kg*	Wgt. of fume/ length of electrode used g/m	Wgt. of fume/ length of electrode used g/in.
81-C3	51	30.5	440	92.71	219	1.08	0.56	0.72	9.04	19.89	7.53	16.73	8.50	0 20	0.01
81-C3	51	30 5	440	92.71	219	1.12	0.57	0.74	9.04	19.89	7.59	16.86	8.69	0.20	0.01
81-C3	52	30 5	445	92.71	219	1.12	0.57	0.75	9.04	19.89	7.55	16.78	8.82	0.20	0.01
Avg						1.11	0.57	0.73			7.56	16.79	8.69		
SD						0.02	0.01	0.01			0.03	0.07	0.13		
91-B3	52	30 5	450	104.14	246	1.15	0.57	0.74	9.42	20.73	8.07	17.97	8.47	0.18	>0.01
91-B3	52	30 5	450	104.14	246	1.14	0.57	0.73	9.42	20.73	8.11	18.03	8.37	0.18	>0.01
91-B3	52	30 5	450	104.14	246	1 15	0.57	0.73	9.42	20.73	8.09	17.97	8.43	0.18	>0.01
Avg						1 15	0.57	0.73			8.09	17.99	8.42		
SD						>0.01	>0.01	>0.01			0.01	0.01	0.05		

*g/kg ÷ 10 = weight of fume expressed as a percentage of the deposit weight

Notes

Electrode polarity: positive

Electrode contact tube-to-work distance: 38 mm (1-1/2 in.)

Welding time: 30 seconds

Shielding gas flow rate: 21-24 liters/min (45-50 ft^3/h) CO_2

Welding speed: 6.77 mm/s (16 in./min)

Electrode weight/unit length: Code No. 51: 27.17 g/m (0.69 g/in.)
Code No. 52: 25.20 g/m (0.64 g/in.)

Metal deposition rate determined for only 1 test per electrode

SD: standard deviation (sigma)

Table B15
Fume generation characteristics of 2.38 mm (3/32 in.) diameter self-shielded stainless steel flux cored electrodes

Electrode	Code	Voltage V avg	Current, A avg	Wire feed rate mm/s	Wire feed rate in./min	Fume generation rate Measured g/min	Fume generation rate Normalized g/min/I²x10⁻⁵	Wgt. of fume/wgt of electrode used x 100	Melting rate kg/h	Melting rate lb/h	Metal deposition rate kg/h	Metal deposition rate lb/h	Wgt. of fume/wgt of deposited metal g/kg*	Wgt of fume/length of electrode used g/m	Wgt of fume/length of electrode used g/in
308L	53	31-33	380-420	147 32	34[illegible]	1 64	1 03	0 77	12 89	28 36	10 65	23 66	9 16	0 18	<0 01
308L	53	32-33	390-410	147 32	34[illegible]	1 63	1 02	0 76	12 89	28 36	10 65	23 66	9.12	0 18	<0 01
308L	53	32-33	380-400	147 32	34[illegible]	1 64	1 07	0 76	12 89	28 36	10 71	23 81	9.06	0 18	<0.01
Avg						1 64	1 04	0 76			10 64	23.71	9 11		
SD						<0 01	0 03	<0 01			0 04	0.09	0 05		
316L	55	28-29	360-390	154 94	36[illegible]	1 34	0 95	0 59	13 58	29 87	11 58	25 48	6 93	0.14	<0 01
316L	55	28-29	350-380	154 94	36[illegible]	1 38	1 03	0 61	13 58	29.87	11 58	25 45	7 13	0.14	<0 01
316L	55	28-29	340-370	153 67	363	1 30	1 03	0 58	13 58	29.87	11 50	25.08	6 86	0 14	<0 01
Avg						1 34	1 00	0 59			11 52	25.34	6 97		
SD						0.04	0.05	0 02			0 10	0.22	0 13		
316L	67	28	390	154 94	36[illegible]	2 44	1 60	1 06	13 76	30 28	12 07	26 56	12 13	0.26	<0.01
316L	67	28 5	405	154 94	36[illegible]	2 49	1 52	1 09	13 76	30.28	12 07	26 56	12 37	0 26	<0 01
316L	67	28 5	400	154 94	36[illegible]	2 51	1 57	1 10	13 76	30.28	12 10	26 62	12.45	0 27	<0 01
Avg						2 48	1 56	1 08			12 08	26 58	12 31		
SD						0 04	0 04	0 02			0 02	0 04	0 17		

*g/kg ÷ 10 = weight of fume expressed as a percentage of the deposit weight

Notes

Electrode polarity positive

Electrode contact tube-to-work distance. 38 mm (1-1/2 in.)

Welding time 30 seconds

Welding speed 6 77 mm/s (16 in./min)

Shielding gas none

Electrode weight/unit length Code No. 53 24 41 g/m (0.62 g/in.)
Code No 55 24.41 g/m (0 62 g/in.)
Code No 67: 24.64 g/m (0 63 g/in.)

Metal deposition rate determined for only 1 test per electrode

SD. standard deviation (sigma)

Table B16
Fume generation characteristics of 1.14 mm (0.045 in.) diameter E70S-3 gas metal arc electrodes

Electrode	Code	Voltage, V avg	Current, A avg	Wire feed rate mm/s	Wire feed rate in./min	Fume generation rate Measured g/min	Fume generation rate Normalized $g/min/I^2 \times 10^{-5}$	Wgt. of fume/wgt. of electrode used x 100	Melting rate kg/h	Melting rate lb/h	Deposition rate kg/h	Deposition rate lb/h	Wgt. of fume/wgt. of deposited metal g/kg*	Wgt of fume/length of electrode used g/m	Wgt of fume/length of electrode used g/in
							Spray transfer with Ar-2 O_2 shielding								
E70S-3	54	35	275	171.87	406	0.41	0.54	0.49	5.01	11.02	4.93	10.84	4.99	0.04	<0.01
E70S-3	54	35	275	171.87	406	0.41	0.55	0.49	5.01	11.02	4.93	10.84	5.02	0.04	<0.01
E70S-3	54	34	260	171.87	406	0.40	0.59	0.48	5.01	11.02	4.86	10.70	4.91	0.04	<0.01
					Avg	0.41	0.56	Avg 0.49			Avg 4.91	10.79	Avg 4.97		
					SD	0.01	0.03	SD 0.01			SD 0.03	0.08	SD 0.06		
							Spray transfer with Ar-9 CO_2 shielding								
E70S-3	54	33	210	132.50	313	0.42	0.95	0.65	3.86	8.50	3.84	8.46	6.52	0.05	<0.01
E70S-3	54	33	215	132.50	313	0.41	0.88	0.63	3.86	8.50	3.86	8.48	6.32	0.05	<0.01
E70S-3	54	33	210	132.50	313	0.41	0.92	0.63	3.86	8.50	3.84	8.46	6.33	0.05	<0.01
					Avg	0.41	0.92	Avg 0.64			Avg 3.85	8.47	Avg 6.39		
					SD	0.01	0.04	SD 0.01			SD 0.01	0.01	SD 0.11		
							Globular transfer with CO_2 shielding								
E70S-3	54	37	328	321.73	760	0.48	0.45	0.31	9.38	20.64	9.30	20.46	3.13	0.03	<0.01
E70S-3	54	36	340	321.73	760	0.52	0.45	0.33	9.38	20.64	9.32	20.51	3.34	0.03	<0.01
E70S-3	54	36	338	321.73	760	0.53	0.46	0.34	9.38	20.64	9.35	20.56	3.38	0.03	<0.01
E70S-3	54	36	335	321.73	760	0.53	0.47	0.34	9.38	20.64	9.34	20.54	3.38	0.03	<0.01
					Avg	0.51	0.46	Avg 0.33			Avg 9.33	20.52	Avg 3.31		
					SD	0.02	0.01	SD 0.02			SD 0.02	0.04	SD 0.12		
							Short circuiting with Ar-25 CO_2 shielding								
E70S-3	54	19.5	195	104.14	246	0.21	0.55	0.43	2.97	6.55	2.96	6.52	4.16	0.03	<0.01
E70S-3	54	19.5	195	104.14	246	0.20	0.53	0.45	2.97	6.55	2.95	6.50	4.07	0.03	<0.01
E70S-3	54	19.5	195	104.14	246	0.20	0.53	0.41	2.97	6.55	2.96	6.52	4.11	0.03	<0.01
					Avg	0.20	0.54	Avg 0.43			Avg 4.96	6.51	Avg 4.11		
					SD	0.01	0.01	SD 0.02			SD 0.01	0.01	SD 0.05		

*g/kg ÷ 10 = weight of fume expressed as a percentage of the deposit weight

Notes

Electrode polarity: positive

Electrode contact tube-to-work distance: 25.4 mm (1 in.)

Welding time: 30 seconds

Welding speed: 6.77 mm/s (16 in./min)

Shielding gas flow rate: 21-24 liters/min (45-50 ft^3/h)

Electrode weight: 7.92 g/m (0.20 g/in.)

Metal deposition rate determined for only 1 test per electrode

SD: standard deviation (sigma)

Table B17

Fume generation characteristics of 1.14 mm (0.045 in.) diameter E70S-3 gas metal arc electrodes

Electrode	Code	Voltage, V avg	Current, A avg	Wire feed rate mm/s	Wire feed rate in./min	Fume generation rate Measured g/min	Fume generation rate Normalized g/min/I ^{2}x10^{-5}	Wgt. of fume/wgt. of electrode used x 100	Melting rate kg/h	Melting rate lb/h	Deposition rate kg/h	Deposition rate lb/h	Wgt. of fume/wgt. of deposited metal g/kg*	Wgt. of fume/length of electrode used g/m	Wgt. of fume/length of electrode used g/in.
							Spray transfer with Ar-2 O_2 shielding								
E70S-3	57	35	290	182.03	430	0.46	0.54	0.54	5.04	11.09	4.87	10.70	5.63	0.04	<0.01
E70S-3	57	35	285	182.03	430	0.46	0.56	0.54	5.04	11.09	4.78	10.52	5.74	0.04	<0.01
E70S-3	57	35	270	182.03	430	0.46	0.64	0.55			4.90	10.78	5.67	0.04	<0.01
						Avg 0.46	0.58	Avg 0.54			Avg 4.88	10.67	Avg 5.68		
						SD 0.00	0.05	SD 0.01			SD 0.06	0.15	SD 0.05		
							Spray transfer with Ar-9 CO_2 shielding								
E70S-3	57	34	210	120.23	284	0.44	1.01	0.80	3.32	7.31	3.18	7.00	8.38	0.06	<0.01
E70S-3	57	34	205	120.23	284	0.44	0.99	0.79	3.32	7.31	3.17	6.98	8.31	0.06	<0.01
E70S-3	57	33	210	120.23	284	0.44	1.00	0.80	3.32	7.31	3.20	7.03	8.33	0.06	<0.01
						Avg 0.44	1.00	Avg 0.80			Avg 3.18	7.00	Avg 8.34		
						SD 0.00	0.01	SD 0.01			SD 0.01	0.03	SD 0.01		
							Globular transfer with CO_2 shielding								
E70S-3	57	35	320	312.42	738	0.45	0.44	0.31	8.65	19.03	8.53	18.77	3.16	0.02	<0.01
E70S-3	57	35	325	312.42	738	0.45	0.43	0.31	8.65	19.03	8.51	18.71	3.18	0.02	<0.01
E70S-3	57	35	325	312.42	738	0.48	0.45	0.33			8.53	18.77	3.37	0.02	<0.01
						Avg 0.46	0.44	Avg 0.32			Avg 8.52	18.75	Avg 3.24		
						SD 0.02	0.01	SD 0.01			SD 0.01	0.02	SD 0.11		
							Short circuiting with Ar-25 CO_2 shielding								
E70S-3	57	19.5	205	110.91	262	0.24	0.58	0.48	3.07	6.75	3.05	6.71	4.80	0.04	<0.01
E70S-3	57	19.5	205	110.91	262	0.25	0.59	0.48	3.07	6.75	3.05	6.71	4.87	0.04	<0.01
E70S-3	57	19.5	195	110.91	262	0.26	0.68	0.50	3.07	6.75	3.05	6.71	5.08	0.04	<0.01
						Avg 0.25	0.62	Avg 0.49			Avg 3.05	6.71	Avg 4.91		
						SD 0.01	0.06	SD 0.01			SD 0.00	0.00	SD 0.15		

*g/kg ÷ 10 = weight of fume expressed as a percentage of the deposit weight

Notes:

Electrode polarity: positive

Electrode contact tube-to-work distance: 25 mm (1 in.)

Welding time: 30 seconds

Welding speed: 6.77 mm/s (16 in./min)

Shielding gas flow rate: 21-24 liters/min (45-50 ft^3/h)

Electrode weight: 7.92 g/m (0.20 g/in.)

Metal deposition rate determined for only 1 test per electrode

SD: standard deviation (sigma)

Table B18
Fume generation characteristics of 1.14 mm (0.045 in.) diameter E70S-3 gas metal arc electrodes

Electrode	Code	Voltage, V avg	Current, A avg	Wire feed rate mm/s	Wire feed rate in./min	Fume generation rate Measured g/min	Fume generation rate Normalized $g/min/l^2 \times 10^{-5}$	Wgt. of fume/wgt of electrode used x 100	Melting rate kg/h	Melting rate lb/h	Deposition rate kg/h	Deposition rate lb/h	Wgt. of fume/wgt of deposited metal g/kg*	Wgt of fume/length of electrode used g/m	Wgt of fume/length of electrode used g/in
							Spray transfer with Ar-2 O_2 shielding								
E70S-3	58	34	290	173.57	410	0.46	0.54	0.57	4.82	10.61	4.81	10.57	5.70	0.05	<0.01
E70S-3	58	35	280	173.57	410	0.46	0.59	0.57	4.82	10.61	4.81	10.59	5.67	0.05	<0.01
E70S-3	58	34	285	173.57	410	0.44	0.54	0.55	4.82	10.61	4.81	10.57	5.48	0.05	<0.01
						Avg 0.45	0.56	Avg 0.56			Avg 4.81	10.58	Avg 5.61		
						SD 0.01	0.03	SD 0.01			SD 0.00	0.01	SD 0.12		
							Spray transfer with Ar-9 CO_2 shielding								
E70S-3	58	33	225	137.80	326	0.48	0.95	0.74	3.93	8.65	3.77	8.30	7.63	0.06	<0.01
E70S-3	58	33	225	137.80	326	0.50	0.98	0.76	3.93	8.65	3.80	8.35	7.90	0.06	<0.01
E70S-3	58	33	220	137.80	326	0.48	0.99	0.74	3.93	8.65	3.84	8.46	7.49	0.06	<0.01
						Avg 0.49	0.97	Avg 0.75			Avg 3.80	8.37	Avg 7.67		
						SD 0.01	0.02	SD 0.01			SD 0.04	0.08	SD 0.21		
							Globular transfer with CO_2 shielding								
E70S-3	58	33	320	312.42	738	0.50	0.49	0.34	8.92	19.63	8.77	19.30	3.41	0.02	<0.01
E70S-3	58	32	330	312.42	738	0.44	0.40	0.29	8.92	19.63	8.85	19.48	2.96	0.02	<0.01
E70S-3	58	32	330	312.42	738	0.43	0.39	0.29	8.92	19.63	8.83	19.43	2.91	0.02	<0.01
						Avg 0.45	0.43	Avg 0.31			Avg 8.82	19.40	Avg 3.09		
						SD 0.04	0.05	SD 0.03			SD 0.04	0.10	SD 0.28		
							Short circuiting with Ar-25 CO_2 shielding								
E70S-3	58	19	205	107.31	253	0.24	0.58	0.48	3.06	6.74	2.98	6.56	4.88	0.04	<0.01
E70S-3	58	19	200	107.31	253	0.24	0.60	0.47	3.06	6.74	2.94	6.48	4.88	0.04	<0.01
E70S-3	58	19.5	200	107.31	253	0.24	0.59	0.46	3.06	6.74	2.94	6.47	4.80	0.04	<0.01
						Avg 0.24	0.59	Avg 0.47			Avg 2.95	6.50	Avg 4.85		
						SD 0.00	0.01	SD 0.01			SD 0.03	0.05	SD 0.05		

*g/kg ÷ 10 = weight of fume expressed as a percentage of the deposit weight

Notes

Electrode polarity positive

Electrode contact tube-to-work distance 25.4 mm (1 in.)

Welding time 30 seconds

Welding speed 6.77 mm/s (16 in./min)

Shielding gas flow rate 21-24 liters/min (45-50 ft^3/h)

Electrode weight 7.92 g/m (0.20 g/in.)

Metal deposition rate determined for only 1 test per electrode

SD: standard deviation (sigma)

Table B19
Fume generation characteristics of 1.14 mm (0.045 in.) diameter E70S-5 gas metal arc electrodes

Electrode	Code	Voltage, V avg	Current, A avg	Wire feed rate mm/s	Wire feed rate in./min	Fume generation rate Measured g/min	Fume generation rate Normalized g/min/I ^{2}x10^{-5}	Wgt. of fume/wgt. of electrode used x 100	Melting rate kg/h	Melting rate lb/h	Deposition rate kg/h	Deposition rate lb/h	Wgt of fume/wgt of deposited metal g/kg*	Wgt of fume/length of electrode used g/m	Wgt of fume/length of electrode used g/in
Spray transfer with Ar-2 O_2 shielding															
E70S-5	59	34	275	158.75	375	0.38	0.50	0.49	4.62	10.15	4.56	10.07	4.98	0.04	<0.01
E70S-5	59	34	275	158.75	375	0.39	0.52	0.51	4.62	10.15	4.60	10.12	5.09	0.04	<0.01
E70S-5	59	34	290	158.75	375	0.38	0.45	0.49	4.62	10.15	4.60	10.11	4.98	0.04	<0.01
Avg						0.38	0.49	0.50			4.59	10.10	5.01		
SD						0.01	0.04	0.01			0.01	0.03	0.01		
Globular transfer with CO_2 shielding															
E70S-5	59	32.5	345	317.92	751	0.40	0.34	0.27	9.22	20.28	9.18	20.19	2.63	0.02	<0.01
E70S-5	59	32.5	325	317.92	751	0.40	0.38	0.26	9.22	20.28	9.19	20.22	2.60	0.02	<0.01
E70S-5	59	32.5	330	317.92	751	0.40	0.36	0.26	9.22	20.28	9.19	20.22	2.59	0.02	<0.01
Avg						0.40	0.36	0.26			9.19	20.21	2.61		
SD						0.00	0.02	0.01			0.01	0.02	0.02		
Short circuiting with Ar-25 CO_2 shielding															
E70S-5	59	19	210	120.65	285	0.25	0.56	0.47	3.50	7.70	3.49	7.59	4.33	0.03	<0.01
E70S-5	59	19	215	120.65	285	0.24	0.52	0.45	3.50	7.70	3.42	7.53	4.23	0.03	<0.01
E70S-5	59	19	210	120.65	285	0.24	0.55	0.45	3.50	7.70	3.42	7.53	4.27	0.03	<0.01
Avg						0.24	0.54	0.46			3.44	7.55	4.28		
SD						0.01	0.02	0.01			0.02	0.04	0.05		

*g/kg ÷ 10 = weight of fume expressed as a percentage of the deposit weight

Notes:

Electrode polarity: positive

Electrode contact tube-to-work distance 25.4 mm (1 in.)

Welding time: 30 seconds

Welding speed 6.77 mm/s (16 in./min)

Shielding gas flow rate 21-24 liters/min (45-50 ft^3/h)

Electrode weight: 7.92 g/m (0.20 g/in

Metal deposition rate determined for only 1 test per electrode

SD: standard deviation (sigma)

Table B20

Fume generation characteristics of 0.89 mm (0.035 in.) diameter stainless steel gas metal arc electrodes

Electrode	Code	Voltage, V avg	Current, A avg	Wire feed rate		Fume generation rate		Wgt. of fume/wgt.of electrode used	Melting rate		Deposition rate		Wgt of fume/wgt of deposited metal	Wgt of fume/length of electrode used	
				mm/s	in./min	Measured g/min	Normalized $g/min/I^2 x 10^{-5}$	x 100	kg/h	lb/h	kg/h	lb/h	g/kg*	g/m	g/in
ER 316	65	31	175	271.78	642	0.05	0.15	0.06	4.70	10.34	4.65	10 24	0 59	<0 01	<0 01
ER 316	65	32	165	271.78	642	0.04	0.16	0.06	4.70	10 34	4.61	10.17	0 56	<0 01	<0 01
ER 316	65	32	170	271.78	642	0.04	0.15	0.06	4.70	10.34	4.63	10 18	0.58	<0.01	<0.01
						Avg 0.04	0.15	Avg 0.06			Avg 4.63	10.20	Avg 0.58		
						SD 0.01	0.01	SD 0.00			SD 0.02	0 04	SD 0.02		

*g/kg ÷ 10 = weight of fume expressed as a percentage of the deposit weight

Notes:

Electrode polarity: positive

Electrode contact tube-to-work distance: 25.4 mm (1 in.)

Welding time: 1 minute

Welding speed: 6.77 mm/s (16 in./min)

Shielding gas: Ar-2 O_2

Shielding gas flow rate. 21-24 liters/min (45-50 ft^3/h)

Electrode weight: Code 65: 4 79 g/m (0.12 g/in.)

Base plate: 304 SS

Metal deposition rate determined for only 1 test per electrode

SD: standard deviation (sigma)

Table B21
Fume generation characteristics of 0.89 mm (0.035 in.) and 1.14 mm (0.045 in.) diameter high alloy gas metal arc electrodes

Electrode	Code	Voltage, V avg	Current, A avg	Wire feed rate mm/s	Wire feed rate in./min	Fume generation rate Measured g/min	Fume generation rate Normalized $g/min/I^2 \times 10^{-5}$	Wgt. of fume/wgt of electrode used x 100	Melting rate kg/h	Melting rate lb/h	Deposition rate kg/h	Deposition rate lb/h	Wgt. of fume/wgt. of deposited metal g/kg*	Wgt of fume/length of electrode used g/m	Wgt of fume/length of electrode used g/in
Inconel 625	63	33.5	195	147.74	349	0.07	0.18	0.09	4.49	9 87	4.36	9 59	0.92	<0.01	<0.01
Inconel 625	63	33.5	190	147.74	349	0.06	0.17	0.08	4.49	9.87	4.35	9.58	0.86	<0.01	<0.01
Inconel 625	63	33.5	190	147.74	349	0.06	0.17	0.08	4.49	9.87	4.36	9.59	0.83	<0.01	<0 01
Avg						0.06	0.17	0.08			4.36	9.59	0.87		
SD						0.01	0.01	0.01			0.01	0.01	0.05		
ERNiCu-7	64	31.5	260	147.74	349	0.15	0.23	0.19	4.82	10.59	4.60	10.12	2.00	0.02	<0.01
ERNiCu-7	64	31.5	250	147.74	349	0.15	0.25	0.20	4.82	10.59	4.57	10.04	2.08	0.02	<0.01
ERNiCu-7	64	31.5	260	147.74	349	0.16	0.23	0.20	4.82	10 59	4.64	10.20	2.04	0.02	<0.01
Avg						0.16	0.24	0.20			4.60	10.20	2.04		
SD						0.01	0.01	0.01			0.03	0.08	0.04		
Haynes 25	71	29.5	200	100.75	238	0.08	0.19	0.13	3.59	7.89	3.48	7.66	1 38	0.01	<0.01
Haynes 25	71	29.5	200	100.75	238	0.08	0.19	0.13	3.59	7.89	3.50	7 69	1.37	0.01	<0.01
Haynes 25	71	29.5	205	100.75	238										
Avg						0.08	0.19	0.13			3.49	7.68	1 38		
SD						0.00	0.00	0.00			0.01	0.01	0.01		
Haynes C-276	68	30	165	166.37	393	0.39	1.43	0.69	3.38	7.44	3.36	7.40	6.96	0.04	<0.01
Haynes C-276	68	30	165	166.37	393	0.39	1.43	0.69	3.38	7 37	3.35	7.37	6.99	0.04	<0 01
Haynes C-276	68	30	165	177.37	393	0.39	1.43	0 69	3.38	7 44	3.35	7 37	6 98	0.04	<0 01
Avg						0.39	1.43	0.69			3.35	7 38	6.98		
SD						0.00	0.00	0 00			0.01	0.02	0.02		

*g/kg - 10 = weight of fume expressed as a percentage of the deposit weight.

Notes:

Electrode polarity: positive

Electrode contact tube-to-work distance: Code 63, 64, and 71 25.4 mm (1 in.), Code 68: 19.1 mm (0.75 in.)

Welding time: 1 minute

Welding speed: Code 63, 64, and 71: 6.77 mm/s (15 in./min); Code 68: 5.93 mm/s (14 in./min)

Shielding gas: argon

Shielding gas flow rate: 21-24 liters/min (45-50 ft^3/h)

Electrode diameter Code 63, 64, and 71. 1.14 mm (0.045 in.); Code 68: 0.89 mm (0.035 in.)

Electrode weight: Code 63: 8.44 g/m (0.21 g/in.); Code 64 9.02 g/m (0.23 g/in.),
Code 71: 9.86 g/m (0.25 g/in.); Code 68 5.64 g/m (0.14 g/in)

Base plate. Code 63: Inconel; Code 64. Monel, Code 68 Haynes C-276, Code 71 Haynes 25

Composition. Inconel 626 61Ni-21.5Cr-2.5Fe-9Mo-3.65 (Cb + Ta)-0.25Mn-0.25Si-0.2Al-0.2Ti
ERNiCu-7: 4Mn-2.5Fe-1.25Si-1.25Al-1.5 to 3.0 Ti-62 to 69 N-Bal.Cu
Haynes C-276. Haynes C-276-2.5Co-16Cr-16Mo-4W-5Fe-0.08Si-1.0Mn-2Cu-2 (Cb + Ta)-Bal.Ni
Haynes 25 Haynes 25-10Ni-20Cr-15W-3Fe-1.0Si-1.0Mn-Bal.Co

Metal deposition rate determined for only 1 test per electrode.

SD: standard deviation (sigma).

Table B22
Fume generation characteristics of 1.19 mm (3/64 in.) diameter 4043 aluminum gas metal arc electrodes

Electrode	Code	Voltage, V avg	Current, A avg	Contact tube-to-work distance mm	Contact tube-to-work distance in	Fume generation rate Measured g/min	Fume generation rate Normalized g/min/I² x 10⁻⁵	Wgt. of fume/wgt. of electrode used x 100	Melting rate kg/h	Melting rate lb/h	Deposition rate kg/h	Deposition rate lb/h	Wgt. of fume/wgt. of deposited metal g/kg*	Wgt. of fume/length of electrode used g/m	Wgt. of fume/length of electrode used g/in.
ER4043	61	23	160	19.1	0.75	0.11	0.43	0.60	1.09	2.40	1.03	2.27	6.33	0.02	<0.01
ER4043	61	22	165	19.1	0.75	0.10	0.37	0.55	1.09	2.40	1.02	2.25	5.88	0.02	<0.01
						Avg 0.11	0.40	Avg 0.57			Avg 1.03	2.26	Avg 5.60		
						SD 0.01	0.03	SD 0.03			SD 0.01	0.01	SD 0.27		
ER4043	61	23	165	25.4	1.00	0.28	1.02	1.45	1.15	2.53	1.05	2.30	15.96	0.04	<0.01
ER4043	61	23	160	25.4	1.00	0.27	1.07	1.44	1.15	2.53	1.06	2.33	15.59	0.04	<0.01
ER4043	61	23	165	25.4	1.00	0.27	1.00	1.43	1.15	2.53	1.05	2.30	15.68	0.04	<0.01
						Avg 0.27	1.03	Avg 1.44			Avg 1.05	2.31	Avg 15.74		
						SD 0.01	0.04	SD 0.01			SD 0.01	0.02	SD 0.19		

*g/kg ÷ 10 = weight of fume expressed as a percentage of the deposit weight.

Notes

Electrode polarity: positive

Wire feed rate: 104.14 mm/s (246 in./min) for 3/4 in. stickout; 109.64 mm/s (259 in./min) for 1 in. stickout.

Welding time: 30 seconds

Welding speed: 6.77 mm/s (16 in./min)

Shielding gas: argon

Shielding gas flow rate: 21-24 liters/min (45-50 ft^3/h)

Electrode weight: 2.90 g/m (0.07 g/in.)

Base plate: 1100 Al

Metal deposition rate determined for only 1 test per electrode.

SD: standard deviation (sigma).

Table B23
Fume generation characteristics of 1.19 mm (3/64 in.) diameter ER5356 aluminum gas metal arc electrodes

Electrode	Code	Voltage, V avg	Current, A avg	Contact tube-to-work distance mm	Contact tube-to-work distance in	Fume generation rate Measured g/min	Fume generation rate Normalized g/min/I^2 x 10^{-5}	Wgt of fume/wgt. of electrode used x 100	Melting rate kg/h	Melting rate lb/h	Deposition rate kg/h	Deposition rate lb/h	Wgt of fume/wgt of deposited metal g/kg*	Wgt. of fume/length of electrode used g/m	Wgt. of fume/length of electrode used g/in.
ER5356	60	22	160	12.7	0.50	1.41	5.50	6.26	1.35	2.96	1.30	2.85	65.16	0.18	<0.01
ER5356	60	22	165	12.7	0.50	1.43	5.24	6.34	1.35	2.96	1.31	2.88	65.40	0.18	<0.01
ER5356	60	22	165	12.7	0.50	1.39	5.10	6.16	1.35	2.96	1.31	2.85	64.27	0.18	<0.01
Avg						1.41	5.28	6.25			1.31	2.86	64.94		
SD						0.02	0.20	0.09			0.01	0.02	0.60		
ER5356	60	23.5	165	19.1	0.75	1.75	6.42	6.60	1.59	3.50	1.33	2.93	78.84	0.19	<0.01
ER5356	60	23.5	160	19.1	0.75	1.74	6.79	6.56	1.59	3.50	1.30	2.85	80.59	0.19	<0.01
Avg						1.75	6.60	6.58			1.31	2.89	79.72		
SD						0.01	0.26	0.02			0.02	0.06	1.23		
ER5356	60	23	150	25.4	1.00	3.75	16.62	14.82	1.52	3.34	1.40	2.93	168.24	0.43	0.01
ER5356	60	23	155	25.4	1.00	3.57	14.87	14.15	1.52	3.34	1.25	2.75	171.46	0.41	0.01
ER5356	60	23	155	25.4	1.00	3.47	14.45	13.76	1.52	3.34	1.35	2.96	154.85	0.40	0.01
Avg						3.59	15.31	14.24			1.31	2.88	164.85	0.41	0.01
SD						0.13	1.15	0.05			0.05	0.12	8.81	0.02	0.00

*g/kg ÷ 10 = weight of fume expressed as a percentage of the deposit weight

Notes

Electrode polarity positive

Wire feed rate 129.54 mm/s (306 in./min) for 1/2 in. stickout, 152.40 mm/s (360 in./min) for 3/4 in. stickout, 145.20 mm/s (343 in./min) for 1 in. stickout.

Welding time 30 seconds

Welding speed 6.77 mm/s (16 in./min)

Shielding gas argon

Shielding gas flow rate: 21-24 liters/min (45-50 ft^3/h)

Electrode weight 2.90 g/m (0.07 g/in.)

Base plate 5083 Al

Metal deposition rate determined for only 1 test per electrode

SD standard deviation (sigma).

Table B24
Fume generation characteristics of 1.14 mm (0.045 in.) diameter copper-base gas metal arc electrodes

Electrode	Code	Voltage, V avg	Current, A avg	Wire feed rate mm/s	Wire feed rate in./min	Fume generation rate Measured g/min	Fume generation rate Normalized $g/min/I^2 \times 10^{-5}$	Wgt. of fume/wgt. of electrode used x 100	Melting rate kg/h	Melting rate lb/h	Deposition rate kg/h	Deposition rate lb/h	Wgt of fume/wgt. of deposited metal g/kg*	Wgt of fume/length of electrode used g/m	Wgt of fume/length of electrode used g/in.
ERCu	76	27	205	120 66	285	0.31	0.75	0.49	3.83	8.42	3.70	8.15	5.09	0.04	<0.01
ERCu	76	27	205	120.66	285	0.29	0.69	0.46	3.83	8.42	3.60	7.93	4.84	0.04	<0.01
ERCu	76	27 5	210	120.66	285	0.29	0.67	0.47	3.83	8.42	3 62	7.96	4.87	0.04	<0.01
Avg						0.30	0.70	0.47			3.64	8.01	4.93		
SD						0.01	0.04	0.02			0.05	0.12	0.13		
ERCuAl-A2	75	27.5	210	129.54	306	0.47	1.08	0.80	3.55	7.80	3.50	7.69	8.14	0.06	<0.01
ERCuAl-A2	75	27.5	215	129.54	306	0.47	1.02	0.80	3.55	7.80	3.50	7.69	8.07	0.06	<0.01
ERCuAl-A2	75	28	215	129.54	306	0 46	1.00	0.78	3.55	7.80	3.40	7.47	8.14	0.06	<0.01
Avg						0.47	1.03	0 79			3.46	7.62	8.12		
SD						0.01	0 04	0.01			0.06	0.13	0.04		

*g kg - 10 = weight of fume expressed as a percentage of the deposit weight.

Notes

Electrode polarity positive

Electrode contact tube-to-work distance: 19.05 mm (0.75 in)

Welding time 1/2 minute

Welding speed Code 75 6.77 mm/s (16 in./min), Code 76 5 08 mm/s (12 in./min)

Shielding gas argon

Shielding gas flow rate: 21-24 liters/min (45-50 ft^3/h)

Electrode weight Code 75 7 59 g/m (0 19 g/in.); Code 76 8.8 g/m (0.22 g/in.)

Base plate Copper for ERCu
Aluminum-bronze for ERCu Al-A2

Metal deposition rate determined for only 1 test per electrode

SD standard deviation (sigma).

Table B25

Fume generation characteristics of 2.38 mm (3/32 in.) diameter E70T-1 flux cored electrodes (75 Ar-25 CO_2) shielding

Code	Voltage, V	Current, A	Wire feed rate mm/s	Wire feed rate in./min		Fume generation rate Measured g/min	Fume generation rate Normalized g/min/I^2 x 10^{-5}		Wgt. of fume/wgt. of electrode used x 100	Melting rate kg/h	Melting rate lb/h		Metal deposition rate kg/h	Metal deposition rate lb/h		Wgt. of fume/wgt. of deposited metal g/kg*	Wgt. of fume/lenght of electrode used g/m	Wgt. of fume/lenght of electrode used g/in.
40	30.5	465	88.90	210		1.01	0.47		0.67	9.08	19.70		7.73	17.18		7.79	0.19	>0.01
40	30.5	460	88.90	210		1.01	0.48		0.67	9.08	19.70		7.74	17.21		7.78	0.19	>0.01
40	30.5	445	88.90	210		1.01	0.51		0.67	9.08	19.70		7.72	17.15		7.77	0.19	>0.01
					Avg	1.01	0.49	Avg	0.67			Avg	7.73	17.18	Avg	7.78		
					SD	>0.01	0.02	SD	>0.01			SD	0.01	0.03	SD	0.01		
42	30.5	445	88.90	210		1.94	0.98		1.29	9.01	19.82		7.73	17.18		14.88	0.36	0.01
42	30.5	445	88.90	210		1.92	0.97		1.28	9.01	19.82		7.71	17.13		14.82	0.36	0.01
42	30.5	445	88.90	210		1.93	0.97		1.29	9.01	19.82		7.62	16.94		15.02	0.36	0.01
					Avg	1.93	0.97	Avg	1.29			Avg	7.69	17.08	Avg	14.91		
					SD	0.01	>0.01	SD	>0.01			SD	0.06	0.13	SD	0.10		

*g/kg ÷ 10 = weight of fume expressed as a percentage of the deposit weight

Notes

Electrode polarity: positive

Electrode contact tube-to-work distance: 38 mm (1-1/2 in.)

Welding time: 30 seconds

Shielding gas flow rate: 21-24 liters/min (45-50 ft^3/h)

Welding speed: 6.77 mm/s (16 in./min)

Electrode weight/unit length: Code No. 40: 28.35 g/m (0.72 g/in.)
Code No. 42: 27.95 g/m (0.71 g/in.)

Metal deposition rate determined for only 1 test per electrode.

SD: standard deviation (sigma).

Table B25a

Comparison of average fume generation characteristics of 2.38 mm (3/32 in.) diameter E70T-1 electrodes as a function of shielding gas

Code	Shielding gas	Voltage, V	Current, A	Average fume generation rate		Average metal deposition rate		Average weight of fumes/ weight of deposited metal
				Measured g/min	Normalized $g/min/I^2 \times 10^{-5}$	kg/h	lb/h	g/kg*
40	CO_2	30.5	475	1.35	0.59	7.73	17.18	10.40
40	Ar-25 CO_2	30.5	465	1.01	0.49	7.73	17.18	7.78
42	CO_2	31	440	2.27	1.17	7.71	17.13	17.51
42	Ar-25 CO_2	30.5	445	1.93	0.97	7.69	17.08	14.91

*g/kg ÷ 10 = weight of fume expressed as a percentage of the deposit weight.

Notes:

Wire feed rate: 87.0-88.9 mm/s

Shielding gas flow rate: 21-24 liters/min

Metal deposition rate determined for only 1 test per electrode.

SD: standard deviation (sigma).

Table B26
Fume generation characteristics of a 3.97 mm (5/32 in.) diameter E7018 electrode as a function of current

Electrode	Code	Voltage, V avg	Current, A avg	Fume generation rate Measured g/min	Fume generation rate Normalized g/min/I^2 x 10^{-5}	Wgt. of fume/ wgt. of electrode used x 100	Melting rate kg/h	Melting rate lb/h	Deposition rate kg/h	Deposition rate lb/h	Wgt. of fume/ wgt. of deposited metal g/kg*	Wgt. of fume/ length of electrode used g/m	Wgt. of fume/ length of electrode used g/in
E7018	10	21	130	0.36	2.16	1.23	1.78	3.91	1.26	2.76	17.45	1.84	0.05
E7018	10	20	150	0.40	1.76	1.23	1.93	4.24	1.37	3.01	17.33	1.84	0.05
E7018	10	21	160	0.42	1.63	1.20	2.09	4.61	1.45	3.18	17.38	1.79	0.05
E7018	10	20	170	0.41	1.42	1.18	2.09	4.60	1.44	3.17	17.03	1.75	0.04
E7018	10	20.5	185	0.50	1.45	1.34	2.22	4.89	1.48	3.26	20.14	2.01	0.05
E7018	10	21.5	185	0.55	1.60	1.38	2.38	5.23	1.74	3.83	18.83	2.05	0.05
E7018	10	21	190	0.60	1.66	1.50	2.40	5.28	1.75	3.85	20.49	2.23	0.06
E7018	10	22	210	0.72	1.62	1.69	2.54	5.58	1.75	3.85	22.69	2.62	0.07
E7018	10	21	210	0.73	1.66	1.76	2.49	5.47	1.87	4.11	23.44	2.63	0.07

*g/kg 10 = weight of fume expressed as a percentage of the deposit weight

Notes:

Electrode polarity positive

Welding speed 3.81 mm/s (9 in./min)

Welding time 1 minute

Electrode weight 148.42 g/m (3.77 g/in.)

Metal deposition rate determined for only 1 test per electrode.

SD standard deviation (sigma).

Table B27
Fume generation characteristics of a 3.97 mm (5/32 in.) diameter E6010 covered electrode as a function of current

Electrode	Code	Voltage, V avg	Current, A avg	Fume generation rate: Measured g/min	Fume generation rate: Normalized $g/min/I^2 \times 10^{-5}$	Wgt. of fume/ wgt. of electrode used x 100	Melting rate kg/h	Melting rate lb/h	Deposition rate kg/h	Deposition rate lb/h	Wgt. of fume/ wgt. of deposited metal g/kg*	Weight of fume/ length of electrode used g/m	Weight of fume/ length of electrode used g/in.
E6010	11	25	100	0.36	3.60	1.57	1.40	3.08	1.20	2.63	18.07	1.79	0.05
E6010	11	24	100	0.36	3.60	1.49	1.45	3.20	1.15	2.54	18.71	1.73	0.04
E6010	11	24.5	110	0.36	2.98	1.53	1.41	3.11	1.15	2.52	18.85	1.79	0.05
Avg				0.36	3.39	1.53	1.42	3.13	1.17	2.56	18.54	1.75	0.05
SD				0.00	0.36	0.04	0.03	0.13	0.03	0.06	0.42	0.04	0.01
E6010	11	23	130	0.47	2.78	1.67	1.69	3.72	1.38	3.04	20.41	1.94	0.05
E6010	11	22	135	0.49	2.69	1.68	1.75	3.85	1.39	3.05	21.21	1.94	0.05
E6010	11	22.5	125	0.48	3.07	1.66	1.74	3.82	1.39	3.05	20.77	1.98	0.05
Avg				0.48	2.85	1.67	1.73	3.80	1.39	3.05	20.80	1.96	0.05
SD				0.01	0.02	0.01	0.03	0.07	0.01	0.01	0.40	0.02	<0.01
E6010	11	22.5	140	0.79	4.03	2.62	1.81	3.99	1.39	3.06	34.08	3.03	0.08
E6010	11	23	145	0.78	3.71	2.60	1.80	3.97	1.40	3.07	33.54	3.00	0.08
E6010	11	22	145	0.79	3.76	2.69	1.76	3.88	1.41	3.11	33.53	3.11	0.08
Avg				0.79	3.83	2.64	1.79	3.95	1.40	3.08	33.71	3.05	0.08
SD				0.01	0.17	0.05	0.03	0.06	0.01	0.03	0.32	0.04	<0.01
E6010	11	23.5	150	0.83	3.69	2.74	1.82	4.00	1.43	3.15	34.78	3.17	0.08
E6010	11	23	150	0.83	3.69	2.70	1.84	4.06	1.42	3.13	35.00	3.13	0.08
E6010	11	23	150	0.83	3.69	2.68	1.86	4.10	1.43	3.15	34.78	3.09	0.08
Avg				0.83	3.69	2.71	1.84	4.05	1.43	3.14	34.85	3.13	0.08
SD				0.00	0.00	0.03	0.02	0.05	0.01	0.01	0.13	0.04	<0.01
E6010	11	22	170	1.18	4.08	3.45	2.05	4.51	1.50	3.30	47.19	4.00	0.10
E6010	11	22	175	1.17	3.82	3.43	2.05	4.51	1.50	3.30	46.79	3.96	0.10
E6010	11	23	170	1.17	4.05	3.43	2.05	4.51	1.50	3.30	46.79	3.98	0.10
Avg				1.17	3.98	3.44	2.05	4.51	1.50	3.30	46.93	3.98	0.10
SD				0.01	0.14	0.01	0.00	0.00	0.00	0.00	0.23	0.02	<0.01

*g/kg ÷ 10 = weight of fume expressed as a percentage of the deposit weight.

Notes:

Electrode polarity: positive

Welding speed: 3.39 to 4.87 mm/s (8 to 11.5 in./min)

Welding time: 1 minute

Electrode weight: 114.57 g/m (2.91 g/in.)

Metal deposition rate determined for only 1 test per electrode.

SD: standard deviation (sigma).

Table B28
Fume generation characteristics of 2.38 mm (3/32 in.) diameter E70T-1 flux cored electrode as a function of current and voltage

Electrode	Code	Voltage, V avg	Current, A avg	Wire feed rate mm/s	Wire feed rate in./min	Fume generation rate Measured g/min	Fume generation rate Normalized $g/min/I^2 \times 10^{-5}$	Wgt. of fume/ wgt. of electrode used x 100	Melting rate kg/h	Melting rate lb/h	Deposition rate kg/h	Deposition rate lb/h	Wgt. of fume/ wgt. of deposited metal g/kg*	Wgt. of fume/ length of electrode used g/m	Wgt. of fume/ length of electrode used g/in.
								Effect of current							
E70T-1	41	28	345	52.49	124	1.14	0.96	1.24	5.56	12.23	4.91	10.80	13.99	0.37	0.01
E70T-1	41	27	345	52.49	124	1.11	0.93	1.20	5.56	12.33	4.70	10.33	14.21	0.35	0.01
E70T-1	41	22	360	55.88	132	1.06	0.82	1.07	5.95	13.08	4.81	10.57	13.26	0.32	0.01
E70T-1	41	25	350	52.49	124	1.08	0.88	1.16	5.59	12.29	4.73	10.41	13 72	0.34	0.01
E70T-1	41	22	330	52.49	124	1.01	0.93	1.09	5.58	12.27	5.02	11.05	12.10	0.32	0.01
E70T-1	41	29	425	72.39	171	0.97	0.54	0.76	7 70	16.95	6.79	14.90	8.58	0.22	0.01
E70T-1	41	31	475	85.09	201	0.99	0.44	0.66	9.06	19.92	8.18	18.00	7.30	0.19	<0.01
E70T-1	41	31	485	85.09	201	1.01	0.43	0.67	9.06	19.82	8.36	18.40	7.21	0.20	0.01
E70T-1	41	34	525	90.59	214	1.33	0.49	0.83	9.64	21.21	9.14	20.11	8.73	0.24	<0.01
E70T-1	41	35	600	113.45	268	2.04	0.57	1.01	12.06	26.54	11.38	25.03	10.74	0.30	0.01
								Effect of voltage							
E70T-1	41	20	460	79.16	187	0.61	0.29	0.44	8.41	18.51	7.40	16.28	4.97	0.13	<0.01
E70T-1	41	30	450	79.16	187	0.91	0.45	0.65	8.41	18.51	7.47	16.44	7.32	0.19	<0.01
E70T-1	41	36	435	76.20	180	1.21	0.64	0.90	8.11	17.84	6.92	15.22	10.49	0.26	0.01
E70T-1	41	25	555	110.06	260	0.64	0.21	0.33	11.69	25.72	10.54	23.18	3.63	0.10	<0.01
E70T-1	41	34	540	109.22	258	1.46	0.50	1.51	11.62	25.57	10.72	23.58	8.20	0.22	0.01
E70T-1	41	38	540	103.72	245	2.01	0.69	2.19	11.13	24.48	9.73	21.41	12.41	0.32	0.01
E70T-1	41	39	325	46.99	111	1.45	1.37	1.65	5.00	10.99	4.08	8.98	21.25	0.51	0.01
E70T-1	41	39	405	63.08	149	1.42	0.87	1.28	6.68	14.71	5.89	12.95	14.48	0.37	0.01
E70T-1	44	20	310	48.26	114	0.86	0.90	2.10	5.13	11.29	4.21	9.25	12.32	0.30	0.01
E70T-1	41	39	315	46.99	111	1.47	1.49	1.77	5.00	10.99	4.08	8.98	21.68	0.50	0.01

*g/kg ÷ 10 = weight of fume expressed as a percentage of the deposit weight.

Notes:

Direct current: electrode positive

Electrode contact tube-to-work distance: 31.8 mm (1.25 in.)

Welding time: 30 seconds

Welding speed: 6.77 mm/s (16 in./min)

Shielding gas: CO_2

Shielding gas flow rate: 21-24 liters/min (45-50 ft^3/h)

Electrode weight: (0.75 g/in.)

Metal deposition rate determined for only 1 test per electrode.

SD: standard deviation (sigma).

Table B29

Fume generation characteristics of a 1.14 mm (0.045 in.) diameter E70S-3 gas metal arc electrode as a function of current

Electrode	Code	Voltage, V avg	Current, A avg	Wire feed rate mm/s	Wire feed rate in./min		Fume generation rate Measured g/min	Fume generation rate Normalized $g/min/I^2 \times 10^{-5}$		Wgt. of fume/wgt. of electrode used x 100	Melting rate kg/h	Melting rate lb/h		Deposition rate kg/h	Deposition rate lb/h		Wgt. of fume/wgt. of deposited metal g/kg*	Wgt. of fume/length of electrode used g/m	Wgt. of fume/length of electrode used g/in.
E70S-3	58	35	240	132.50	313		0.25	0.43		0.80	3.77	8.30		3.75	8.25		4.00	0.03	<0.01
E70S-3	58	35.5	245	132.50	313		0.24	0.40		0.77	3.77	8.30		3.52	7.74		4.14	0.03	<0.01
E70S-3	58	35	240	132.50	313		0.25	0.44		0.81	3.77	8.30		3.65	8.04		4.17	0.03	<0.01
E70S-3	58	35	240	132.50	313		0.26	0.45		0.83	3.77	8.30		3.60	7.93		4.32	0.03	<0.01
						Avg	0.25	0.43	Avg	0.80			Avg	3.63	7.99	Avg	4.15		
						SD	0.01	0.02	SD	0.03			SD	0.10	0.21	SD	0.13		
E70S-3	58	34.5	260	162.56	384		0.48	0.71		0.62	4.64	10.20		4.54	9.99		6.32	0.05	<0.01
E70S-3	58	35	270	162.56	384		0.47	0.64		0.61	4.64	10.20		4.53	9.96		6.21	0.05	<0.01
E70S-3	58	34.5	270	162.56	384		0.47	0.62		0.60	4.64	10.20		4.54	9.99		6.16	0.05	<0.01
						Avg	0.47	0.66	Avg	0.61			Avg	4.54	9.98	Avg	6.22		
						SD	0.01	0.05	SD	0.01			SD	0.01	0.02	SD	0.08		
E70S-3	58	34	290	173.57	410		0.46	0.54		0.57	4.82	10.61		4.81	10.57		5.70	0.05	<0.01
E70S-3	58	35	280	173.57	410		0.46	0.59		0.57	4.82	10.61		4.81	10.59		5.67	0.05	<0.01
E70S-3	58	34	285	173.57	410		0.44	0.54		0.55	4.82	10.61		4.81	10.57		5.48	0.05	<0.01
						Avg	0.45	0.56	Avg	0.56			Avg	4.81	10.58	Avg	5.61		
						SD	0.01	0.03	SD	0.01			SD	0.00	0.01	SD	0.12		
E70S-3	58	35.5	295	189.65	448		0.45	0.51		0.50	5.41	11.90		5.20	11.44		5.12	0.04	<0.01
E70S-3	58	35	300	189.65	448		0.54	0.60		0.60	5.41	11.90		5.17	11.39		6.25	0.05	<0.01
E70S-3	58	35	300	189.65	448		0.53	0.59		0.59	5.41	11.90		5.19	11.42		6.13	0.05	<0.01
						Avg	0.51	0.57	Avg	0.56			Avg	5.19	11.42	Avg	5.85		
						SD	0.05	0.05	SD	0.06			SD	0.01	0.03	SD	0.57		

*g/kg - 10 = weight of fume expressed as a percentage of the deposit weight.

Notes

Electrode polarity positive

Electrode contact tube-to-work distance: 25.4 mm (1 in.)

Welding time 30 seconds

Welding speed 6.77 mm/s (16 in./min)

Shielding gas: A-$2O_2$

Shielding gas flow rate: 21-24 liters/min (45-50 ft^3/h)

Electrode weight: 7.92 g/m (0.20 g/in.)

Metal deposition rate determined for only 1 test per electrode.

SD: standard deviation (sigma).

Table B30

Fume generation rate of E6013 electrodes with different diameters (electrode negative)

Electrode	Code	Voltage, V avg	Current, A avg	Electrode diameter mm	Electrode diameter in	Fume generation rate Measured g/min	Fume generation rate Normalized g/min/I^2 x 10^{-5}	Wgt of fume/wgt. of electrode used x 100	Melting rate kg/h	Melting rate lb/h	Deposition rate kg/h	Deposition rate lb/h	Wgt. of fume/wgt. of deposited metal g/kg*	Wgt. of fume/length of electrode used g/m	Wgt. of fume/length of electrode used g/in.
E6013	4	25	155	3.18	1/8	0.83	3.47	2.24	2.22	4.90	1.13	2.48	44.34	1.68	0.04
E6013	4	26	155	3.18	1/8	0.85	3.53	2.26	2.26	4.98	1.03	2.27	49.45	1.70	0.04
E6013	4	24	155	3.18	1/8	0.84	3.50	2.24	2.25	4.96	1.14	2.51	44.18	1.69	0.04
Avg						0.84	3.50	2.25	2.25	4.95	1.10	2.42	45.98	1.69	0.04
SD						0.01	0.03	0.01	0.02	0.04	0.06	0.13	3.01	0.01	0.00
E6013	4	21	145	3.97	5/32	0.30	1.42	1.02	1.76	3.87	1.39	3.05	12.93	1.20	0.03
E6013	4	22	150	3.97	5/32	0.30	1.32	1.04	1.71	3.77	1.37	3.01	12.98	1.22	0.03
E6013	4	22	150	3.97	5/32	0.29	1.30	1.00	1.76	3.87	1.38	3.03	12.73	1.21	0.03
Avg						0.30	1.35	1.02	1.74	3.84	1.38	3.03	12.88	1.21	0.03
SD						0.01	0.06	0.02	0.03	0.06	0.01	0.02	0.13	0.01	0.00
E6013	4	19	140	4.76	3/16	0.14	0.74	0.51	1.73	3.79	1.44	3.16	5.91	0.81	0.02
E6013	4	19	140	4.76	3/16	0.15	0.77	0.54	1.70	3.75	1.48	3.25	6.14	0.88	0.02
E6013	4	18	150	4.76	3/16	0.15	0.69	0.53	1.74	3.83	1.46	3.21	6.36	0.88	0.02
Avg						0.15	0.73	0.53	1.72	3.79	1.46	3.21	6.14	0.86	0.02
SD						0.01	0.04	0.02	0.02	0.04	0.02	0.04	0.20	0.04	0.00

*g/kg ÷ 10 = weight of fume expressed as a percentage of the deposit weight

Notes

Electrode polarity negative

Welding speed 1/8 in diameter electrode 4.02 mm/s (9.5 in./min),
5/32 in. diameter electrode 3.55 mm/s (8.38 in./min),
3/16 in diameter electrode 2.75 mm/s (6.5 in./min)

Electrode weight 1/8 in. diameter electrode 75.15 g/m (1.91 g/in.)
5/32 in. diameter electrode 117.50 g/m (2.98 g/in.)
3/16 in. diameter electrode 164.74 g/m (4.18 g/in.)

Code 4: All electrodes made by same manufacturer.

Metal deposition rate determined for only 1 test per electrode.

SD: standard deviation (sigma).

Table B31
Fume generation rate of E6013 electrodes with different diameters (electrode positive)

Electrode	Code	Voltage, V avg	Current, A avg	Electrode diameter mm	Electrode diameter in.	Fume generation rate Measured g/min	Fume generation rate Normalized g/min/I² x 10⁻⁵	Wgt. of fume/ wgt of electrode used x 100	Melting rate kg/h	Melting rate lb/h	Deposition rate kg/h	Deposition rate lb/h	Wgt. of fume/ wgt. of deposited metal g/kg*	Wgt. of fume/ length of electrode used g/m	Wgt. of fume/ length of electrode used g/in.
E6013	4	25	150	3.18	1/8	1.12	4.99	3.42	1.97	4.34	1.06	2.34	63.45	2.57	0.07
E6013	4	24	155	3.18	1/8	1.16	4.84	3.43	2.04	4.48	1.07	2.36	65.02	2.58	0.07
E6013	4	24	160	3.18	1/8	1.14	4.46	3.38	2.03	4.46	1.06	2.33	64.63	2.54	0.06
Avg						1.14	4.76	3.41	2.01	4.43	1.06	2.34	64.37	2.56	0.07
SD						0.02	0.27	0.03	0.03	0.08	0.01	0.02	0.82	0.02	0.01
E6013	4	23	145	3.97	5/32	0.45	2.12	1.54	1.73	3.81	1.14	2.51	23.42	1.81	0.05
E6013	4	22	145	3.97	5/32	0.46	2.17	1.58	1.73	3.81	1.15	2.52	23.90	1.86	0.05
E6013	4	21	150	3.97	5/32	0.47	2.08	1.62	1.73	3.80	1.14	2.51	24.62	1.91	0.05
Avg						0.46	2.12	1.58	1.73	3.81	1.14	2.51	23.97	1.86	0.05
SD						0.01	0.05	0.04	0.00	0.01	0.01	0.01	0.61	0.05	0.00
E6013	4	19	150	4.76	3/16	0.21	0.94	0.70	1.82	4.00	1.51	3.33	8.36	1.15	0.03
E6013	4	19	150	4.76	3/16	0.24	1.06	0.81	1.78	3.93	1.53	3.36	9.44	1.33	0.03
E6013	4	20	150	4.76	3/16	0.24	1.08	0.82	1.78	3.91	1.53	3.36	9.55	1.35	0.03
Avg						0.23	1.03	0.77	1.79	3.95	1.52	3.35	9.11	1.28	0.03
SD						0.02	0.07	0.07	0.02	0.05	0.01	0.02	0.66	0.11	0.00

*g/kg – 10 = weight of fume expressed as a percentage of the deposit weight.

Notes

Electrode polarity positive

Welding speed 3.91 mm/s (9 25 in./min) with 1/8 in. diameter electrode,
3.81 mm/s (9 in./min) with 5/32 in. diameter electrode,
3.12 mm/s (7.38 in./min) with 3/16 in. diameter electrodes

Electrode weight 1/8 in. diameter electrode: 75.15 g/m (1.91 g/in.)
5/32 in. diameter electrode: 117.50 g/m (2.98 g/in.)
3/16 in. diameter electrode 164.74 g/m (4.18 g/in.)

Code 4 All electrodes made by same manufacturer.

Metal deposition rate determined for only 1 test per electrode.

SD standard deviation (sigma).

Table B32
Fume generation characteristics of E7018 covered electrodes with different diameters

Electrode	Code	Voltage, V avg	Current, A avg	Electrode diameter mm	Electrode diameter in.	Fume generation rate: Measured g/min	Fume generation rate: Normalized g/min/I² x 10⁻⁵	Wgt. of fume/wgt. of electrode used x 100	Melting rate kg/h	Melting rate lb/h	Deposition rate kg/h	Deposition rate lb/h	Wgt. of fume/wgt. of deposited metal g/kg*	Wgt. of fume/length of electrode used g/m	Wgt. of fume/length of electrode used g/in.
E7018	10	21	65	2.38	3/32	0.19	4.58	1.30	0.90	1.97	0.59	1.29	19.69	0.77	0.02
E7018	10	22	63	2.38	3/32	0.19	4.87	1.28	0.90	1.97	0.58	1.25	19.61	0.76	0.02
E7018	10	21	65	2.38	3/32	0.19	4.45	1.31	0.86	1.89	0.57	1.25	19.79	0.79	0.02
Avg						0.19	4.64	1.30	0.88	1.94	0.58	1.28	19.69	0.77	0.02
SD						>0.01	0.21	0.02	0.02	0.04	0.01	0.02	0.09	0.01	>0.01
E7018	10	21	100	3.18	1/8	0.33	3.34	1.63	1.23	2.71	0.90	1.98	22.26	1.62	0.04
E7018	10	23	100	3.18	1/8	0.32	3.19	1.54	1.24	2.73	0.85	1.88	22.42	1.54	0.04
E7018	10	22	100	3.18	1/8	0.33	3.34	1.62	1.24	2.72	0.91	2.00	22.07	1.62	0.04
Avg						0.33	3.29	1.30	1.24	2.72	0.89	1.95	22.25	1.59	0.04
SD						0.01	0.09	0.05	0.01	0.01	0.03	0.06	0.18	0.04	>0.01
E7018	10	22	160	3.97	5/32	0.50	1.95	1.53	1.96	4.32	1.45	3.20	20.64	2.27	0.06
E7018	10	22	155	3.97	5/32	0.49	2.10	1.52	1.98	4.35	1.49	3.28	20.13	2.25	0.06
E7018	10	22	155	3.97	5/32	0.50	2.06	1.50	1.98	4.35	1.48	3.25	20.07	2.22	0.06
Avg						0.50	2.03	1.52	1.97	4.34	1.47	3.24	20.28	2.25	0.06
SD						>0.01	0.07	0.01	0.01	0.02	0.02	0.04	0.31	0.02	>0.01
E7018	10	22	230	4.76	3/16	0.73	1.37	1.53	2.84	6.25	2.16	4.74	20.21	2.97	0.07
E7018	10	22	230	4.76	3/16	0.73	1.34	1.55	2.85	6.28	2.17	4.78	20.36	3.00	0.08
E7018	10	21	220	4.76	3/16	0.72	1.56	1.56	2.79	6.13	2.13	4.68	20.40	2.98	0.08
Avg						0.73	1.42	1.55	2.83	6.22	2.15	4.74	20.32	2.98	0.08
SD						0.01	0.12	0.01	0.04	0.08	0.02	0.05	0.10	0.01	>0.01

*g/kg ÷ 10 = weight of fume expressed as a percentage of the deposit weight

Notes

Direct current electrode positive

Welding speed
- 3/32 in. diam. 2.54 mm/s (6 in./min).
- 1/8 in. diam. 3.07 mm/s (7.25 in./min).
- 5/32 in. diam. 3.81 mm/s (9 in./min).
- 3/16 in. diam. 3.81 mm/s (9 in./min)

Electrode weight
- 3/32 in. diam. 59.46 g/m (1.5103 g/in.).
- 1/8 in. diam. 99.01 g/m (2.5149 g/in.).
- 5/32 in. diam. 147.98 g/m (3.7587 g/in.).
- 3/16 in. diam. 194.40 g/m (4.9379 g/in.)

Metal deposition rate determined for only 1 test per electrode.

SD standard deviation (sigma).

Table B33
Fume generation characteristics of E6010 covered electrodes with different diameters

Electrode	Code	Voltage, V avg	Current, A avg	Electrode diameter mm	Electrode diameter in	Fume generation rate Measured g/min	Fume generation rate Normalized g/min/I^2 x 10^{-5}	Wgt. of fume/wgt. of electrode used x 100	Melting rate kg/h	Melting rate lb/h	Deposition rate kg/h	Deposition rate lb/h	Wgt. of fume/wgt. of deposited metal g/kg*	Wgt of fume/length of electrode used g/m	Wgt of fume/length of electrode used g/in
E6010	1	24	55	2.38	3/32	0.15	4 83	1.46	0 60	1.32	0 50	1.10	17.57	0 59	0 01
E6010	1	- - -	55	2.38	3/32	0.14	4.66	1.41	0.60	1.32	0 47	1.04	17.84	0.58	0 01
E6010	1	26	55	2.38	3/32	0.15	4.86	1.46	0 60	1.33	0.50	1.10	17.69	0.59	0 01
						Avg 0.15	4.78	Avg 1.44	Avg 0 60	1.32	Avg 0.49	1.08	Avg 17 70	Avg 0.59	0.01
						SD >0.01	0.10	SD 0.03	SD 0 00	>0.01	SD 0.01	0.03	SD 0.13	SD 0 01	>0.01
E6010	1	29	85	3.18	1/8	0.31	4.24	1.72	1 07	2.36	0.76	1 66	24 28	1 31	0.03
E6010	1	28	90	3.18	1/8	0.33	4.08	1.80	1 10	2.43	0 77	1 70	25.60	1 41	0.04
E6010	1	28	90	3.18	1/8	0.35	4.30	1.88	1 12	2 46	0 78	1 72	26.79	1 32	0.03
						Avg 0.33	4.24	Avg 1.72	Avg 1.10	2 41	Avg 0.77	1.70	Avg 25 56	Avg 1 31	0.03
						SD 0.02	0.11	SD 0.08	SD 0.03	0.05	SD 0.01	0.03	SD 1.26	SD 0.10	>0.01
E6010	1	27	145	3.97	5/32	0.45	2.15	1.60	1 70	3 73	1.29	2.84	20.98	1.72	0.04
E6010	1	27	145	3.97	5/32	0.46	2.20	1.64	1.70	3.73	1 30	2.85	21.44	1.77	0 04
E6010	1	26	145	3.97	5/32	0 43	2.06	1.46	1.78	3.92	1.50	3.30	17.32	1 63	0.04
E6010	1	26	145	3.97	5/32	0.44	2.09	1.54	1.72,	3 79	1.52	3.34	17.39	1.66	0 04
E6010	1	26	145	3.97	5/32	0.44	2 09	1 48	1 79	3.94	1 51	3.32	17 49	1.63	0 04
						Avg 0.45	2.12	Avg 1 54	Avg 1 74	3.82	Avg 1.42	3.13	Avg 18.92	Avg 1.68	0 04
						SD 0.01	0.06	SD 0.08	SD 0.05	0.10	SD 0 12	0 26	SD 2.09	SD 0 06	>0 01
E6010	1	29	180	4.76	3/16	0.70	2.16	1 73	2 42	5.33	1 92	4.22	21 87	2.72	0 07
E6010	1	30	190	4.76	3/16	0.72	1.99	1 77	2.44	5.37	1.94	4.27	22.21	2 76	0 07
E6010	1	29	190	4.76	3/16	0.73	2.02	1.75	2.50	5.50	1.96	4.31	22.37	2 72	0 07
						Avg 0.72	2.06	Avg 1.75	Avg 2.45	5.40	Avg 1.94	4.26	Avg 22 15	Avg 2.73	0 07
						SD 0.02	0 09	SD 0.02	SD 0.02	0.08	SD 0.02	0 05	SD 0.25	SD 0.02	>0 01

*g/kg ÷ 10 = weight of fume expressed as a percentage of the deposit weight.

Notes

Direct current: electrode positive

Welding speed. 3/32 in. diam.: 2.65 mm/s (6.25 in./min),
1/8 in. diam.: 3.18 mm/s (7.5 in./min),
5/32 in. diam. 3.81 mm/s (9 in./min);
3/16 in. diam. 4.02 mm/s (9.5 in./min)

Electrode weight 3/32 in diam 40.36 g/m (1.0252 g/in.),
1/8 in. diam. 70.37 g/m (1 7875 g/in.),
5/32 in. diam 107 73 g/m (2 7363 g/in.),
3/16 in. diam 154.48 g/m (3.9237 g/in.)

Metal deposition rate determined for only 1 test per electrode

SD standard deviation (sigma)

Table B34

Fume generation characteristics of E6013 covered electrodes with different diameters

Electrode	Code	Voltage, V avg	Current, A avg	Electrode diameter mm	Electrode diameter in.		Fume generation rate Measured g/min	Fume generation rate Normalized g/min/I^2 x 10^{-5}	Wgt. of fume/ wgt. of electrode used x 100	Melting rate kg/h	Melting rate lb/h	Deposition rate kg/h	Deposition rate lb/h	Wgt. of fume/ wgt of deposited metal g/kg*	Wgt. of fume/ length of electrode used g/m	Wgt. of fume/ length of electrode used g/in.
E6013	4	24	50	2.38	3/32		0.13	5.25	1.17	0.67	1.48	0.49	1.08	15.99	0.55	0.01
E6013	4	26	53	2.38	3/32		0.13	4.70	1.17	0.67	1.47	0.47	1.04	16.38	0.52	0.01
						Avg	0.13	4.98	1.17	0.67	1.47	0.48	1.06	16.19	0.53	0.01
						SD	>0.01	0.39	>0.01	0.00	>0.01	0.01	0.03	0.28	0.02	>0.01
E6013	4	24	83	3.18	1/8		0.17	2.53	0.08	0.96	2.11	0.65	1.43	15.93	0.81	0.02
E6013	4	21	83	3.18	1/8		0.18	2.65	1.21	0.90	1.98	0.62	1.36	17.49	0.86	0.02
E6013	4	22	83	3.18	1/8		0.17	2.57	1.08	0.96	2.11	0.57	1.48	15.40	0.81	0.02
						Avg	0.17	2.57	1.12	0.94	2.07	0.65	1.42	16.27	0.83	0.02
						SD	>0.01	0.07	0.07	0.03	0.08	0.03	0.06	1.09	0.03	>0.01
E6013	4	22	135	3.97	5/32		0.34	1.88	1.32	1.56	3.43	1.09	2.40	18.79	2.51	0.06
E6013	4	25	137	3.97	5/32		0.35	1.87	1.37	1.56	3.44	1.09	2.39	19.56	2.59	0.07
E6013	4	23	150	3.97	5/32		0.36	1.61	1.33	1.64	3.60	1.20	2.64	18.05	2.65	0.07
						Avg	0.35	1.79	1.34	1.59	3.49	1.13	2.48	18.80	2.58	0.07
						SD	0.01	0.16	0.02	0.04	0.10	0.06	0.14	0.75	0.07	>0.01
E6013	4	21	205	4.76	3/16		0.41	0.99	1.09	2.29	5.04	1.84	4.04	13.53	1.79	0.04
E6013	4	20	205	4.76	3/16		0.41	0.98	1.06	2.35	5.18	1.86	4.10	13.34	1.69	0.04
E6013	4	20	200	4.76	3/16		0.42	1.04	1.07	2.40	5.29	1.78	3.91	14.11	1.79	0.04
						Avg	0.41	1.01	1.07	2.35	5.17	1.83	4.02	13.66	1.76	0.04
						SD	>0.01	0.03	0.02	0.06	0.12	0.04	0.10	0.40	0.06	>0.01

*g/kg ÷ 10 = weight of fume expressed as a percentage of the deposit weight

Notes

Direct current electrode positive

Welding speed: 3/32 in. diam.: 2.54 mm/s (6 in./min),
1/8 in. diam.: 2.96 mm/s (7 in./min),
5/32 in. diam: 3.81 mm/s (9 in./min);
3/16 in. diam.: 3.81 mm/s (9 in./min)

Electrode weight 3/32 in. diam 44.04 g/m (1.1186 g/in.),
1/8 in diam. 75.15 g/m (1.9088 g/in.);
5/32 in diam. 117.50 g/m (2.98 g/in.),
3/16 in diam 164.74 g/m (4.18 g/in.)

Metal deposition rate determined for only 1 test per electrode

SD standard deviation (sigma)

Table B35
Fume generation characteristics of a 3.97 mm (5/32 in.) diameter E6010 covered electrode as a function of voltage

Electrode	Code	Voltage, V avg	Current, A avg		Fume generation rate: Measured g/min	Fume generation rate: Normalized $g/min/I^2 \times 10^{-5}$		Wgt of fume/ wgt. of electrode used x 100		Melting rate kg/h	Melting rate lb/h		Deposition rate kg/h	Deposition rate lb/h		Wgt. of fume/ wgt. of deposited metal g/kg*		Wgt. of fume/ length of electrode used g/m	Wgt. of fume/ length of electrode used g/in
E6010	11	21	140		0.78	3.98		2.53		1.85	4.08		1.43	3.14		32.78		2.91	0.07
E6010	11	20.5	140		0.78	3.98		2.52		1.86	4.09		1.42	3.12		33.42		2.98	0.08
E6010	11	20.5	140		0.79	4.02		2.75		1.72	3.79		1.40	3.09		33.75		3.11	0.08
				Avg	0.78	3.99	Avg	2.60	Avg	1.81	4.00	Avg	1.42	3.12	Avg	33.31	Avg	3.00	0.08
				SD	0.01	0.02	SD	0.13	SD	0.08	0.17	SD	0.01	0.03	SD	0.49	SD	0.10	0.01
E6010	11	23.5	150		0.83	3.69		2.74		1.82	4.00		1.43	3.15		34.78		3.17	0.08
E6010	11	23	150		0.83	3.69		2.70		1.84	4.06		1.42	3.13		35.00		3.13	0.08
E6010	11	23	150		0.83	3.69		2.68		1.84	4.10		1.43	3.15		34.78		3.09	0.08
				Avg	0.83	3.69	Avg	2.71	Avg	1.83	4.05	Avg	1.43	3.14	Avg	34.88	Avg	3.13	0.08
				SD	0.00	0.00	SD	0.03	SD	0.01	0.05	SD	0.01	0.01	SD	0.13	SD	0.04	0.00
E6010	11	27	160		1.20	4.97		3.65		1.96	4.32		---	---		---		4.17	0.11
E6010	11	27	155		1.20	4.70		3.68		1.96	4.32		---	---		---		4.17	0.11
E6010	11	28	155		1.20	4.68		3.67		1.96	4.32		---	---		---		4.17	0.11
E6010	11	28	160		1.20	4.99		3.79		1.90	4.18		1.32	2.91		54.36			
				Avg	1.20	4.84	Avg	3.70	Avg	1.95	4.29								
				SD	0.00	0.17	SD	0.06	SD	0.03	0.07								
E6010	11	38.5	145		1.53	7.28		5.38		1.71	3.76		1.40	3.09		65.36		6.22	0.16
E6010	11	37.5	150		1.64	7.29		5.49		1.79	3.95		1.43	3.15		68.72		6.34	0.16
E6010	11	37.5	150		1.56	6.93		5.36		1.75	3.85		1.40	3.09		66.64		6.22	0.16
				Avg	1.58	7.17	Avg	5.41	Avg	1.75	3.85	Avg	1.41	3.11	Avg	66.91	Avg	6.26	0.16
				SD	0.06	0.02	SD	0.07	SD	0.04	0.10	SD	0.01	0.03	SD	1.70	SD	0.07	0.00

*g/kg ÷ 10 = weight of fume expressed as a percentage of the deposit weight.

Notes

Electrode polarity: positive

Welding speed: 3.39 to 4.45 mm/s (8 to 10.5 in./min)

Welding time: 1 minute

Electrode weight: 114.57 g/m (2.91 g/in.)

Metal deposition rate determined for only 1 test per electrode.

SD: standard deviation (sigma).

Table B36
Fume generation rates of 3.97 mm (5/32 in.) diameter E6010, E6013, and E7018 electrodes as a function of arc length

Electrode	Code	Arc length		Voltage, V	Current, A	Fume generation rate, g/m	
		mm	in.			Measured	Normalized*
E6010	1	0	0	23	140	0.53	0.61
E6010	1	0	0	24	140	0.47	0.54
E6010	1	1.59	1/16	29	135	0.68	0.84
E6010	1	3.18	1/8	33	140	0.77	0.88
E6010	1	4.76	3/16	35	140	0.68	0.78
E6010	1	6.35	1/4	41	150	0.98	0.98
E6010	1	6.35	1/4	43	150	1.08	1.08
E6013	3	0	0	23	140	0.49	0.56
E6013	3	1.59	1/16	26	150	0.63	0.63
E6013	3	3.18	1/8	29	140	0.53	0.62
E6013	3	4.76	3/16	32	135	0.67	0.83
E6013	3	6.35	1/4	42	165	1.35	1.12
E6013	3	6.35	1/4	43	155	1.19	1.11
E6013	3	7.94	5/16	43	155	1.08	1.07
E6013	3	15.88	5/8	50	135	1.12	1.39
E7018	10	3.18	1/8	23.5	145	0.47	0.51
E7018	10	6.35	1/4	26	145	0.46	0.50
E7018	10	7.94	5/16	31	130	0.45	0.60
E7018	10	9.53	3/8	33	140	0.57	0.65
E7018	10	11.11	7/16	43	150	0.75	0.75

*All fume generation rates normalized to current of 150 amperes.

Notes:

Electrode polarity: positive

Welding time: 30 seconds (nominal).

Table B37
Fume generation characteristics of some 3.97 mm (5/32 in.) diameter high titania mild steel covered electrodes with iron powder additions

Electrode	Code	Voltage, V avg	Current, A avg	Fume generation rate: Measured g/min	Fume generation rate: Normalized g/min/I^2 x 10^{-5}	Wgt of fume/wgt of electrode used x 100	Melting rate kg/h	Melting rate lb/h	Deposition rate kg/h	Deposition rate lb/h	Wgt. of fume/wgt. of deposited metal g/kg*	Weight of fume/length of electrode used g/m	Weight of fume/length of electrode used g/in.
E6013	34	25	165	0.74	2.70	2.03	2.17	4.78	1.47	3.24	30.00	2.27	0.06
E6013	34	24	165	0.73	2.67	2.00	2.18	4.80	1.46	3.21	29.88	2.25	0.06
E6013	34	24	170	0.72	2.51	2.00	2.18	4.79	1.46	3.22	29.69	2.25	0.06
Avg				0.73	2.63	2.01	2.18	4.79	1.46	3.22	29.86	2.26	0.06
SD				0.01	0.01	0.02	0.01	0.01	0.01	0.02	0.15	0.01	0.00
E7014	30	30	175	0.72	2.34	1.63	2.65	5.83	1.83	4.03	23.53	2.42	0.06
E7014	30	29	175	0.72	2.35	1.62	2.67	5.87	1.81	3.99	23.86	2.43	0.06
E7014	30	29	175	0.72	2.35	1.63	2.66	5.85	1.81	3.99	23.81	2.41	0.06
Avg				0.72	2.35	1.63	2.66	5.85	1.82	4.00	23.72	2.42	0.06
SD				>0.01	>0.01	>0.01	0.01	0.02	0.01	0.02	0.18	0.01	0.00
E7024	35	41	210	0.69	1.56	0.82	5.01	11.03	3.35	7.37	12.34	2.13	0.05
E7024	35	41	210	0.68	1.54	0.81	5.01	11.02	3.41	7.50	11.93	2.10	0.05
E7024	35	41	210	0.68	1.55	0.82	5.01	11.02	3.38	7.44	12.12	2.11	0.05
Avg				0.68	1.54	0.82	5.01	11.02	3.38	7.44	12.13	2.11	0.05
SD				0.01	0.01	0.01	0.00	0.01	0.03	0.06	0.22	0.02	0.00

*g/kg ÷ 10 = weight of fume expressed as percentage of the deposit weight

Notes

Electrode polarity positive

Welding speed Code 30 and 34 electrodes 4.23 mm/s (10 in./min), Code 35 electrode 4.45 mm/s (10.5 in./min).

Electrode weight Code 30 electrode 147.64 g/m (3.75 g/in.)
Code 34 electrode 115.75 g/m (2.94 g/in.)
Code 35 electrode 146.85 g/m (3.73 g/in.)

Metal deposition rate determined for only 1 test per electrode.

SD. standard deviation (sigma).

Table B38
Fume generation characteristics of some 3.97 mm (5/32 in.) diameter low hydrogen mild steel covered electrodes with iron powder additions

Electrode	Code	Voltage, V avg	Current, A avg	Fume generation rate: Measured g/min	Fume generation rate: Normalized g/min/I^2 x 10^{-5}	Wgt. of fume/wgt of electrode used x 100	Melting rate kg/h	Melting rate lb/h	Deposition rate kg/h	Deposition rate lb/h	Wgt. of fume/wgt of deposited metal g/kg*	Weight of fume/length of electrode used g/m	Weight of fume/length of electrode used g/in.
E7016	29	23	170	0.48	1.67	1.37	2.12	4.66	1.40	3.07	20.78	1.80	0.05
E7016	29	23	175	0.51	1.67	1.44	2.14	4.71	1.48	3.26	20.77	1.79	0.05
E7016	29	23	170	0.49	1.70	1.38	2.14	4.70	1.40	3.09	20.99	1.81	0.05
				Avg 0.50	1.68	Avg 1.40	Avg 2.13	4.69	Avg 1.43	3.14	Avg 20.84	Avg 1.80	0.05
				SD 0.01	0.01	SD 0.04	SD 0.01	0.03	SD 0.05	0.10	SD 0.13	SD 0.01	0.00
E7018	36	24	175	0.51	1.66	1.33	2.30	5.06	1.68	3.70	18.16	1.98	0.05
E7018	36	24	180	0.51	1.59	1.36	2.27	5.00	1.65	3.63	18.69	2.02	0.05
E7018	36	24	175	0.51	1.68	1.36	2.28	5.01	1.68	3.70	18.33	2.02	0.05
				Avg 0.51	1.64	Avg 1.35	Avg 2.28	5.02	Avg 1.67	3.68	Avg 18.39	Avg 2.00	0.05
				SD <0.01	0.05	SD 0.02	SD 0.01	0.03	SD 0.02	0.04	SD 0.28	SD 0.02	0.00
E7028	32	33	235	1.04	1.89	1.48	4.24	9.34	2.96	6.52	21.11	3.53	0.09
E7028	32	32	235	1.01	1.82	1.43	4.23	9.30	2.88	6.34	20.96	3.45	0.09
E7028	32	34	225	1.03	2.03	1.46	4.24	9.32	2.85	6.26	21.71	3.52	0.09
				Avg 1.03	1.91	Avg 1.45	Avg 4.24	9.32	Avg 2.90	6.37	Avg 21.27	Avg 3.50	0.09
				SD 0.02	0.01	SD 0.02	SD 0.01	0.02	SD 0.06	0.13	SD 0.39	SD 0.04	0.00

*g/kg ÷ 10 = weight of fume expressed as percentage of the deposit weight.

Notes

Electrode polarity positive

Welding speed Code 32 and 36 electrodes, 4.23 mm/s (10 in./min), Code 29 electrode 3.81 mm/s (9 in./min).

Electrode weight Code 29 electrode 130.71 g/m (3.32 g/in.)
Code 32 electrode 196.46 g/m (4.99 g/in.)
Code 36 electrode 146.85 g/m (3.73 g/in.)

Metal deposition rate determined for only 1 test per electrode.

SD standard deviation (sigma).

Table B39

Fume generation characteristics of 3.97 mm (5/32 in.) diameter E6010, E6013, and E7018 covered electrodes as a function of electrode angle

Electrode	Code	Voltage, V avg	Current, A avg	Electrode angle, °	Fume generation rate: Measured g/min	Fume generation rate: Normalized g/min/I^2 x 10^{-5}	Wgt. of fume/ wgt. of electrode used x 100	Melting rate kg/h	Melting rate lb/h	Deposition rate kg/h	Deposition rate lb/h	Wgt of fume/ wgt. of deposited metal g/kg*	Weight of fume/ length of electrode used g/m	Weight of fume/ length of electrode used g/in.
E6010	1	25.5	150	90	0.79	3.51	2.46	1.92	4.23	1.45	3.20	32.59	2.73	0.07
E6010	1	27	155	60	0.95	3.95	3.13	1.83	4.02	1.33	2.93	42.80	3.45	0.09
E6010	1	27.5	155	30	1.02	4.24	2.97	2.19	4.82	1.59	3.49	38.58	3.30	0.08
E6010	1	29	165	90	1.16	4.26	3.80	1.85	4.06	1.19	2.62	58.44	4.38	0.11
E6010	1	29	150	60	1.25	5.55	4.13	1.75	3.86	1.19	2.61	63.22	5.07	0.13
E6010	1	28	150	30	1.36	6.04	4.48	1.82	4.00	1.22	2.69	66.74	4.87	0.12
E6013	3	25	150	90	0.41	1.82	2.14	2.30	5.06	1.46	3.21	16.84	2.36	0.06
E6013	3	25	145	60	0.47	2.19	2.70	2.08	4.59	1.34	2.95	21.01	3.29	0.08
E6013	3	26	145	30	0.55	2.44	3.17	2.08	4.59	1.29	2.83	25.67	3.94	0.10
E7018	10	23	165	90	0.45	1.65	1.14	1.92	4.21	1.44	3.17	18.74	4.17	0.11
E7018	10	25	165	60	0.58	2.13	1.91	1.82	4.00	1.23	2.70	28.35	5.53	0.14
E7018	10	23.5	170	30	0.68	2.35	2.15	1.90	4.18	1.22	2.69	33.37	5.65	0.15

*g/kg ÷ 10 = weight of fume expressed as percentage of the deposit weight

Notes

Electrode polarity positive

Welding speed 3.39 to 4.23 mm/s (8 to 10 in./min).

Electrode weight Code 1 111.02 g/m (2.82 g/in.)
Code 3 122.05 g/m (3.10 g/in.)
Code 10 148.42 g/m (3.77 g/in.)

Metal deposition rate determined for only 1 test per electrode.

SD standard deviation (sigma).

Table B40
Fume generation characteristics of 6.35 mm (1/4 in.) diameter E7018 and E7028 covered electrodes at high current levels

Electrode	Code	Voltage, V avg	Current, A avg	Fume generation rate: Measured g/min	Fume generation rate: Normalized g/min/I^2 x 10^{-5}	Wgt. of fume/wgt. of electrode used x 100	Melting rate kg/h	Melting rate lb/h	Deposition rate kg/h	Deposition rate lb/h	Wgt. of fume/wgt. of deposited metal g/kg*	Weight of fume/length of electrode used g/m	Weight of fume/length of electrode used g/in.
E7018	72	24	395	1.59	1.02	2.16	4.42	9.71	3.46	7.61	27.57	6.96	0.18
E7018	72	24.5	390	1.61	1.06	1.94	4.97	10.94	3.45	7.59	28.00	7.00	0.18
E7018	72	22.5	400	1.60	1.00	1.93	4.97	10.93	3.41	7.51	28.15	7.00	0.18
				Avg 1.60	1.03	Avg 2.01	Avg 4.78	10.52	Avg 3.44	7.50	Avg 27.91	Avg 6.99	0.18
				SD 0.01	0.03	SD 0.13	SD 0.32	0.71	SD 0.02	0.05	SD 0.30	SD 0.01	0.00
E7024	73	44	385	2.10	1.41	1.35	9.36	20.60	6.62	14.56	19.04	6.89	0.17
E7024	73	43	390	2.14	1.41	1.37	9.40	20.69	6.63	14.59	19.36	7.02	0.18
E7024	73	42.5	395	2.15	1.38	1.37	9.41	20.70	6.63	14.59	19.44	7.05	0.18
				Avg 2.13	1.40	Avg 1.36	Avg 9.39	20.66	Avg 6.63	14.58	Avg 19.28	Avg 6.99	0.18
				SD 0.02	0.02	SD 0.02	SD 0.03	0.06	SD 0.01	0.02	SD 0.02	SD 0.04	0.01

*g/kg ÷ 10 = weight of fume expressed as percentage of the deposit weight.

Notes

Direct current electrode positive

Welding speed Code 72 electrode 4.23 mm/s (10 in./min), Code 73 electrode 5.08 mm/s (12 in./min).

Electrode weight Code 72 electrode: 357.87 g/m (9.09 g/in.)
Code 73 electrode 514.46 g/m (13.07 g/in.)

Metal deposition rate determined for only 1 test per electrode.

SD standard deviation (sigma).

Table B41

Comparison of fume generation characteristics of 1.59 mm (1/16 in.) diameter and 2.39 mm (3/32 in.) diameter E70T-1 flux cored electrode (CO_2 shielding)

Code	Voltage, V avg	Current, A avg	Wire feed rate mm/s	Wire feed rate in/min	Fume generation rate Measured g/min	Fume generation rate Normalized g/min/I^2 x 10^{-5}	Wgt. of fume/wgt. of electrode used x 100	Melting rate kg/h	Melting rate lb/h	Deposition rate kg/h	Deposition rate lb/h	Wgt of fume/wgt. of deposited metal g/kg*	Wgt. of fume/length of electrode used g/m	Wgt. of fume/length of electrode used g/in
40	30	485	87.63	207	1.36	0.58	0.91	8.95	19.70	7.72	17.19	10.44	0.26	0.01
40	31	485	87.63	207	1.34	0.57	0.90	8.95	19.70	7.70	17.10	10.37	0.26	0.01
40	31	470	87.63	207	1.36	0.61	0.91	8.95	19.70	7.77	17.27	10.38	0.26	0.01
					Avg 1.36	0.59	Avg 0.91			Avg 7.73	17.17	Avg 10.40		
					SD 0.01	0.02	SD 0.01			SD 0.04	0.09	SD 0.04		
45	34	335	139.07	328.5	1.01	0.90	0.87	6.99	15.39	6.27	13.93	9.57	0.12	>0.01
45	35	325	139.07	328.5	1.01	0.96	0.87	6.99	15.39	6.33	14.06	9.53	0.12	>0.01
45	34	335	139.07	328.5	1.01	0.90	0.87	6.99	15.39	6.29	13.98	9.58	0.12	>0.01
					Avg 1.01	0.92	Avg 0.87			Avg 6.30	13.98	Avg 9.56		
					SD 0.00	0.03	SD >0.01			SD 0.03	>0.01	SD 0.02		

*g/kg − 10 = weight of fume expressed as a percentage of the deposit weight.

Notes

Electrode polarity positive

Electrode contact tube-to-work distance 31.8 mm (1-1/4 in.)

Welding time 30 seconds

Shielding gas flow rate 21-24 liters/min (45-50 ft^3/h)

Welding speed 6.77 mm/s (16 in./min)

Electrode weight/unit length Code No. 45 13.78 (0.35 g/in.)
Code No. 40 28.35 (0.72 g/in.)

Electrode diameter Code No 45 1.59 mm (1/16 in.)
Code No 40. 2.38 mm (3/32 in.)

Metal deposition rate determined for only 1 test per electrode

SD standard deviation (sigma)

Table B42

Fume generation characteristics of E70T-1 flux cored electrodes as a function of contact tube-to-work distance when operating at 450 amperes

Code	Voltage, V avg	Current, A avg	Contact tube-to-work distance mm	Contact tube-to-work distance in		Fume generation rate Measured g/min	Fume generation rate Normalized g/min/I^2 x 10^{-5}		Wgt. of fume/wgt. of electrode used x 100	Melting rate kg/h	Melting rate lb/h		Deposition rate kg/h	Deposition rate lb/h		Wgt. of fume/wgt. of deposited metal g/kg*	Wgt. of fume/length of electrode used g/m	Wgt. of fume/length of electrode used g/in.
41	30	450	19.1	0.75		0.92	0.46		0.80	6.89	15.16		6.12	13.59		8.97	0.24	<0.01
41	31	455	19.1	0.75		0.92	0.45		0.81	6.89	15.16		6.07	13.48		9.05	0.24	<0.01
41	29	455	19.1	0.75		0.91	0.44		0.80	6.89	15.16		6.07	13.48		8.96	0.24	<0.01
					Avg	0.92	0.45	Avg	0.80			Avg	6.09	13.50	Avg	8.99		
					SD	0.01	0.01	SD	0.01			SD	0.03	0.08	SD	0.05		
41	31	450	25.4	1.00		1.04	0.52		0.76	8.17	17.98		7.30	16.23		8.42	0.22	<0.01
41	30	450	25.4	1.00		1.02	0.50		0.75	8.17	17.98		7.12	15.83		8.52	0.22	<0.01
41	29	445	25.4	1.00		1.02	0.51		0.75	8.17	17.98		7.31	16.25		8.27	0.22	<0.01
					Avg	1.03	0.51	Avg	0.75			Avg	7.24	16.10	Avg	8.40		
					SD	0.01	0.01	SD	0.01			SD	0.10	0.24	SD	0.13		
41	30	450	31.8	1.25		0.94	0.46		0.63	8.92	19.62		7.20	15.99		7.76	0.19	<0.01
41	31	450	31.8	1.25		0.95	0.47		0.64	8.92	19.62		7.32	16.26		7.72	0.19	<0.01
41	30	440	31.8	1.25		0.95	0.49		0.64	8.92	19.62		7.34	16.31		7.75	0.19	<0.01
					Avg	0.95	0.47	Avg	0.64			Avg	7.29	16.19	Avg	7.74		
					SD	0.01	0.02	SD	0.01			SD	0.08	0.17	SD	0.02		
41	30	450	38.1	1.50		1.03	0.51		0.64	9.19	20.21		8.28	18.40		7.38	0.20	<0.01
41	31	450	38.1	1.50		1.03	0.51		0.68	9.19	20.21		8.27	18.37		7.41	0.20	<0.01
41	30	450	38.1	1.50		1.03	0.51		0.68	9.19	20.21		8.28	18.40		7.42	0.20	<0.01
					Avg	1.03	0.51	Avg	0.67			Avg	8.28	18.39	Avg	7.40		
					SD	0.00	0.00	SD	0.01			SD	0.01	0.02	SD	0.02		

*g/kg ÷ 10 = weight of fume expressed as a percentage of the deposit weight.

Notes

Electrode polarity positive

Welding time 30 seconds

Shielding gas flow rate: 21-24 liters/min (45-50 ft^3/h) CO_2

Welding speed 6.77 mm/s (16 in./min)

Electrode weight/unit length 29.53 g/m (0.75 g/in.)

Metal deposition rate determined for only 1 test per electrode

SD standard deviation (sigma).

Table B43
Fume generation characteristics of E70T-1 flux cored electrodes as a function of contact tube-to-work distance when operating at 520 amperes

Code	Voltage, V avg	Current, A avg	Contact tube-to-work distance mm	Contact tube-to-work distance in.	Fume generation rate Measured g/min	Fume generation rate Normalized $g/min/I^2 \times 10^{-5}$	Wgt. of fume/wgt. of electrode used x 100	Melting rate kg/h	Melting rate lb/h	Deposition rate kg/h	Deposition rate lb/h	Wgt. of fume/wgt. of deposited metal g/kg*	Wgt. of fume/length of electrode used g/m	Wgt. of fume/length of electrode used g/in.
41	30	515	19.1	0.75	1.31	0.49	0.96	8.17	17.98	7.84	17.42	9.92	0.28	<0.01
41	29	520	19.1	0.75	1.32	0.49	0.97	8.17	17.98	7.86	17.47	9.95	0.28	<0.01
41	29	510	19.1	0.75	1.31	0.50	0.96	8.17	17.98	7.84	17.42	9.93	0.28	<0.01
Avg					1.31	0.49	0.96			7.85	17.44	9.93		
SD					0.01	0.01	0.01			0.01	0.03	0.02		
41	28	515	25.4	1.00	1.32	0.49	0.79	10.00	21.99	8.95	19.88	8.75	0.23	<0.01
41	29	520	25.4	1.00	1.32	0.50	0.79	10.00	21.99	8.98	19.96	8.73	0.23	<0.01
42	29	525	25.4	1.00	1.32	0.49	0.80	10.00	21.99	8.92	19.82	8.82	0.23	<0.01
Avg					1.32	0.49	0.79			8.95	19.89	8.77		
SD					0.00	0.01	0.01			0.03	0.07	0.05		
41	29	510	31.8	1.25	1.33	0.51	0.77	10.38	22.84	9.43	20.96	8.35	0.23	<0.01
41	31	520	31.8	1.25	1.33	0.49	0.77	10.38	22.84	9.45	29.99	8.42	0.23	<0.01
41	30	525	31.8	1.25	1.32	0.48	0.77	10.38	22.84	9.35	20.78	8.40	0.22	<0.01
Avg					1.33	0.49	0.77			9.41	20.91	8.39		
SD					0.01	0.02	0.00			0.05	0.11	0.04		
41	29	510	38.1	1.50	1.34	0.52	0.77	10.56	23.23	9.77	21.70	8.17	0.22	<0.01
41	29	525	38.1	1.50	1.34	0.49	0.76	10.56	23.23	9.88	21.96	8.05	0.22	<0.01
41	28	525	38.1	1.50	1.33	0.49	0.76	10.56	23.23	9.79	21.75	8.12	0.22	<0.01
Avg					1.34	0.50	0.76			9.81	21.80	8.11		
SD					0.01	0.02	0.01			0.06	0.14	0.06		

*g/kg ÷ 10 = weight of fume expressed as a percentage of the deposit weight.

Notes:

Electrode polarity: positive

Welding time: 30 seconds

Shielding gas flow rate: 21-24 liters/min (45-50 ft^3/h) CO_2

Welding speed: 6.77 mm/s (16 in./min)

Electrode weight/unit length: 29.53 g/m (0.75 g/in.)

Metal deposition rate determined for only 1 test per electrode.

SD: standard deviation (sigma).

Appendix C: Tables C1 to C4
(Summary of Original Data, Part III)

Table C1
Brazing fume generation by flux covered BAg-1 filler metal

	1200° F (649° C)				
Test number	1	7	15	20	27
Filter type	G	A	A	G	A
Relative humidity, %	43	49	45	48	51
Time molten, min	3	6	5	5	6
Sampling time, min	39	28	23	24	24
Ingot loss, g	0.0037	0.0495	0.0380	0.0412	0.0392
Fume weight, g	0.0069	0.0821	0.0453	0.0043	0.0777
	1350° F (732° C)				
Test number	3	11	18	24	28
Filter type	G	A	A	G	A
Relative humidity, %	- - -	47	45	- - -	53
Time molten, min	9	8	9	9	8
Sampling time, min	29	26	26	26	31
Ingot loss, g	0.0900	0.0487	0.0608	0.0529	0.0486
Fume weight, g	0.0183	0.0802	0.0645	0.0090	0.0865
	1500° F (816° C)				
Test number	5	9	17	19	26
Filter type	A	A	G	A	A
Relative humidity, %	48	49	- - -	47	53
Time molten, min	13	7	13	9	13
Sampling time, min	36	26	35	25	29
Ingot loss, g	0.1125	1.2877	0.1431	0.1236	0.0923
Fume weight, g	0.2809	0.1077	0.0751	0.0831	0.1371

A = membrane filter

G = glass fiber

Table C2
Brazing fume generation by flux covered low cadmium filler metal

	1250° F (677° C)					
Test number	36	42	48	52		
Filter type	A	A	A	A		
Relative humidity, %	49	45	38	47		
Time molten, min	7	9	7	7		
Sampling time, min	29	25	27	28		
Ingot loss, g	0.0125	0.0028	0.0004	0.0075		
Fume weight, g	0.1035	0.0366	0.0198	0.0601		
	1350° F (732° C)					
Test number	34	40	47	53		
Filter type	A	A	A	A		
Relative humidity, %	53	53	39	48		
Time molten, min	9	6	9	9		
Sampling time, min	22	30	26	26		
Ingot loss, g	0.0062	0.0413	0.0245	0.0287		
Fume weight, g	0.1057	0.0332	0.0365	0.0652		
	1500° F (816° C)					
Test number	35	41	41A	46	56	58
Filter type	A	A	A	A	A	A
Relative humidity, %	54	52	42	42	45	45
Time molten, min	14	16	16	17	14	16
Sampling time, min	29	32	31	32	30	33
Ingot loss, g	0.2275	0.1021	0.2255	0.3350	0.1175	0.1298
Fume weight, g	0.1269	0.1354	0 0963	0.0687	0.0679	0.0643

A = membrane filter

Table C3
Brazing fume generation by CO_2 covered BAg-1 filler metal

	1200° F (649 C)							
Test number	4	10	14	25	30			
Filter type	G	A	G	A	A			
Relative humidity, %	49	47	---	51	53			
Time molten, min	7	7	7	6	5			
Sampling time, min	29	28	24	31	25			
Ingot loss, g	0.0136	0.0301	0.0024	0.0061	0.0041			
Fume weight, g	0.0092	0.0639	0.0030	0.1595	0.0871			
	1350° F (732° C)							
Test number	6	8	16	23	29			
Filter type	A	G	G	A	A			
Relative humidity, %	49	49	45	---	53			
Time molten, min	9	7	8	7	8			
Sampling time, min	29	26	26	24	27			
Ingot loss, g	0.0495	0.0173	0.0374	0.0408	0.0436			
Fume weight, g	0.0805	0.0027	0.0032	0.1045	0.0923			
	1500° F (816° C)							
Test number	2	12	13	21	22	31	32	33
Filter type	A	G	G	G	G	A	A	A
Relative humidity, %	53	46	46	49	52	49	48	48
Time molten, min	11	11	12	13	12	12	11	12
Sampling time, min	43	36	29	35	28	31	27	31
Ingot loss, g	0.0795	0.0806	0.0616	0.1259	0.0631	0.0233	0.0537	0.0940
Fume weight, g	0.5127	0.0033	0.0000	0.0055	0.0014	0.1207	0.1065	0.1108

A = membrane filter

G = glass fiber

Table C4
Brazing fume generation by CO_2 covered low cadmium filler metal

	1250° F (677° C)				
Test number	38	44	50	57	39*
Filter type	A	A	A	A	A
Relative humidity, %	46	42	36	45	44
Time molten, min	9	6	7	7	---
Sampling time, min	34	25	36	27	21
Ingot loss, g	0.0011	0.0048	0.0119	0.0138	0.0250
Fume weight, g	0.0696	0.0230	0.0483	0.0330	0.0117
	1350° F (732° C)				
Test number	45	39A	49	55	
Filter type	A	A	A	A	
Relative humidity, %	42	42	36	45	
Time molten, min	10	8	8	9	
Sampling time, min	31	30	36	32	
Ingot loss, g	0.0243	0.0027	0.0709	0.0265	
Fume weight, g	0.0177	0.0463	0.0370	0.0416	
	1500° F (816° C)				
Test number	37	43	51	54	
Filter type	A	A	A	A	
Relative humidity, %	49	42	42	45	
Time molten, min	13	17	15	14	
Sampling time, min	33	39	35	33	
Ingot loss, g	0.1083	0.1586	0.0496	0.1411	
Fume weight, g	0.1154	0.0876	0.0488	0.0021	

* = Data questionable: furnace failure during test, data not used.

A = membrane filter

Appendix D: Tables D1 and D2
(Laboratory Data Sheets for Part IV)

Table D1
Fume generation characteristics of gas combustion sprayed metals

Metal wire sprayed	Fume generation rate, g/min	Melting rate		Deposition rate		Weight of fumes/ weight of metal sprayed		Weight of fumes/ weight of metal deposited
		kg/hr	lbs/hr	kg/hr	lbs/hr	g/kg	g/lb	g/kg*
Aluminum	0.23	3.14	6.90	2.81	6.18	4.40	2.20	4.91
	0.22	3.69	8.07	3.42	7.53	3.60	1.65	3.86
	0.38	2.87	6.31	2.10	4.62	7.99	3.63	10.85
	Avg 0.28	Avg 3.22	7.09	Avg 2.78	6.11	Avg 5.34	2.45	Avg 6.43
Zinc	2.22	6.99	15.39	4.71	10.36	19.04	8.65	28.28
	1.99	8.53	18.76	4.95	10.88	14.07	6.39	24.14
	2.71	7.85	17.26	3.96	8.72	20.72	9.42	41.02
	Avg 2.31	Avg 7.79	17.14	Avg 4.54	9.99	Avg 17.94	8.15	Avg 31.14
Stainless steel	0.44	1.70	3.75	1.17	2.56	15.49	7.04	22.68
	0.44	1.61	3.54	1.13	2.48	16.41	7.46	23.42
	0.48	1.61	3.54	1.25	2.75	17.91	8.14	23.04
	Avg 0.45	Avg 1.64	3.61	Avg 1.18	2.60	Avg 16.60	7.55	Avg 23.04

*g/kg ÷ 10 = weight of fume expressed as a percentage cf the deposit weight.

Table D2
Fume generation characteristics of electric arc sprayed metals

Metal wire sprayed	Fume generation rate, g/min	Melting rate		Deposition rate		Weight of fumes/ weight of metal sprayed		Weight of fumes/ weight of metal deposited
		kg/hr	lbs/hr	kg/hr	lbs/hr	g/kg	g/lb	g/kg*
Aluminum	0.97	1.32	2.90	0.59	1.30	44.15	20.07	98.49
	0.90	1.32	2.90	0.60	1.32	40.97	18.62	88.00
	0.88	1.32	2.90	0.61	1.35	40.05	18.20	86.04
	Avg 0.92	Avg 1.32	2.90	Avg 0.60	1.32	Avg 41.72	18.96	Avg 90.84
Zinc	1.42	4.45	9.78	2.05	4.52	19.16	8.71	41.47
	1.48	4.00	8.80	1.91	4.20	22.20	10.09	46.51
	1.42	3.91	8.61	2.04	4.49	21.77	9.90	41.75
	Avg 1.44	Avg 4.10	9.06	Avg 2.00	4.40	Avg 21.05	9.57	Avg 43.24
Stainless steel	1.60	5.65	12.42	2.46	5.42	17.00	7.73	38.96
	1.36	4.01	8.82	2.28	5.02	20.35	9.25	35.76
	1.58	4.33	9.52	2.29	5.05	21.91	9.96	41.30
	Avg 1.51	Avg 4.66	10.25	Avg 2.35	5.16	Avg 19.75	8.98	Avg 38.67
Aluminum-bronze	1.39	3.82	8.40	1.79	3.94	21.84	9.93	46.57
	1.33	4.00	8.88	1.78	3.91	19.77	8.97	44.90
	1.35	3.86	8.50	1.80	3.96	20.96	9.53	45.00
	Avg 1.34	Avg 3.89	8.57	Avg 1.79	3.94	Avg 20.85	9.48	Avg 45.49

*g/kg ÷ 10 = weight of fume expressed as a percentage of the deposit weight.

Appendix E: Tables E1 to E6
(Laboratory Data Sheets for Part V)

Table E1
Characteristics of fumes produced by oxyacetylene cutting of carbon steel plate at recommended conditions

Cutting tip number[1]	Oxygen pressure		Acetylene pressure		Plate thickness		Cutting speed			Fume generation rate, g/min		Weight of fumes/ length of plate cut	
	kPa	psi	kPa	psi	mm	in.	mm/s	in./min				g/m	g/in.
4	241.3	35	34.5	5	12.7	0.50	5.50	13		0.45		1.38	0.035
4	241.3	35	34.5	5	12.7	0.50	5.50	13		0.42		1.26	0.032
4	241.3	35	34.5	5	12.7	0.50	5.50	13		0.38		1.14	0.029
									Avg	0.42	Avg	1.26	0.032
6	275.8	40	34.5	5	25.4	1.00	4.66	11		0.53		1.89	0.048
6	275.8	40	34.5	5	25.4	1.00	4.66	11		0.53		1.89	0.048
6	275.8	40	34.5	5	25.4	1.00	4.66	11		0.56		2.01	0.051
									Avg	0.54	Avg	1.93	0.049
8	275.8	40	41.4	6	50.8	2.00	2.75	6.5		0.67		4.06	0.103
8	275.8	40	41.4	6	50.8	2.00	2.75	6.5		0.67		4.06	0.103
8	275.8	40	41.4	6	50.8	2.00	2.75	6.5		0.66		4.02	0.102
									Avg	0.67	Avg	4.05	0.103

1. Cutting orifice diameter: No. 4: 0.04 in. (1.0 mm); No. 6: 0.06 in. (1.5 mm); No. 8: 0.08 in. (2.0 mm).

Table E2
Characteristics of fumes produced by oxyacetylene cutting of carbon steel plate at various oxygen and acetylene pressures

Cutting tip number[1]	Oxygen pressure		Acetylene pressure		Plate thickness		Cutting speed		Fume generation rate, g/min	Weight of fumes/ length of plate cut	
	kPa	psi	kPa	psi	mm	in.	mm/s	in./min		g/m	g/in.
					High gas pressures						
4	275.8	40	48.3	7	13.7	0.50	5.50	13	0.46	1.38	0.035
4	275.8	40	48.3	7	13.7	0.50	5.50	13	0.47	1.42	0.036
4	275.8	40	48.3	7	13.7	0.50	5.50	13	0.46	1.38	0.035
Avg									0.46	1.39	0.035
6	310.3	45	48.3	7	25.4	1.00	4.66	11	0.63	2.24	0.057
6	310.3	45	48.3	7	25.4	1.00	4.66	11	0.65	2.23	0.059
6	310.3	45	48.3	7	25.4	1.00	4.66	11	0.63	2.24	0.057
Avg									0.64	2.24	0.058
8	310.3	45	55.2	8	50.8	2.00	2.75	6.5	0.80	4.84	0.123
8	310.3	45	55.2	8	50.8	2.00	2.75	6.5	0.81	4.92	0.125
8	310.3	45	55.2	8	50.8	2.00	2.75	6.5	0.81	4.92	0.125
Avg									0.81	4.89	0.124
					Low gas pressures						
4	206.8	30	20.7	3	13.7	0.50	5.50	13	0.33	0.98	0.025
4	206.8	30	20.7	3	13.7	0.50	5.50	13	0.34	1.02	0.026
4	206.8	30	20.7	3	13.7	0.50	5.50	13	0.34	1.02	0.026
Avg									0.34	1.01	0.026
6	241.3	35	20.7	3	25.4	1.00	4.66	11	0.34	1.22	0.031
6	241.3	35	20.7	3	25.4	1.00	4.66	11	0.35	1.26	0.032
6	241.3	35	20.7	3	25.4	1.00	4.66	11	0.35	1.26	0.032
Avg									0.35	1.25	0.032
8	241.3	35	27.6	4	50.8	2.00	2.75	6.5	0.49	2.95	0.075
8	241.3	35	27.6	4	50.8	2.00	2.75	6.5	0.46	2.80	0.071
8	241.3	35	27.6	4	50.8	2.00	2.75	6.5	0.47	2.83	0.072
Avg									0.48	2.86	0.073

1. **Cutting orifice diameter**: No. 4: 0.04 in. (1.0 mm); No. 6: 0.06 in. (1.5 mm); No. 8: 0.08 in. (2.0 mm).

Table E3
Characteristics of fumes produced by oxyacetylene cutting of carbon steel plates at various speeds

Cutting tip number[1]	Oxygen pressure		Acetylene pressure		Plate thickness		Cutting speed		Fume generation rate, g/min	Weight of fumes/ length of plate cut	
	kPa	psi	kPa	psi	mm	in.	mm/s	in./min		g/m	g/in.
					High cutting speeds						
4	241.3	35	34.5	5	13.7	0.50	6.77	16	0.48	1.18	0.030
4	241.3	35	34.5	5	13.7	0.50	6.77	16	0.49	1.22	0.031
4	241.3	35	34.5	5	13.7	0.50	6.77	16	0.49	1.22	0.031
									Avg 0.49	Avg 1.21	0.031
6	275.8	40	34.5	5	25.4	1.00	5.93	14	0.58	1.61	0.041
6	275.8	40	34.5	5	25.4	1.06	5.93	14	0.58	1.61	0.041
6	275.8	40	34.5	5	25.4	1.00	5.93	14	0.56	1.57	0.040
									Avg 0.57	Avg 1.60	0.041
8	275.8	40	41.4	6	50.8	2.00	4.02	9.5	0.71	2.95	0.075
8	275.8	40	41.4	6	50.8	2.00	4.02	9.5	0.72	2.99	0.076
8	275.8	40	41.4	6	50.8	2.00	4.02	9.5	0.72	2.99	0.076
									Avg 0.72	Avg 2.98	0.076
					Low cutting speeds						
4	241.3	35	34.5	5	13.7	0.50	4.23	10	0.45	1.77	0.045
4	241.3	35	34.5	5	13.7	0.50	4.23	10	0.45	1.77	0.045
4	241.3	35	34.5	5	13.7	0.50	4.23	10	0.45	1.77	0.045
									Avg 0.45	Avg 1.77	0.045
6	275.8	40	34.5	5	25.4	1.00	3.39	8	0.55	2.72	0.069
6	275.8	40	34.5	5	25.4	1.00	3.39	8	0.55	2.72	0.069
6	275.8	40	34.5	5	25.4	1.00	3.39	8	0.53	2.60	0.066
									Avg 0.54	Avg 2.68	0.068
8	275.8	40	41.4	6	50.8	2.00	1.48	3.5	0.52	5.87	0.149
8	275.8	40	41.4	6	50.8	2.00	1.48	3.5	0.54	6.06	0.154
8	275.8	40	41.4	6	50.8	2.00	1.48	3.5	0.53	5.94	0.151
									Avg 0.53	Avg 5.96	0.151

1. Cutting orifice diameter: No. 4: 0.04 in. (1.0 mm); No. 6: 0.06 in. (1.5 mm); No. 8: 0.08 in. (2.0 mm).

Table E4
Characteristics of fumes produced by oxymethane cutting of carbon steel plate at recommended conditions

Cutting tip number[1]	Oxygen pressure		Methane pressure		Plate thickness		Cutting speed		Fume generation rate, g/min	Weight of fumes/length of plate cut	
	kPa	psi	kPa	psi	mm	in.	mm/s	in./min		g/m	g/in.
00VVC	517.1	75	6.9	1	12.7	0.50	7.83	18.5	0.51	1.10	0.028
00VVC	517.1	75	6.9	1	12.7	0.50	8.09	19.1	0.71	1.46	0.037
00VVC	517.1	75	6.9	1	12.7	0.50	7.79	18.4	0.58	1.26	0.032
00VVC	517.1	75	6.9	1	12.7	0.50	7.45	17.6	0.59	1.33	0.034
							Avg 7.79	18.4	Avg 0.60	Avg 1.28	0.032
0-1/2VVC	689.7	100	6.9	1	25.4	1.00	6.22	14.7	0.53	1.42	0.036
0-1/2VVC	689.7	100	6.9	1	25.4	1.00	5.80	13.7	0.52	1.50	0.038
0-1/2VVC	689.7	100	6.9	1	25.4	1.00	6.01	14.2	0.55	1.54	0.039
							Avg 6.01	14.2	Avg 0.53	Avg 1.49	0.038
1VVC	723.9	105	6.9	1	50.8	2.00	5.08	12.0	0.59	1.93	0.049
1VVC	723.9	105	6.9	1	50.8	2.00	5.59	13.2	0.55	1.65	0.042
1VVC	723.9	105	6.9	1	50.8	2.00	5.16	12.2	0.48	1.54	0.039
							Avg 5.28	12.47	Avg 0.54	Avg 1.71	0.043

1. Cutting orifice diameter No. 00VVC: 0.036 in. (0.9 mm); No. 0-1/2 VVC: 0.040 in. (1.0 mm); No. 1 VVC: 0.046 in. (1.2 mm).

Table E5
Characteristics of fumes produced by oxymethane cutting of carbon steel plate at various oxygen and acetylene pressures

Cutting tip number[1]	Oxygen pressure		Methane pressure		Plate thickness		Cutting speed		Fume generation rate, g/min	Weight of fumes/length of plate cut	
	kPa	psi	kPa	psi	mm	in.	mm/s	in./min		g/m	g/in.
						High gas pressure					
00VVC	689.4	100	6.9	1	12.7	0.50	8.17	19.3	0.66	1.38	0.035
00VVC	689.4	100	6.9	1	12.7	0.50	7.24	17.1	0.66	1.53	0.039
00VVC	689.4	100	6.9	1	12.7	0.50	8.17	19.3	0.73	1.50	0.038
						Avg	7.86	18.6	Avg 0.68	Avg 1.47	0.037
						Low gas pressure					
00VVC	344.7	50	6.9	1	12.7	0.50	7.11	16.8	0.19	0.43	0.011
00VVC	344.7	50	6.9	1	12.7	0.50	6.86	16.2	0.30	0.71	0.018
00VVC	344.7	50	6.9	1	12.7	0.50	7.11	16.8	0.37	0.87	0.022
00VVC	344.7	50	6.9	1	12.7	0.50	6.73	15.9	0.33	0.83	0.021
						Avg	6.95	16.4	Avg 0.30	Avg 0.71	0.018
0-1/2VVC	517.1	75	20.7	3	25.4	1.00	4.23	10.0	0.61	2.40	0.061
1VVC	517.1	35	27.6	4	50.8	2.00	1.48	3.5	0.15	1.69	0.043
1VVC	517.1	35	27.6	4	50.8	2.00	1.48	3.5	0.12	1.65	0.042
1VVC	517.1	35	27.6	4	50.8	2.00	1.48	3.5	0.15	1.93	0.049
						Avg	1.48	3.5	Avg 0.14	Avg 1.58	0.045

1. Cutting orifice diameter: No. 00VVC: 0.036 in. (0.9 mm); No. 0-1/2 VVC: 0.040 in. (1.0 mm); No. 1 VVC: 0.046 in. (1.2 mm).

Table E6
Characteristics of fumes produced by oxymethane cutting of carbon steel plate at various speeds

Cutting tip number[1]	Oxygen pressure kPa	Oxygen pressure psi	Methane pressure kPa	Methane pressure psi	Plate thickness mm	Plate thickness in.	Cutting speed mm/s	Cutting speed in./min	Fume generation rate, g/min	Weight of fumes/length of plate cut g/m	Weight of fumes/length of plate cut g/in.
High cutting speed											
00VVC	689.4	100	6.9	1	12.7	0.50	8.93	21.1	0.27	0.51	0.013
0-1/2VVC	689.4	100	6.9	1	25.4	1.00	8.66	20.5	0.33	0.63	0.016
0-1/2VVC	689.4	100	6.9	1	25.4	1.00	8.66	20.5	0.37	0.71	0.018
Avg							8.66	20.5	0.35	0.67	0.017
Slow cutting speed											
00VVC	689.4	100	6.9	1	12.7	0.50	6.69	15.8	0.69	1.73	0.044
00VVC	689.4	100	6.9	1	12.7	0.50	7.11	16.8	0.49	1.41	0.029
Avg							6.90	16.3	0.59	1.57	0.037
0-1/2VVC	689.4	100	6.9	1	25.4	1.00	2.96	7.0	0.36	2.05	0.052
0-1/2VVC	689.4	100	6.9	1	25.4	1.00	2.96	7.0	0.36	2.01	0.051
0-1/2VVC	689.4	100	6.9	1	25.4	1.00	2.75	6.5	0.36	2.17	0.055
Avg							2.89	6.8	0.36	2.08	0.053
1VVC	723.9	105	6.9	1	50.8	2.00	4.78	11.3	0.41	1.43	0.036

1. Cutting orifice diameter: No. 00VVC: 0.036 in. (0.9 mm); No. 0-1/2 VVC: 0.040 in. (1.0 mm); No. 1 VVC: 0.046 in. (1.2 mm).

References and Literature Surveyed

Introduction

1. Arc Welding and Cutting Noise, Project Committee VI on Noise of the AWS Committee on Safety and Health and Battelle-Columbus Laboratories, American Welding Society, Miami, Florida (1979).

2. Lyon, T.L., Marshall, W.J., Sliney, D. H., Krial, N.P., and Del Valle, P.F., "Nonionizing Radiation Protection Special Study No. 42-0053-77, Evaluation of the Potential Hazards from Actimic Ultraviolet Radiation Generated by Electric Welding and Cutting Arcs," December 1975 to September 1976, U.S. Army Environmental Hygiene Agency, Aberdeen Proving Grounds, Md. 21010.

3. Marshall, W.J., Sliney, D.H., Lyon, T.L., Krial, N.P., and Del Valle, P.F., "Nonionizing Radiation Protection Special Study No. 42-0312-77, Evaluation of the Potential Retinal Hazards from Optical Radiation Generated by Electric Welding and Cutting Arcs," December 1975 to April 1977, U.S. Army Environmental Hygiene Agency, Aberdeen Proving Grounds, Md. 21010.

4. Hinrichs, J.F., "Project Committee on Radiation —Summary Report," Welding Journal *57*(1), 62, 65 (1978).

5. The Welding Environment, Edited by the AWS Technical Department, American Welding Society, Miami, Florida (1973).

6. *Threshold Limit Values for Chemical Substances and Physical Agents in the Workroom Environment*, American Conference of Governmental Industrial Hygienists, Cincinnati, Ohio (updated and issued yearly).

Part I

1.1 AWS Standard Method for Sampling Airborne Particulates Generated by Welding and Allied Processes, March 9, 1976.

1.2 Millipore Corporation Catalogue, Bedford, Massachusetts, 01730 (1976).

1.3 Liu, B.Y.H., and Lee, K.W., "Efficiency of Membrane and Nuclepore Filters for Submicrometer Aerosols," Environmental Science and Technology, Vol. 10 (1976), p 345.

1.4 Industrial Ventilation, A Manual of Recommended Practice, 12th Edition, American Conference of Government Industrial Hygienists, Committee on Industrial Ventilation, P.O. Box 453, Lansing, Michigan 48902 (1972).

Part II

2.1 The Welding Environment, Edited by the AWS Technical Department, American Welding Society, Miami, Florida (1973).

2.2 Altshuler, B., Yarmus, L., Palmes, E.D., and Nelson, N., "Aerosol Deposition in the Human Respiratory Tract: I. Experimental Hygiene," Vol. 15, p. 293 (1957).

2.3 Heile, R.F., and Hill, D.C., "Particulate Fume Generation in Arc Welding Processes," Welding Journal, *54*, (7), 201s-210s (1975).

2.4 Dahlberg, J.A., Andersson, H., and Wettstrom, R., "On the Decomposition of Trichloroethylene, Perchloroethylene, and Methyl Chloroform in Welding Works," International Institute of Welding, Document VIII-585-74 (1974).

2.5 Kobayashi, M., Maki, S., and Ohe, J., "Factors Affecting the Amount of Fumes Generated by Manual Metal Arc Welding," International Institute of Welding, Document VIII-670-76 (1976).

2.6 Standard for Grouping Manual Arc Welding Electrodes into Fume Classes, International Institute of Welding, Document VIII-543-73 (1973).

2.7 Fume Classification of Coated Electrodes for Manual Welding, International Institute of Welding, Document VIII-749-77 (1977).

2.8 Kobayashi, M., Maki, S., Hashimoto, Y., and Suga, T., "Some Considerations About Formation Mechanism of Welding Fumes," International Institute of Welding, Document VIII-715-77 (1977).

2.9 Kimura, S., Kobayashi, M., and Maki, S., "Some Considerations about Fumes Generated in Arc Welding Processes," International Institute of Welding, Document VIII-687-76 (1976).

2.10 Morita, N., and Tanigaki, T., "Investigation on Welding Fumes from Covered Electrodes and Development of Low Fume Electrodes," International Institute of Welding, Document VIII-724-77 (1977).

2.11 Katrenko, V.T., Kassov, D.S., and Karpenko, V.M., "A Comparative Assessment of Slag Bases with Regard to Dust Generation in Welding and Self-Shielding, Flux Cored Electrodes," Welding Production, *21* (9), 53-54 (1974).

2.12 Erman, M.J., and Kabasin, S.G., "Comparative Health Appraisal of Semiautomatic Welding with Shielding Gases Consisting of a Mixture of Argon and Oxygen, CO_2, and Mixtures with CO_2," Automatic Welding, *24* (4), 69-70 (1971).

2.13 Vorontsova, E.J., and Mosolov, N.J., "Influence of Wire Composition and Equipment Settings on the Health Aspects of CO_2 Welding," Welding Production, *16* (11), 47-49 (1969).

2.14 Psaras, G.G., and Popova, Z.M., "Health and Hygiene Aspects of The Automatic Welding of Aluminum with a Consumable Electrode in a Mixture of Argon and Oxygen," Welding Production, *18* (3), 48 (1971).

2.15 Erman, M.J., "Some Ways of Reducing Dust Emission During Welding with a Consumable Electrode in Shielding Gases," Welding Production, *19* (3), 51-52 (1972).

2.16 Pantucek, M., "Influence of Filler Materials on Air Contamination in Manual Electric Arc Welding," American Industrial Hygiene Association Journal, *32* (10), 687-692 (1971).

2.17 Smith, D.C., "Flux Cored Electrodes-Their Composition and Use," Welding Journal, *49* (7), 535-547 (1970).

2.18 Smith, D.C., and Rhinehart, W.G., "Application and Development of Modern Heavy-Coated Arc Welding Electrodes," Welding Journal, *24* (6), 541-548 (1945).

2.19 Smith, D.C., "Development Properties, and Usability of Low Hydrogen Electrodes," Welding Journal, *38* (9), 377-s-392-s (1959).

2.20 Magnusson, E.J., "Environment and Productivity," Svetsaren, No. 1, 1977.

2.21 Hazards from Arc Welding: A Tentative Standard Method for the Evaluation of Potential Health Hazard of Fumes from the Use of Coated Mild Steel Welding Electrodes, International Institute of Welding, Document VIII-525-73 (1973).

2.22 Frant, R., "Formation of Ozone in Gas-Shielded Welding," Annals of Occupational Hygiene, *6*, 113-125 (1963).

2.23 Lunau, F.W., "Ozone in Arc Welding," Annals of Occupational Hygiene, *10* (7), 175-188 (1967).

2.24 Alpaugh, E.L., Phillipo, K.A., and Pulsifier, H.C., "Ventilation Requirements for Gas Metal Arc Welding Versus Covered Electrode Welding," American Industrial Hygiene Association Journal, *29* (6), 551-557 (1968).

Part III

3.1 Dumphy, B., "Acute Occupational Cadmium Poisoning—A Critical Review of the Literature," J. Occup. Med., Vol. 9, no. 1, January 1967.

3.2 Fishburn, C.W., and Zenz, C., "Metal Fume Fever, A Report of a Case," J. Occup. Med., Vol. 11, no. 3, April 1969.

3.3 Mangold, C.A., and Beckett, R.R., "Combined Occupational Exposure of Silver Brazers to Cadmium Oxide, Nitrogen Dioxide, and Fluorides at a Naval Shipyard," Am. Ind. Hygiene Assoc. J., Vol. 32, no. 2, 1971.

3.4 Vance, G.H., "Cadmium Exposure in Silver Soldering," Am. Ind. Hygiene Assoc. J., Vol. 21, no. 4, 1960.

3.5 Blejer, H.F., "Death Due to Cadmium Oxide Fume," Ind. Med. and Surg., Vol. 35, May 1966.

3.6 Winston, R.M., "Cadmium Fume Poisoning," Brit. Med. J., May 15, 1971.

3.7 *Working With Silver Solder.* U.S. Department of Health, Education, and Welfare, Public Health Service, Division of Occupational Health, PHS Publication No. 1518, 4 pp. October 1966.

3.8 *Braze Safely,* Am. Weld. Soc., Miami, Florida.

3.9 Hara, N., Homma, K., and Koshi, S., "Metallic Fume Generated in the Welding Using Silver Solder," Ind. Health, Vol. 6, 1968.

3.10 Private communication, May 1976.

3.11 Aluminum Brazing Handbook, The Aluminum Association, New York, 1971.

3.12 McCann, J.K., "Health Hazard from Flux Used in Joining Aluminum Electric Cables," Ann. Occup. Hyg., Vol. 7, no. 3, 1964.

3.13 Ruza, Viliam, (Vyzk. Vstay Zvarnic, Bratislava, Czech.) "New Data on Brazing Alloys and Fluxes," Zvarnie, Vol. 23, no. 2, pp. 56-62, 1974.

3.14 Steenbeek, J., "Safety Measures Before, During, and After Brazing," De Veiligheid, Vol. 44, no. 7-8, pp. 227-229, July-August 1968 (Dutch).

3.15 Silver Soldering Hazards and Control, *Michigan's Occupational Health*, Vol. 11, no. 4, pp. 6-7, 1966.

3.16 Oliver, T.P., and Molyneux, M.D.B., "Calculation of Risks in Burning and Welding," Ann. Occup. Hyg., Vol. 17, pp. 295-302, February 1975.

3.17 Zavon, M.R. and Meadows, C.C., "Vascular Sequelae to Cadmium Fume Exposure," Am. Ind. Hyg. Ass. J., Vol. 31, no. 2, pp. 180-182, 1970.

3.18 Townshend, R.H., "A Case of Acute Cadmium Pneumonitis: Lung Function Tests During a Four-Year Follow-Up," Brit. J. Ind. Med., Vol. 25, no. 1, pp. 68-71, 1968.

3.19 Humperdinck, K., "Kadmium and Lungenkrebs. (Cadmium and lung cancer)," Med. Klin., Vol. 63, no. 24, pp. 948-951, 1968.

3.20 Triffett, A.C.K., "Hazards in the Use of Cadmium-Containing Silver Brazing Alloys," Proc. Roy. Aust. Chem. Inst., Vol. 35, no. 5, pp. 124-125, 1968.

3.21 Belknapp, E.L., "Hidden Dangers in Silver Solder," Health, Quarterly Bulletin of the Wisconsin State Board of Health, pp. 17-18, 1964.

3.22 Favorite, F.G., Roslinski, L.M., and Wands, R.C., "Guides for Short-Term Exposures of the Public to Air Pollutants," Aerospace Medical Research Laboratory, Wright-Patterson AFB, Ohio, AVRL-TR-71-120-Paper 16.

3.23 Market, H.L., Sr., et al., "Health Hazard Evaluation/Toxicity Determination, Rheem Manufacturing Company," Nat'l. Inst. for Occupational Safety and Health, Report NIOSH TR HHE 73 138 156, 13 pp. November 1974.

3.24 Gunter, Bobby J., et al., "Health Hazard Evaluation/Toxicity Determination, Empire Detroit Steel Company, New Boston, Ohio," National Institute for Occupational Safety and Health, Report NIOSH TR 045 74, 21 pp. June 1973.

3.25 Boogaerdy, C.A., and De Mooij, H., "Air Pollution From Soldering Fumes," Ind. Med. and Surg., Vol. 29, no. 1, pp. 36-38, 1960.

3.26 Gordts, J., "Determination of Lead Content in the Air During Soft Soldering," T. Soc. Geneesk, Vol. 49, no. 25, pp. 912-914, 1971.

3.27 Agafonova, L.I., et al., "Lead Aerosol Content in the Air of the Working Zone of Bulk-Soldering Shops," Vodosnabzh Sanit. Tekh., Vol. 10, no. 35, 1973 (Russian).

3.28 Bruno, L., "Lead Poisoning Hazard in Workers Employed on Car Bodies," Bulletin of Hygiene, Vol. 36, no. 24, 1961.

3.29 Kharlashko, N.K., and Atanas eva, N.A., "Sanitary Hygienic Evaluation of Working Conditions in the Assembly Shop of a Telephone Plant," Tr. Permsk. Gos. Med. Inst., Vol. 110, pp. 78-9, 1972 (Russian).

3.30 Feldman, R.G., and Sedman, T., "Hobbyists Working With Lead," New Engl. J. Med. Vol. 292, pp. 929, April 24, 1975.

3.31 Okawa, M.T., and Rodner, A., "Health Hazard Evaluation Report 72-74-51 Western Electric Company, Dublin, California," Hazard Evaluation Services Branch, NIOSH, Cincinnati, Ohio, Report 72-74-51, 1973.

3.32 Hehir, R.M., et al., "Toxicity of Zinc Soldering Compounds," Food and Drug Administration, Division of Pharmacology, Washington, D.C.

3.33 Stahl, Quade R., "Air Pollution Aspects of Hydrochloric Acid," Litton Systems, Inc., Bethesda, Md., Environmental Systems Div., Report PH 22 68 25, September 1969.

3.34 "Exposure of Eyes to Soldering Fumes," Journal of the American Medical Association, Vol. 165, pp. 764-765, 1957.

3.35 Snell, A.C., Birren, F., Wirt, S.E., and Kuhn, H.S., "Occupational Opthalmology," Transactions American Academy of Ophthalmology and Otolaryngology, November-December, 1947.

3.36 Ivanova, L.V., and Lyubimova, M.I., "Flux For Tin Solders," U.S.S.R. Patent 215,020 (Cl. B. 23 k), granted March 29, 1968, Izobret. Prom. Obrazlsy, Tovarnye Znaki, Vol. 45, no. 12, p. 154, 1968.

3.37 Cheffers, R.R., "Factors Affecting the Selection of Fluxes and Solders," Sub-Assembly Components, Fastening, Vol. 10, pp. 39-41, January 1972.

3.38 Wheeler, C.E., Jr., Penn, S.R., and Cawley, E.P., "Dermatitis From Hydrazine Hydrobromide Solder Flux," Arch. Dermat., Vol. 91, pp. 235-238, March 1965.

3.39 McCann, J.K., "Health Hazard from Flux Used in Joining Aluminum Electricity Cables," in: Conference on health and safety aspects of welding. London, 1963. Ann. Occup. Hyg., Vol. 7, no. 3, pp. 261-268, 1968.

3.40 Sterling, G.M., "Asthma due to Aluminum Soldering Flux," Thorax, Vol. 22, pp. 533-537, 1967.

3.41 Gunther, Bobby J., et al., "Health Hazard Evaluation Toxicity Determination, Docutel Corporation, Irving, Texas," National Institute for Occupational Safety and Health, Cincinnati, Ohio, Report NIOSH TR 73 180 183 159, 21 pp., November 1974.

3.42 Pepys, J., and Pickering, C.A.C., "Asthma Due

to Inhaled Chemical Fumes, Amino-ethyl Ethanolamine in Aluminum Soldering Flux," Clinallergy, Vol. 2, no. 2, pp. 197-204, 1972.

3.43 Greenburg, L., "Safety Measures for Use of Chlorinated Naphthalenes and Diphenyls in Industry," Industrial Bulletin, Vol. 22, no. 10, 5 pp., 1943.

3.44 Mayers, M.R., and Silverberg, M.G., "Effects Upon the Skin Due to Exposure to Some Chlorinated Hydrocarbons," Industrial Bulletin, New York State Department of Labor, Vol. 17, nos. 8-9, pp. 29-43, 1938.

Part V

5.1 Steel, J. "Respiratory Hazards in Shipbuilding and Shiprepairing," Annals of Occupational Hygiene, Vol. 11, pp. 155-121, 1968.

5.2 Silk, S.J., "Fumes from Welding and Flame Cutting," International Institute of Welding Document VIII-414-70(d), 1970.

Index

A

B

C

D

E

F

G

H